NUCLEIC ACID-PROTEIN INTERACTIONS

NUCLEIC ACID SYNTHESIS IN VIRAL INFECTION

MIAMI WINTER SYMPOSIA, VOLUME 2, 1971

nucleic acid-protein interactions

nucleic acid synthesis in viral infection

Proceedings of the Miami Winter Symposia, January 18–22, 1971
Organized by the Department of Biochemistry, University of Miami,
and the Papanicolaou Cancer Research Institute

D.W. RIBBONS, J.F. WOESSNER and J. SCHULTZ, Editors

NORTH-HOLLAND PUBLISHING COMPANY – AMSTERDAM · LONDON

© 1971 North-Holland Publishing Company

All rights reserved. No part of this publication may be reproduced, stored in a retrieval system, or transmitted, in any form or by any means, electronic, mechanical, photocopying, recording or otherwise, without the prior permission of the copyright owner.

Library of Congress Catalog Card Number: 79-161446
North Holland ISBN: 0 7204 4099 8
American Elsevier ISBN: 0 444 10084 9

Publishers:

NORTH-HOLLAND PUBLISHING COMPANY – AMSTERDAM
NORTH-HOLLAND PUBLISHING COMPANY, LTD. – LONDON

SOLE DISTRIBUTORS FOR THE WESTERN HEMISPHERE:
American Elsevier Publishing Company, Inc.
52 Vanderbilt Avenue
New York, N.Y. 10017

Printed in The Netherlands

FOREWORD

Symposia on biochemical topics have been arranged by the Department of Biochemistry and the Program in Cellular and Molecular Biology of the University of Miami for a number of years. In January 1969 the Department of Biochemistry joined with the University-affiliated Papanicolaou Cancer Research Institute to continue this tradition by presenting pairs of Symposia on closely related biochemical topics which now attracted national interest. In 1970, the two Symposia were entitled "Homologies in Enzymes and Metabolic Pathways" and "Metabolic Alterations in Cancer". The full report of this meeting was published as the first volume of a continuing series under the title: *Miami Winter Symposia.* The present volume, the second in this series, contains the report of the January 1971 Symposia which were entitled "Nucleic Acid-Protein Interactions" and "Nucleic Acid Synthesis in Viral Infection". This year, it was decided to include all discussions as well as the full text of the reports. It is hoped that this will enhance the usefulness and timeliness of this volume.

Associated with the symposia is an opening lecture, now named in honor of the University of Miami's distinguished Visiting Professor, Professor Feodor Lynen. The first Lynen Lecture was given by Dr. George Wald. This year the Lynen Lecture was delivered by Dr. Arthur Kornberg. This lecture forms the opening chapter of the present volume.

It is hoped that the method of organization of the symposia will ensure their publication as rapidly as possible. The speakers are strictly enjoined to present their finished manuscripts at the time of the meeting, the discussants edit their comments before the symposia adjourn, and our publishers use a rapid, highly attractive photo-offset printing process. We thank them, the speakers, and the many local helpers, faculty and administrative staff, coordinated by Dr. W.J. Whelan, Dr. K. Savard, Dr. B.J. Catley and Dr. Z. Brada, who have made this venture possible. We also acknowledge with gratitude the financial assistance of the National Science Foundation, the Howard Hughes Medical Institute, the Departments of Anesthesiology, Dermatology, Ophthalmology and the Institute of Molecular Evolution of the University of Miami, Coulter Electronics, Inc., Eli Lilly and Company, and M.C.B. Manufacturing Chemists.

Previous Volumes in This Series

1969 — "Biochemistry of the Phagocytic Process." Ed. by J. Schultz, North-Holland Publishing Company, Amsterdam. (Selected papers)

1970 — "Homologies in Enzymes and Metabolic Pathways" and "Metabolic Alterations in Cancer". *Miami Winter Symposia, Volume 1.* Ed. by W.J. Whelan and J. Schultz, North-Holland Publishing Company, Amsterdam.

CONTENTS

Foreword

Part 1. Nucleic acid-protein interactions

The second Feodor Lynen Lecture: Enzymes, DNA and Membranes
   A. Kornberg   3

On the mechanism of DNA synthesis in bacteria
   R. Werner   24

Discussion: Alberts, Cairns, Firshein, Greer, A. Kornberg, Littauer, Tomasz

Enzymatic mechanisms in DNA replication
   R.E. Moses, J.L. Campbell, R.A. Fleischman, and C.C. Richardson   48

Discussion: Cairns, Chang, Firshein, Hsiang, Novak, Rabinowitz, Tomasz, Weiss, Werner, Willis

Studies on termination during *in vitro* DNA-dependent RNA synthesis
   A.R. Goldberg and J. Hurwitz   70

Discussion: Cohen, Huijing, Novak, Qasba

Repair of damaged DNA in human and other eukaryotic cells
   J.E. Cleaver   87

Discussion: Greer, Levi, Michaelis, Penneys, Warren

Interactions of thermal proteinoids with polynucleotides
   S.W. Fox, J.C. Lacey, Jr., and T. Nakashima   113

Discussion: Novelli

On the structure of the replication apparatus
   B.M. Alberts   128

Discussion: Allfrey

Changes in chromosomal proteins associated with gene activation
V.G. Allfrey, C.S. Teng, and C.T. Teng 144

Discussion: Alberts, Borovsky, Byvoet, Friedman, Horecker, Novelli, Warren, Weiss

Quaternary structure and substrate binding of the aminoacyl-tRNA synthetases, particularly tryptophanyl-tRNA synthetase of *Escherichia coli*
K.H. Muench and D.R. Joseph 172

Discussion: Jacobson, A. Kornberg, Littauer, Weiss

Aminoacyl-tRNA synthetases: Their biological stability and regulation in *Escherichia coli*
F.C. Neidhardt and J.J. Anderson 184

Discussion: Jacobson, Muench, Novelli, Skarstedt, Stebbing

The role of a specific isoacceptor tRNA in genetic suppression and in enzyme regulation in Drosophila
K.B. Jacobson and E.H. Grell 204

Discussion: Armstrong, Horecker, Skarstedt

Regulation of gene expression in *Escherichia coli* by cyclic AMP
I. Pastan, B. de Crombrugghe, B. Chen, W. Anderson, J. Parks, P. Nissley, M. Straub, M. Gottesman and R.L. Perlman 215

Discussion: Lipscomb, Novak, Shaw, Whelan

DNA-RNA polymerase interaction: The effects of the formation of a single phosphodiester bond
A.G. So, K.M. Downey and B.S. Jurmark 221

Discussion: Friedman, Horecker, Michaelis

Part 2. Nucleic acid synthesis in viral infection

Translational control mechanisms in bacteriophage development
S.B. Weiss 237

Discussion: Armstrong, Cohen, Greer, Mans, Neidhardt, Novelli, Schultz

CONTENTS

Transcription *in vitro* of the *Escherichia coli* tRNA^{Tyr} gene carried by the transducing bacteriophage $\phi 80\text{psu}_3^+$
U.Z. Littauer, V. Daniel, J.S. Beckmann, S. Sarid ... 248

Discussion: Cohen, Novelli, Weiss

Polyamines and the multiplication of T-even phages
S.S. Cohen, A. Dion ... 266

Discussion: Alberts, Greer, Littauer, Neidhardt, Rabinovitz, Schultz, Sigel, Werner

DNA polymerases and other enzymes of RNA tumor viruses
H.M. Temin, S. Mizutani, J. Coffin ... 291

Discussion: Baltimore, Duesberg, Greer

DNA polymerase of avian tumor viruses
P. Duesberg, E. Canaani, K. v.d. Helm ... 311

Discusssion: Baltimore, Hurwitz, Temin

Template and primer requirements for the avian myeloblastosis DNA polymerase
D. Baltimore, D. Smoler ... 328

Replication of reovirus
A.F. Graham, S. Millward ... 333

Discussion: Schultz, Sigel, Temin

Some observations on DNA polymerases of human normal and leukemic cells
R.C. Gallo, S.S. Yang, R.G. Smith, F. Herrera, R.C. Ting, S. Fujioka ... 353

Discussion: Chang, Novelli, Schultz, Spiegelman

Transcription of viral genes in cells transformed by DNA and RNA tumor viruses
M. Green ... 380

Discussion: Baltimore, Kit, Temin

Structure and synthesis of adenovirus capsid proteins
H.S. Ginsberg ... 419

Discussion: Baltimore, Jacobson, Raskas, Sigel

The role of herpesvirus glycoproteins in the modification of membranes of infected cells
B. Roizman, P.G. Spear ... 435

Discussion: Cohen, Sigel, Temin

SV40 DNA replication in normal and transformed cells
 S. Kit, D.R. Dubbs   461

Discussion: Baltimore

Discussion on: Viral oncology, S. Spiegelman,
 *Alberts, Hurwitz, Kit, Littauer, Novelli, Parks, Pieczenik, Schultz,
 Tomasz, Warren, Willis*   490

LIST OF SPEAKERS, CHAIRMEN AND DISCUSSANTS

B.M. Alberts, Department of Biochemistry, Princeton University, Princeton, New Jersey.
V.G. Allfrey, Department of Cell Biology, Rockefeller University, New York, New York.
R. Armstrong, Department of Biological Chemistry, University of Michigan, Ann Arbor, Michigan, 48104.
D. Baltimore (Session Chairman), Department of Biology, Massachusetts Institute of Technology, Cambridge, Massachusetts, 02139.
J.S. Beckman, Department of Biochemistry, The Weizmann Institute of Science, Rehovot, Israel.
D. Borovsky, Department of Biochemistry, University of Miami School of Medicine, Miami, Florida, 33152.
P. Byvoet, Molecular and Cellular Division, Biological Sciences, University of Florida, Gainesville, Florida.
J. Cairns (Session Chairman), Cold Spring Harbor Laboratory, Cold Spring Harbor, New York, 11724.
E. Canaani, Department of Molecular Biology and Virus Laboratory, University of California, Berkeley, California, 94720.
S.H. Chang, Department of Biochemistry, Louisiana State University, Baton Rouge, Louisiana, 70803.
J.E. Cleaver, Department of Radiology, Laboratory of Radiobiology, University of California, San Francisco Medical Center, San Francisco, California.
J. Coffin, McArdle Laboratories for Cancer Research, University of Wisconsin, Madison, Wisconsin.
G. Cohen, Albert Einstein College of Medicine, Bronx, New York, 10461.
S.S. Cohen (Session Chairman), Department of Therapeutic Research, University of Pennsylvania School of Medicine, Philadelphia, Pennsylvania.
V. Daniel, Department of Biochemistry, The Weizmann Institute of Science, Rehovot, Israel.
A. Dion, Department of Therapeutic Research, University of Pennsylvania School of Medicine, Philadelphia, Pennsylvania.
Kathleen Downey, Departments of Medicine and Biochemistry, University of Miami School of Medicine, Miami, Florida, 33152.

LIST OF PARTICIPANTS

D.R. Dubbs, Division of Biochemical and Virology, Baylor College of Medicine, Houston, Texas.

P. Duesberg, Department of Molecular Biology, Virology Laboratory, University of California, Berkeley, California, 94720.

W. Firshein, Shanklin Laboratory for Biology, Wesleyan University, Middletown, Connecticut, 06457.

R.M. Franklin, Department of Molecular Biophysics, The Public Health Research Institute of the City New York, New York, New York.

S. Friedman, Department of Biochemistry, Faculty of Science, Laval University, Quebec, Canada.

S. Fujioka, Section on Cellular Control Mechanisms, National Cancer Institute, and Bionetics Research Laboratories, Bethesda, Maryland.

R.C. Gallo, Section on Cellular Control Mechanisms, National Cancer Institute, and Bionetics Research Laboratories, Bethesda, Maryland.

H.S. Ginsberg, Department of Microbiology, University of Pennsylvania School of Medicine, Philadelphia, Pennsylvania.

A.F. Graham, Department of Biochemistry, McGill University Medical School, Montreal, Canada.

M. Green, Institute of Molecular Virology, St. Louis University School of Medicine, Philadelphia, Pennsylvania.

S.B. Greer, Department of Microbiology and Biochemistry, University of Miami School of Medicine, Miami, Florida, 33152.

K. v.d. Helm, Department of Molecular Biology and Virus Laboratory, University of California, Berkeley, California 94720.

F. Herrera, Section on Cellular Control Mechanisms, National Cancer Institute and Bionetics Research Laboratoires, Bethesda, Maryland.

B.L. Horecker (Session Chairman), Department of Molecular Biology, Albert Einstein College of Medicine, New York, 10461.

W. Hsiang, Department of Microbiology, University of Miami School of Medicine, Miami, Florida, 33152.

F. Huijing, Departments of Biochemistry and Medicine, University of Miami School of Medicine, Miami, Florida, 33152.

J. Hurwitz, Department of Developmental Biology and Cancer, Albert Einstein College of Medicine, Bronx, New York, 10461.

K.B. Jacobson, Biology Division, Oak Ridge National Laboratory, Oak Ridge, Tennessee, 37830.

S. Kit, Division of Biochemistry and Virology, Baylor College of Medicine, Houston, Texas.

A. Kornberg, Department of Biochemistry, Stanford University School of Medicine, Stanford, California.

LIST OF PARTICIPANTS

H.L. Kornberg (Session Chairman), Department of Biochemistry, University of Leicester, England.

J.C. Lacey, Jr., Department of Biochemistry, University of Miami School of Medicine and Institute of Molecular Evolution.

J. Levi, Department of Biochemistry-CMB Program, University of Miami School of Medicine, Miami, Florida, 33152.

M. Lipscomb, Department of Biochemistry, University of Miami School of Medicine, Miami, Florida, 33152.

U.Z. Littauer, Department of Biochemistry, Weizmann Institute of Science, Rehovot, Israel.

R.J. Mans, Departments of Microbiology and Radiation Biology, University of Florida, Gainesville, Florida, 32601.

M. Michaelis, Ophthalmology Research Laboratory, University of Maryland School of Medicine, Baltimore, Maryland, 21201.

S. Millward, Department of Biochemistry, McGill University School of Medicine, Montreal, P.Q. Canada.

S. Mizutani, McArdle Laboratories for Cancer Research, University of Wisconsin, Madison, Wisconsin.

K.H. Muench, Departments of Medicine and Biochemistry, University of Miami School of Medicine, Miami, Florida, 33152.

F.C. Neidhardt, Department of Microbiology, The University of Michigan Medical School, Ann Arbor, Michigan.

R. Novak, Department of Chemistry, De Paul University, Chicago, Illinois.

G.D. Novelli (Session Chairman), Biology Division, Oak Ridge National Laboratory, Oak Ridge, Tennessee, 37830.

R.E. Parks, Medical School, Brown University, Providence, Rhode Island.

I. Pastan, Molecular Biology Section, National Cancer Institute, National Institutes of Health, Bethesda, Maryland.

N. Penneys, Department of Biochemistry, University of Miami School of Medicine, Miami, Florida, 33152.

G. Pieczenik, New York University, New York, New York.

P.K. Qasba, Departments of Cell Biology and Pharmacology, Baltimore, Maryland.

M. Rabinowitz, Laboratory of Physiology, National Cancer Institute, National Institutes of Health, Bethesda, Maryland, 20014.

H. Raskas, Institute of Molecular Virology, St. Louis University, St. Louis, Missouri.

C.C. Richardson, Department of Biological Chemistry, Harvard Medical School Boston, Massachusetts.

B. Roizman, Department of Microbiology, University of Chicago, Chicago, Illinois.

S. Sarid, Department of Biochemistry, The Weizmann Institute of Science, Rehovot, Israel.

J. Schultz, Papanicolaou Cancer Research Institute, Miami, Florida, 33136.

W.V. Shaw, Departments of Medicine and Biochemistry, University of Miami School of Medicine, Miami, Florida, 33152.

M.M. Sigel (Session Chairman), Department of Microbiology, University of Miami School of Medicine, Miami, Florida, 33155.

M.T. Skarstedt, Department of Biochemistry, University of Miami School of Medicine, Miami, Florida, 33152.

R.G. Smith, Section of Cellular Control Mechanisms, National Cancer Institute and Bionetics Research Laboratories, Bethesda, Maryland.

D. Smoler, Department of Biology, Massachusetts Institute of Technology, Cambridge, Massachusetts.

P.G. Spear, Department of Microbiology, University of Chicago, Chicago, Illinois.

S. Spiegelman, Institute for Cancer Research, College of Physicians and Surgeons, Columbia University, New York, New York.

H.K. Stanford (Session Chairman), University of Miami, Coral Gables, Florida.

N. Stebbing, Searle Research Laboratories, England.

H.M. Temin, McArdle Laboratories for Cancer Research, University of Wisconsin, Madison, Wisconsin.

R.C. Ting, Section on Cellular Control Mechanisms, National Cancer Institute and Bionetics Research Laboratories, Bethesda, Maryland.

A. Tomasz, The Rockefeller University, New York, New York, 10021.

J. Warren, Life Science Center, Nova University, Fort Lauderdale, Florida.

R. Warren, Department of Pediatrics, University of Miami School of Medicine, Miami, Florida, 33152.

S.B. Weiss, Department of Biochemistry, University of Chicago, Chicago, Illinois.

R.K. Werner, Department of Biochemistry, University of Miami School of Medicine, Miami, Florida, 33152.

W.J. Whelan, Department of Biochemistry, University of Miami School of Medicine, Miami, Florida, 33152.

J. Willis, P-L Biochemicals, Milwaukee, Wisconsin, 53205.

S.S. Yang, Section on Cellular Control Mechanisms, National Cancer Institute and Bionetics Research Laboratories, Bethesda, Maryland.

PART 1

NUCLEIC ACID-PROTEIN INTERACTIONS

Editors: D.W. Ribbons and J.F. Woessner

THE SECOND FEODOR LYNEN LECTURE:

ENZYMES, DNA AND MEMBRANES

Arthur Kornberg
Department of Biochemistry
Stanford University School of Medicine

The tradition for this lecture is for it to be anecdotal and autobiographical. It is the kind of assignment that a scientist would be wise to avoid. To paraphrase Satchell Page: a scientist should never look back because he knows they're gaining on him. But this is an opportunity to honor Fitzie Lynen. He is an old friend and a great scientist. The effort is well worth the risks.

Enzymes, DNA and membranes describe my past, present and future interests and those of many other biochemists, as well. My formal training taught me very little about these subjects. The course in biochemistry I took in medical school thirty-three years ago reflected the emphasis in American biochemistry of that era: nutrition and the analysis of body tissues and fluids.

Enzymes were submerged somewhere in the physiology of digestion. DNA had been isolated from thymus gland; it was defined as the animal nucleic acid; RNA was isolated from yeast and defined as the plant nucleic acid. This distribution of nucleic acids, we were taught, served as a basic distinction between plants and animals. The route of biosynthesis of nucleic acids, had this question ever come up, would have had to be inferred from what dogs, fed purines and pyrimidines, excreted in their urine. As for membranes, it is hardly a proper biochemical subject, even today.

When I came to Ochoa's laboratory in 1946, eight years later, I hardly knew the difference between ADP and ATP. I knew even less about enzymes. At twenty-eight years of age, I was a very retarded biochemist and I have always been grateful to Severo Ochoa for his generosity in taking me in. During that year, the first after the war, reports filtered back about the work of a young German biochemist, named Lynen. He had, like Ochoa, made considerable headway in describing some of the enzymes of the Krebs cycle. The two-carbon fragment, the mythical compound that condensed with oxalacetate to form citrate, was just coming into focus as a prize of major dimensions. A few years later, Lynen was the one who identified acetyl CoA as the two-carbon fragment.

Enzyme Purification

From the very first assay, I fell in love with enzymes. I have always been in awe of enzymes, even the so-called dull ones. With such affection, purifying an enzyme is always rewarding. I doubt that any effort in purifying an enzyme was ever wasted. The purified enzyme may blaze a new pathway, or it may disclose unimagined subtleties in a reaction, or, at the very least, it will provide a unique and precious analytical reagent. I recall a visit from Hans Krebs about 1950 after I had returned to the N.I.H. I was working on the enzymatic synthesis of coenzymes and needed an assay for ATP. I was therefore purifying glucose 6-phosphate dehydrogenase (Zwischenferment). When Krebs asked me what I was doing I told him: "I am driven to purifying Zwischenferment." "Who's driving you?", he asked. I have realized since that I was really attracted to this job rather than driven to it.

Nucleotidyl transfer in coenzyme biosynthesis

The coenzymes, DPN, TPN and FAD, are the simplest of the nucleotide condensation products. Work on the biosynthesis of these coenzymes got me interested in the biosynthesis of the most complex of the nucleotide condensation products, the nucleic acids. The synthesis of DPN involves a nucleophilic attack by nicotinamide mononucleotide on the α-phosphate of ATP and the elimination of PPi (Fig. 1). This transfer of one 5'-nucleotide to another produces a pyrophosphate bond between them. This reaction is the prototype of a large number of such nucleotidyl transfers to other phosphate compounds (Fig. 2). These include transfers to various sugar phosphates to form the nucleoside diphosphate sugar coenzymes, to choline phosphate to form cytidine diphosphate choline, and to phosphatidic acid to form cytidine diphosphate diglyceride. This nucleotidyl transfer is the prototype also for transfers of nucleotides to produce mixed acid anhydrides with fatty acids, amino acids, and sulfates.

However, the nucleotides of nucleic acids are linked by a phosphodiester rather than a pyrophosphate bond!

Phospholipids, a poor model for nucleic acid synthesis

At about this time, Gerhard Schmidt found large quantities of glycerophosphorylcholine in liver (Fig. 3). Glycerophosphorylcholine is a simple phosphodiester and so I became interested in its biosynthesis. I was mistaken first in thinking that glycerophosphorylcholine had accumulated as an intermedi-

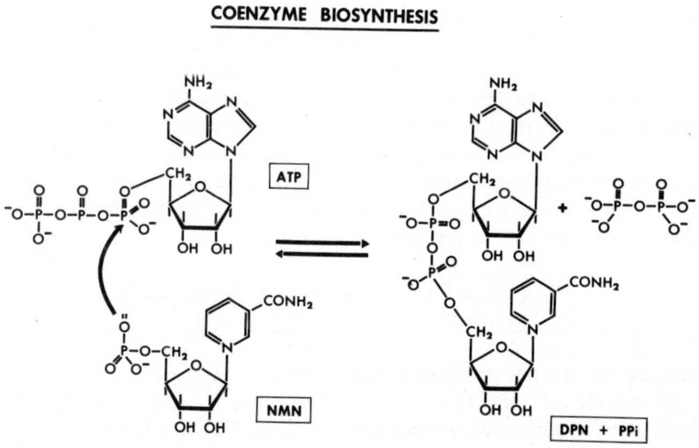

COENZYME BIOSYNTHESIS

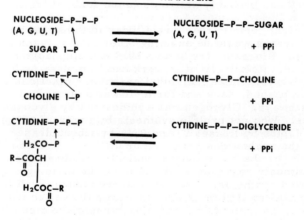

Figure 1

NUCLEOTIDYL TRANSFERS

```
NUCLEOSIDE-P-P-P          NUCLEOSIDE-P-P-SUGAR
(A, G, U, T)      ⇌       (A, G, U, T)
  SUGAR 1-P                    + PPi


CYTIDINE-P-P-P            CYTIDINE-P-P-CHOLINE
                  ⇌
  CHOLINE 1-P                  + PPi


CYTIDINE-P-P-P            CYTIDINE-P-P-DIGLYCERIDE
  H₂CO-P          ⇌
  R-COCH                       + PPi
     ‖
     O
  H₂COC-R
     ‖
     O
```

Figure 2

ate on the biosynthetic route to phospholipids. It was also
naive to think that the biosynthesis of this very simple compound would prove to be a useful model for nucleic acids. Our
search for enzymes that would synthesize glycerophosphorylcholine from plausible precursors was futile. Instead, we
found particulate fractions that condensed L-α-glycerophosphate with long-chain fatty acyl CoA to make phosphatidic acid,
and converted phosphoryl choline into a choline phospholipid
(Fig. 3). Eugene Kennedy went on in elegant style to delineate
the pathways of phospholipid metabolism but I was happy to
escape from these greasy compounds and particulate enzymes
back into the cool clear aqueous phase.

Nucleoside 5'-phosphate as the valid nucleotide isomer in biosynthesis

Back in water-soluble systems, we could find the biosynthetic pathway of pyrimidine nucleotides. A key step involved
5'-phosphoribosylpyrophosphate (PRPP). Orotic acid condensed with PRPP to form a nucleoside 5'-phosphate (Fig. 4).
John Buchanan found in purine nucleotide biosynthesis that
PRPP was also involved in the first step of that pathway. Thus,
at the very outset there was a commitment to the synthesis of
the 5'-nucleotide isomer.

Purine and pyrimidine rings are assembled from simple
precursors or salvaged from the diet. In every case, the
nucleotide formed is the 5' isomer. These facts convinced me
that the building blocks of nucleic acids must be nucleoside
5'-phosphates, probably activated in the form of triphosphates.

It might seem unreasonable today to think that nucleic
acid precursors would be anything other than 5' isomers. Not
so in 1950. Biochemistry before 1950 had emphasized energy
metabolism. Relatively little work had been done on biosynthetic pathways. Since individual enzymatic reactions could be
pulled and pushed, back and forth, it was assumed so could
entire pathways. Glycogen was synthesized by glycogen phosphorylase. Why not protein synthesis by reversal of peptidases? Lipids by lipases? And RNA by ribonuclease? Inasmuch as the nucleotides produced or utilized by ribonuclease
were the 3' or the 2', 3'-cyclic nucleotides, these nucleotides
were commonly regarded as likely to be the building blocks in
nucleic acid synthesis. What we learned during the fifties
radically altered this thinking. We came to realize that pathways of energy metabolism and biosynthesis are distinct. They
are dual and divided highways. Now that we know how reasonable and essential it is to separate the pathways of biosynthesis
of proteins, lipids and nucleic acids from their degradation, it

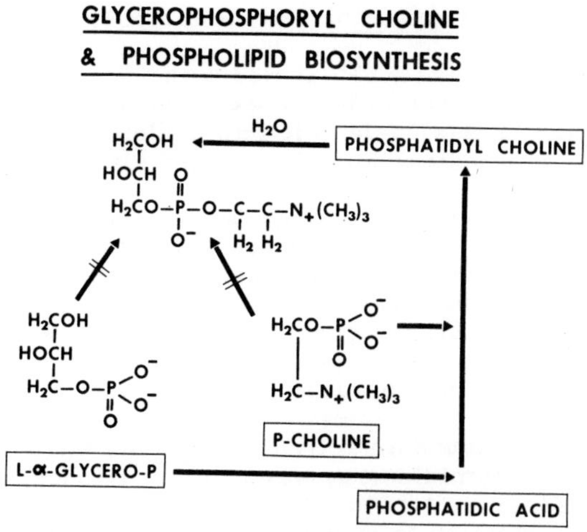

ENZYMES, DNA AND MEMBRANES

GLYCEROPHOSPHORYL CHOLINE & PHOSPHOLIPID BIOSYNTHESIS

Figure 3

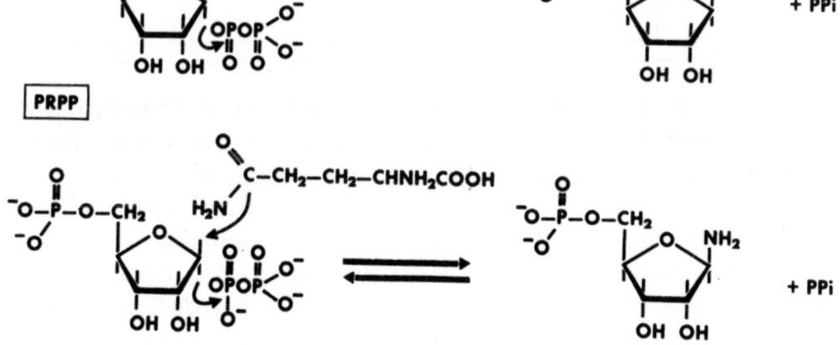

NUCLEOTIDE BIOSYNTHESIS

Figure 4

is hard to imagine why this wasn't obvious from the beginning.

Early attempts at nucleic acid biosynthesis

In 1954, we used ATP labeled with ^{32}P in the adenyl group to look in various cell-free extracts for an activity that would incorporate this nucleotide into nucleic acids. We found such an activity in extracts of E. coli and started to purify it.

One fine day we learned that Ochoa's laboratory had done it. They had discovered an enzyme, in Azotobacter, that made RNA. Polynucleotide phosphorylase polymerized ADP and other nucleoside diphosphates into RNA. So we switched from using ATP to using ADP in our E. coli system. As a result, we purified a polynucleotide phosphorylase from E. coli instead of discovering RNA polymerase. What a blunder!

We had also found an activity in E. coli extracts that incorporated ^{14}C-thymidine, by way of its 5' triphosphate, into a nucleic acid sensitive to DNase. So as a form of psychotherapy, we paid more attention to purifying the DNA-synthesizing activity.

Last year, in Cambridge, when Francis Crick, Gobind Khorana and I happened to be together, Crick asked us what prompted our working on DNA. With Khorana, his success in synthesizing ATP led him to the synthesis of coenzyme A. This, in turn, led him to more and more difficult condensations of chains of nucleotides. Finally, this trail brought him to the synthesis of a gene for a transfer RNA. With him, as with me, one thing led to another. We were aware of the importance of DNA, but it seems to me we were as fascinated by the journey as by the destination. Crick's style, by contrast, has been to search for grand designs and to his great credit, he has found them.

DNA polymerase, a template-directed enzyme

As we purified the activity responsible for the polymerization of deoxyribonucleotides, it became apparent that we had to supply preformed DNA and the four commonly occurring deoxyribonucleotides. The purified enzyme, called DNA polymerase, used the preformed DNA as a template. The DNA provided directions to DNA polymerase in assembling the DNA chain and it was clear at once why all four nucleotides were absolutely required. This template direction of an enzyme function was unique and unanticipated in enzymology. Some biochemists found template-direction of an enzyme very hard to believe, even after a number of years.

Not so the biologists. Although the Watson-Crick proposal for the structure and replication of DNA had not predicted an enzyme for assembling a DNA chain, yet the functions of DNA polymerase were easy for biologists to accept immediately.

Enzymological work gains its full significance when it is related to the structure and function of the cell.

DNA polymerase binds at a nick

The DNA polymerase purified from E. coli is a single polypeptide chain of about 1000 amino acids and includes one disulfide and one sulfhydryl group (1). The molecular weight is 109,000. The enzyme is globular, about 65 Å across. When Hg^{++} is added to the enzyme, one atom can bind two molecules through their sulfhydryl groups to form a dimer, without altering the enzyme's function; ^{203}Hg also provides a sensitive and convenient tag for observing the enzyme in physico-chemical studies, such as studies of binding of the enzyme to DNA.

How does the enzyme bind DNA? It does not bind to the complete, double-stranded circles of viral DNA. However when such circles have phosphodiester cleavages (nicks) in one of the strands, enzyme molecules are bound in exact proportion to the number of nicks. The binding is tight; the K_D is $<10^{-9}M$. T7 phage DNA, which is a linear, double-stranded helix of 40,000 base pairs binds only two molecules of enzyme, one at each end of the DNA.

We can see this binding of DNA polymerase to DNA in the electron microscope with a new technique developed by Jack Griffith (2). The standard Kleinschmidt method uses a protein film to complex and visualize the DNA, and so must be avoided (Fig. 5). The DNA we used was a homopolymer pair consisting of a chain of deoxyriboadenylates (4000 units long) with 20 to 30 chains of deoxyribothymidylates (150 units, 500Å long) annealed to it; this double-stranded helix should have nicks, where the thymidylate chains abut each other, spaced about 500Å apart. DNA polymerase molecules when bound to this homopolymer showed exactly this spacing (Fig. 6). With Hg-enzyme dimers, one molecule of the enzyme pair was bound to the DNA while the other assumed a variety of positions on the grid (Fig. 7). We infer that there is considerable flexibility at the Hg-enzyme dimer joint.

Since intact helical DNA does not bind DNA polymerase, it is not surprising that such DNA is inert as a template. What did surprise us at first was that the enzyme molecules bound at the ends of a T7 DNA are also inactive. For an enzyme molecule to be active it must be located at a nick. In fact, the very

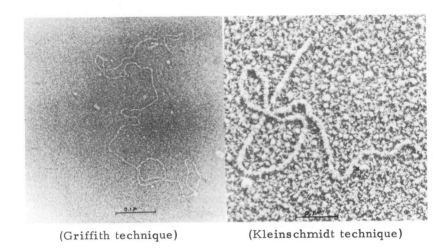

(Griffith technique)　　(Kleinschmidt technique)

Figure 5

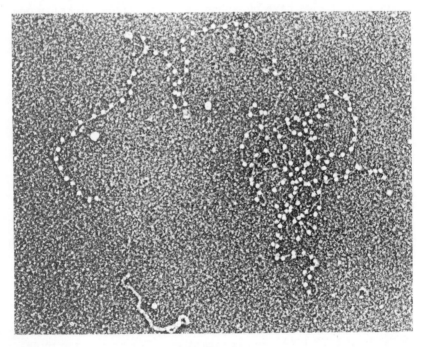

Figure 6

best rates of polymerization are observed when a polymerase molecule acts at a nick which has been enlarged to a gap as the result of prior action by an exonuclease (i.e., exonuclease III).

The polymerizing function of DNA polymerase

DNA polymerase requires three components to achieve chain growth (Fig. 8): (1) a chain with a 3'-hydroxyl terminus (primer), (2) a complimentary strand to which the primer is annealed and which extends beyond the primer terminus (template), and (3) a deoxyribonucleoside 5'-triphosphate which can form a base pair with the exposed template. Nucleophilic attack by the 3'-hydroxyl on the 5'-phosphate of the triphosphate results in a 3'-5' phosphodiester bond, the elimination of inorganic pyrophosphate and the appearance of a fresh primer terminus. The chain grows in a 5'⟶3' direction.

If a DNA chain is to grow at a rapid rate, it seems reasonable that the DNA should not dissociate from the enzyme after each polymerization step. Is there in fact a concerted polar movement of the chain in the 3'⟶5' direction, along the enzyme, as the diester bond is forged and the incoming triphosphate is becoming the new primer terminus? This question about the chain movement applies to most enzymes which catalyze polymerization and depolymerization reactions of biopolymers; the nucleic acid-protein interactions of DNA synthesis provide an intriguing example.

Binding sites for polymerization

The enzyme has two distinctive binding sites which seem well-suited for making the 3'-5' phosphodiester bond between a 5'-triphosphate and the 3'-hydroxyl end of a growing chain. One site accommodates any one of the four deoxyribonucleoside triphosphates and also a large variety of their analogues. It is essential that the molecule be a triphosphate (3). The other site specifies a 3'-hydroxyl nucleotide in the ribo configuration; xylo, lyxo and arabino analogues are inactive (4). We believe this site to be related to the one occupied by the primer terminus. We picture the active center of the enzyme as some specially adapted polypeptide surface that recognizes and accommodates these nucleotide structures (Fig. 9) and is designed to carry out the polymerization reaction (Fig. 10).

The proximity of the triphosphate site to the presumed primer terminus site was tested by Tom Krugh using electron spin resonance and nuclear magnetic resonance techniques (5). A spin-labeled derivative of ATP was prepared in which a nitroxide, bearing a free electron, was attached to the 6-amino

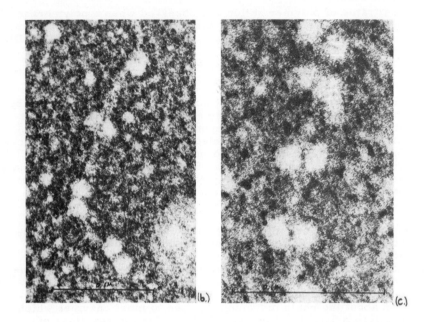

Figure 7

Figure 8

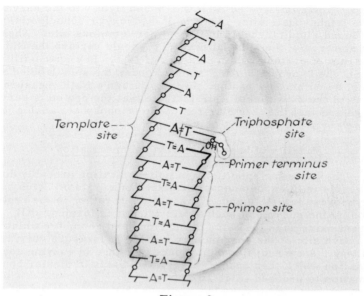

Figure 9

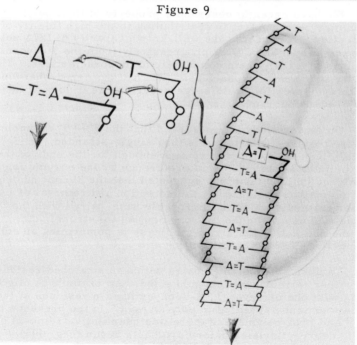

Figure 10

group. The spin-labeled ATP was bound tightly to the enzyme in the triphosphate site. AMP (a 3'-hydroxyl ribonucleotide) was bound simultaneously in the primer terminus site. The free electron of the spin-labeled ATP should perturb the NMR spectrum of protons within a radius of 10 Å. In a space-filling model, the 2-proton of AMP stacked under the spin-labeled ATP should be within this radius. In fact, Krugh's NMR measurements of the 2-proton of AMP indicate that the proton is perturbed and that the free electron of the triphosphate analog is 7.1 Å away from it.

Fidelity of base-pairing in polymerization

The basis for specificity in polymerization probably does not reside in the spontaneous formation of a Watson-Crick base pair between template and triphosphate but rather in the recognition by the enzyme of such a pair when it is formed. All correct pairs have the same dimensions and geometry which the active site of the enzyme may be able to measure very accurately. A correct fit might induce a change in enzyme conformation which then enables a series of steps in diester bond formation to proceed (Fig. 10).

We have recently observed a property of DNA polymerase which may be important in contributing to the high fidelity of replication. It concerns the well-known capacity of DNA polymerase to hydrolyze DNA progressively from the 3'-hydroxyl terminus of a DNA chain completely to mononucleotides. This activity is a 3'→5' exonuclease. We had wondered what function this degradation could serve occurring as it does at the very growing point of the chain.

On closer study, we observed that the 3'→5' exonuclease activity requires DNA that is either single-stranded, or if helical, then at a temperature high enough for the ends of the DNA helix to be frayed. This preference by the enzyme for an unmatched primer terminus suggested that the 3'→5' nuclease function may be designed for the detection and removal of a primer terminus which is improperly base-paired with the template. Recent studies by Doug Brutlag (6) strongly support the notion that the 3'→5' nuclease is performing an editing function in polymerization.

Various homopolymer pairs with a mismatched residue in the primer terminus were synthesized. An example is oligo dT_{260} with one or several ^{3}H-deoxycytidylate residues at the primer terminus annealed to poly dA_{4000}. In the presence of dTTP and DNA polymerase there was quantitative removal of all the deoxycytidylates before synthesis began (Fig. 11); deoxythymidylate residues subjacent to deoxycytidylate were

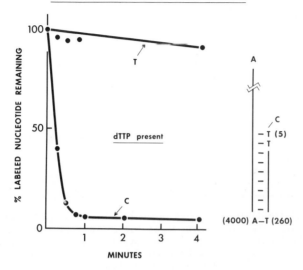

Figure 11

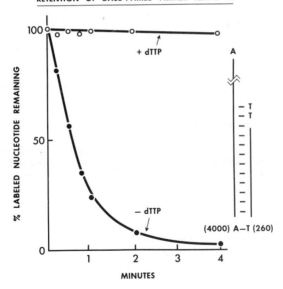

Figure 12

quantitatively retained. Analogous experiments with other polymer pairs establish that the enzyme invariably removes mismatched residues at the primer terminus before proceeding with polymerization. With a properly matched terminus, and with the triphosphate supplied that is appropriate for basepairing with the template, hydrolysis was not detectable (Fig. 12). Thus, the polymerization complex preempts the primer terminus making it unavailable for hydrolysis.

Thus it seems that the extraordinary accuracy of DNA replication may be a consequence of base pair recognition by the polymerase not only at the step of adding a nucleotide to the primer chain but also at a proof-reading step right after the primer terminus has been extended.

The 5'→3' nuclease, an entirely distinct "repair" function of DNA polymerase

Our attention up to this point has focused quite reasonably on how DNA polymerase manages the growth of a DNA chain. We did not realize until recently that the single polypeptide chain of DNA polymerase is shaped into two distinct enzymes with two separate functions and that these functions can be nicely coordinated. So far as I am aware, such an architectural design of a polypeptide chain is unique among enzymes and suggests an ancient fusion of two gene products.

The distinct second function of DNA polymerase is not concerned with the primer terminus. It degrades DNA progressively from the 5' end of a chain and can therefore function as a 5'→3' nuclease (Fig. 13). The additional distinguishing features of 5'→3' nuclease action are that: (1) the cleavage while predominantly at the terminal bond also occurs at subterminal bonds, (2) the DNA must be double stranded, and (3) concomitant DNA synthesis enhances the rate ten-fold and increases the percentage of oligonucleotides among the products.

Of what use could this degradative property of the enzyme be? The properties of the 5'→3' nuclease suggested two possible functions. One might be to facilitate the start of replication at a nick by clearing away the 5' strand that occupies the template above the triphosphate site (Fig. 13). Upon examining the kinetics of early replication at a nick, we observed a burst of nuclease action which is exclusively 5'→3' and which exactly matches the extent of synthesis. In effect, there is in the first few minutes of in vitro replication, translation of the nick along the template. Thereafter, by a mechanism which still is not clear to us, the 5'-strand is displaced and conserved.

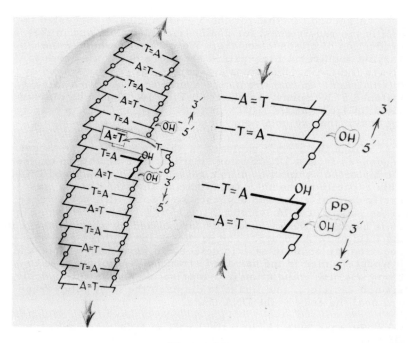

Figure 13

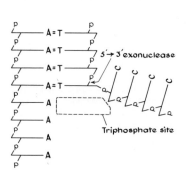

Figure 14

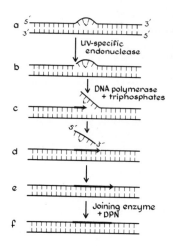

Figure 15

A second function for the 5'⟶3' nuclease suggested itself in the requirement for double-stranded DNA and in the production of oligonucleotides. We wondered whether the enzyme requires a base-paired structure but is also able to cleave the chain subterminally when a chain end is frayed or mismatched. In that event, the synthetic DNA with a mismatched 5' terminus, as shown in Fig. 14, should be cleaved to yield $(pC)_4pT$ and $(pC)_4(pT)_2$ but not $(pC)_{1-4}$. This is exactly the result that we obtained.

Since the 5'⟶3' nuclease is able to excise mismatched sequences from a DNA duplex, it might then be able to excise the distorted pyrimidine dimer regions in UV-irradiated DNA. This capacity of the nuclease to excise thymine dimers was easily demonstrated and suggests a role for DNA polymerase in the repair of DNA lesions (Fig. 15)(7). The scheme in Fig. 15 is basically the same as that formulated by many workers in the field of DNA repair. The participation of three enzymes is required: an endonuclease to detect a mismatched regions and introduce a nick in the damaged strand, DNA polymerase to carry out nick translation (polymerization and 5'⟶3' hydrolysis and excision), and ligase to displace the DNA polymerase and seal the ends of the DNA chain.

Support for a physiological role of DNA polymerase in DNA repair has come recently from studies of an E. coli mutant isolated by De Lucia and Cairns (8) and called pol A_1. This mutant which appears to lack DNA polymerase also shows an increased susceptibility to damage by UV light, X-rays and DNA-alkylating agents, such as methyl methanesulfonate. Among additional mutants isolated on the basis of their sensitivity to methyl methanesulfonate, deficiency or absence of DNA polymerase is common.

<u>One polypeptide chain - two enzymes;
two polypeptide chains - one enzyme</u>

Over a year ago when the late Maurice Atkinson was in our laboratory, he analyzed a sample of DNA polymerase on SDS-polyacrylamide gel. There was a single band as expected, but the molecular weight was 75,000. This was startling. He ran another sample freshly thawed from liquid nitrogen. This time, the molecular weight was properly 109,000. The first sample was an active preparation. But it had been thawed for a few weeks and had developed a slight turbidity. The mystery of the light DNA polymerase was solved easily. Proteolytic cleavage by a bacterial contaminant had split the enzyme. The polymerase activity was there, but the 5'⟶3' nuclease was gone.

Controlled cleavage of DNA polymerase by proteases, including subtilisin and trypsin, produces two products: a large fragment of 75,000 and a small fragment of 36,000 (9). Hans Klenow of Copenhagen had made these observations independently (10). The fragments have been isolated in pure form by Peter Setlow (11) and have entirely distinct activities.

The large fragment contains the single disulfide and the single sulfhydryl group of the enzyme. It contains the triphosphate binding site and the primer terminus binding site. It retains the polymerase and the $3' \longrightarrow 5'$ nuclease functions associated with the primer terminus (Fig. 16). What the large fragment lacks is the $5' \longrightarrow 3'$ nuclease activity. As a result, polymerization proceeds without detectable degradation. In primed, poly dAT synthesis, the triphosphate substrates are converted quantitatively to the polymer; synthesis of dAT, de novo, does not take place.

The small fragment shows none of the primer terminus functions, but it retains the $5' \longrightarrow 3'$ nuclease activity (Fig. 16). However, there is no stimulation of this activity by substrates which support polymerization because the polymerase function is absent.

Of importance to all of us working with DNA is enlarging our arsenal of nucleases to do specific jobs. The separated nuclease activities of polymerase promise to be useful reagents. The large fragment degrades DNA completely to mononucleotides and specifically from the 3' end. Because it acts on both single- and double-stranded DNA, the large fragment offers an advantage over the nucleolytic reagents currently used for total hydrolysis of DNA from the 3' end. Exonuclease I is blocked by any secondary structure, whereas exonuclease III is blocked at denatured regions in DNA. The large fragment can also be specifically inhibited in the presence of components that support polymerization. The small fragment, on the other hand, degrades DNA specifically from the 5' end. It requires a region of base-pairing and can excise mismatched regions as oligonucleotides. It is inert on single strands (Fig. 17).

When the small and large fragments are mixed together, there is no obvious physical interaction. But in the presence of nicked DNA, the fragments bind to the DNA and form a functional complex which resembles the intact enzyme in all respects.

DNA polymerase thus contains within <u>one polypeptide chain, two discrete enzymes.</u> They are linked by a peptide region susceptible to cleavage by proteases. For lack of any

Properties of the Intact Enzyme and the Proteolytic Fragments

Reaction*	Intact enzyme	Large fragment	Small fragment
DNA synthesis	148	396	<3
3'→5' nuclease	16	20	<0.2
5'→3' nuclease	27	<1	37
5'→3' nuclease (concom. synthesis)	260	-	37

* Moles nucleotide polymerized or hydrolyzed/mole enzyme/min

Figure 16

Distinction between the Nuclease Activities of DNA Polymerase

	3'→5' nuclease	5'→3' nuclease
Start of action	3' End (primer terminus)	5' End (OH, P, PPP)
Nucleotide products	Mono - 100%	Mono - 80% Di, tri - 20%
Excision of oligomers	No	Yes
DNA structure	Frayed, single strand	Double strand
Influence of concom. polymerization	Blocked	Enhanced 10-fold
T4 DNA polymerase	Yes	No

Figure 17

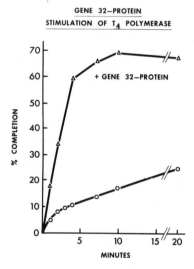

GENE 32–PROTEIN
STIMULATION OF T$_4$ POLYMERASE

+ GENE 32–PROTEIN

Figure 18

detailed knowledge about the three-dimensional structure of the enzyme, we cannot visualize how these two components are related. Nevertheless, the two polypeptide fragments have the capacity to form, at a DNA nick, a functional complex in which they must be closely related. This suggests physical proximity of the multiple functions and sites of the <u>two polypeptide chains</u> to form the active center of <u>one enzyme</u>.

Physiological role of DNA polymerase

The versatility of DNA polymerase <u>in vitro</u> is impressive. How are these resources of the enzyme exploited in the cell? The pol A_1 mutant of <u>E. coli</u>, which appears to lack DNA polymerase offers an opportunity for such an evaluation. As mentioned earlier, the deficiencies of this mutant in coping with agents that damage DNA suggest that DNA polymerase serves in excising lesions and replacing defective DNA, perhaps as suggested earlier (Fig. 15). Another role for the enzyme is indicated by Helinski's (12) recent experiments. The circular DNA plasmids which carry the genes for colicins are not sustained in pol A_1 mutants and would seem to require DNA polymerase for their replication. DNA polymerase may also share in chromosomal replication to an extent that can be replaced by another replicating system. For example, a recombination-deficient mutant, called <u>rec A</u>, can be made nonviable when it is also made pol A_1 deficient; presumably some vital function is either shared or can be performed by the products of these two genes and this cannot occur in the double mutant.

If we assume that the pol A_1 mutant has no functional DNA polymerase <u>in vivo</u>, then some other DNA replicating system must be serving this organism. Recently a DNA polymerizing enzyme which may be membrane bound has been found in the mutant (13). The enzyme has been solubilized, partially purified and can be distinguished from DNA polymerase by its susceptibility to sulfhydryl blocking reagents, to inhibition by salt and by its insensitivity to <u>E. coli</u> DNA polymerase antiserum. The resemblance to DNA polymerase in its basic requirements for template, for deoxyribonucleoside triphosphates and the mechanism of chain growth has prompted a designation of DNA polymerase II for this enzyme. The level of this new enzyme activity in extracts of mutant or wild type cells is only about 10% of that measured for DNA polymerase I. We are eager to hear more about the properties of this new polymerase and what contribution it may make to replication <u>in vivo</u>.

DNA is a dynamic molecule. It undergoes replication and transcription. It is also modified, nicked, sealed, excised, degraded, patched and rearranged. No wonder there are 19 genes in T_4 phage involved in aspects of DNA function. No wonder there are about 20 genes for DNA synthesis mapped in B. subtilis and surely as many in the E. coli chromosome.

It has been clear for some time that there must be a unique complex of several enzymes, a ribosomal equivalent for DNA replication. Such a complex would resemble the fatty acid synthetase in E. coli in being readily dissociated and dispersed in extracts. It was a boon having soluble polymerases, nucleases and ligases, but the search now for pieces to put the "replication ribosome" back together has been difficult. One of the pieces may be similar to the product of gene 32 in phage T4. The gene 32 protein, isolated and studied by Bruce Alberts (14) binds to single-stranded DNA (1). By extending the random coil of DNA, gene 32 protein greatly facilitates recomination. Gene 32 protein also stimulates replication by T_4 DNA polymerase (Fig. 18)(15). The rate of DNA replication was increased up to 10-fold and the replacement of a large section of λ DNA was completed in 4 minutes as compared with 200 minutes in its absence.

Membranes, a biochemical frontier

I want to close this travelogue with a brief statement about membranes. They represent to me one of the most intriguing frontiers of biochemistry.

Enzymes have three faces. Most familiar is their catalytic face. Lately we have come to recognize that many enzymes have a second aspect, their control or regulatory face. It seems certain to me now that most enzymes also have a third face. In this period of social concern for ecology, community and our neighbor, we might call this third aspect of an enzyme, the social face. This third face is the one that recognizes other proteins and certain of the lipids among the variety that form membranes.

Hydrophobic interactions dominate the third face and social behavior of membrane proteins. Our problem is to isolate these proteins in order to see and appreciate their catalytic and control faces. Clearly, the refined methods available for separating soluble proteins are inadequate for this job. Many of us were trained under the Warburg dictum: "Where structure begins, biochemistry ends". As biochemists, we obediently discarded the material that sedimented to the bottom of a tube. We called it debris or worse. Yet it has been obvious for many years that the most fascinating machinery of the cell is contained in that debris. The debris contains cell envelopes, the plasma membrane that surrounds the cell and the membranous organelles found inside the cell: mitochondria and microsomes. Quite often the chromosome and enzymes associated with it are in the sediment.

To do biochemistry on membranes, we must devise new methods that can separate membrane proteins. I wish I could describe significant progress to you but the best I can do now is to indicate our attitudes and aspirations. Put simply: purify enzymes, trust primary structure, and be patient.

First of all, we assume that membrane proteins have catalytic functions until proven otherwise. The history of protein chemistry records a succession of mistakes in lumping proteins into functionless groups: the albumins, the globulins, the histones, and now the structural membrane proteins. We must purify membrane proteins by assaying for their functions. It is simplest to let function lead us to structure; it is exceedingly difficult when given a structure to divine its function.

Second of all, when we resort to fractionation procedures that denature proteins, we must be careful to keep the primary structure intact. We must then assume that the native functional state of the protein can be restored from its primary structure. The use of naturally occurring lipids and other membrane components may be essential to achieve renaturation. Careful control of temperature, time, ionic conditions and sequence of interactions must be exercised.

Finally we must be patient. In the past 25 years the time required to purify a soluble protein has been reduced by a factor of ten. Today we must be prepared to invest 10 man years in the purification of membrane protein just as we did 25 years ago for soluble protein.

I will close by paraphrasing an old saw of Efraim Racker: Don't waste clean thinking on dirty enzymes, even if they're in membranes.

REFERENCES

1. A. Kornberg, Science, 163 (1969) 1410.

2. J. Griffith, J.A. Huberman and A. Kornberg, J. Mol. Biol., In press.

3. P.T. Englund, J.A. Huberman, T.M. Jovin and A. Kornberg, J. Biol. Chem., 244 (1969) 3038.

4. J.A. Huberman and A. Kornberg, J. Biol. Chem., 245 (1970) 5326.

5. T. Krugh. In preparation.

6. D. Brutlag and A. Kornberg. In preparation.

7. R.B. Kelly, M.R. Atkinson, J.A. Huberman and A. Kornberg, Nature, 224 (1969) 495.

8. P. De Lucia and J. Carins, Nature, 224 (1969) 1164.

9. D. Brutlag, M.R. Atkinson, P. Setlow and A. Kornberg, Biochem. Biophys. Res. Comm., 37 (1969) 982.

10. H. Klenow and I. Henningsen, P.N.A.S., 65 (1970) 168; H. Klenow and K. Overgaard-Hansen, FEBS Letters, 6 (1970) 25.

11. P. Setlow, D. Brutlag and A. Kornberg. In preparation.

12. D.T. Kingsbury and D.R. Helinski, Biochem. Biophys. Res. Comm., in press.

13. T. Kornberg and M.L. Gefter, Biochem. Biophys. Res. Comm., 40 (1970) 1348; R. Knippers, Nature, 228 (1970) 1050; D. Smith, H. Schaller and F. Bonhoeffer, Nature, 226 (1970) 1313.

14. B.M. Alberts and L. Frey, Nature, 227 (1970) 1313.

15. J.A. Huberman, A. Kornberg and B.M. Alberts. In preparation.

ON THE MECHANISM OF DNA SYNTHESIS IN BACTERIA

R. WERNER
Department of Biochemistry
University of Miami School of Medicine

Abstract: Meaningful information on the in vivo DNA replication process can only be obtained from steady-state incorporation studies using thymine rather than thymidine as radioactive precursor. The addition of thymidine to E. coli causes a tremendous expansion of the intracellular deoxyribonucleotide pools thus leading to the results obtained by Okazaki and coworkers. Very short pulse-labeling with ^{3}H-thymine produces only large DNA whereas in a similar pulse with ^{3}H-thymidine given to the same cells about 70% of the radioactivity can be found in small DNA fragments. Longer pulses with thymine produce up to 50% small DNA even under steady-state conditions. The results indicate that Okazaki fragments are not intermediates in DNA replication but possibly arise from single-strand interruptions in the daughter helices that might be required for rotation. Thymidine is very rapidly converted to TTP. Its addition to bacteria presumably stimulates a repair reaction which normally occurs at a much more reduced level. Studies on the labeling of the various thymidine nucleotide pools indicate that most of the intracellular deoxyribonucleoside triphosphates are not directly used for DNA replication which apparently uses a different precursor.

Despite the efforts of many laboratories to elucidate the mechanism of DNA synthesis, little progress has been made in understanding the basic replication process. With the recent isolation of a mutant of E. coli that lacks the Kornberg DNA polymerase (1),-- the only enzyme previously known to synthesize DNA--, it is no longer necessary to postulate models of discontinuous DNA replication that would allow the Kornberg polymerase or similar enzymes the simultaneous elongation of both 3'- and 5'-terminated DNA strands (2,3). Compared with repair synthesis or RNA transcription, --both processes that synthesize only one polymer chain at a time--, the replication of DNA with its two-strandedness and unwinding problems is a completely different reaction. We should therefore not discard the possibility of a fundamentally different reaction mechanism, including perhaps a different precursor.

Okazaki's observation that short pulses of ^{3}H-thymidine are incorporated almost exclusively into short polynucleotide chains, which only later are joined to long DNA, has led to the widely held belief that DNA synthesis is discontinuous (4). However, more recent work on the synthesis and properties of Okazaki pieces has revealed two significant paradoxes. The first paradox is that clearly more than half the label incorporated by a bacterium such as <u>Bacillus subtilis</u>, during a very short pulse of ^{3}H-thymidine, is found in small pieces, yet the same pieces have been reported to be complementary to only one of the two parental strands (5) and so, if both parental chains are being copied at the same rate, should not have received more than half the incorporated label. The second paradox concerns the number of pieces at each growing point: when the number is calculated from the relation between the overall rate of incorporation of label and the maximum amount of label that can be found in short pieces (unpublished calculations) or from comparing the time required for pieces to accumulate with the rate at which the replication fork progresses (6), it appears that there must be very few pieces per growing point that are engaged in DNA synthesis; however, if calculated from the step time for the creation of TMP residues in DNA, it seems as if the chain growth rate is so much less than the rate of movement of the growing point that there must be many growing pieces at each fork (7).

Most published experiments on the labeling of Okazaki pieces have used labeled thymidine because it is quickly incorporated into DNA by wild-type cells and is apparently a preferred source of DNA-TMP for thymine-requiring cells growing in the presence of thymine. Despite the preference for thymidine the incorporation of thymine is not affected by the presence of thymidine. In fact, the addition of thymidine can actually boost the incorporation of ^{3}H-thymine. From this it seems that thymidine and thymine go two different pathways into DNA and that the incorporation of thymidine may not be representative of normal DNA synthesis. For this reason I have studied the labeling of Okazaki pieces under steady-state conditions using thymine rather than thymidine as label. The results are rather different from the results of labeling with ^{3}H-thymidine and lead inexorably to a new model for the relation between Okazaki pieces and DNA synthesis. This model sees the creation of Okazaki pieces as an event that follows, rather than accompanies, replication; it demands that replication of DNA and the repair synthesis associated with the sealing of the Okazaki pieces occur by distinct mechanisms, and the model incidentally resolves the paradoxes mentioned earlier.

In the second half of my talk, I will present some studies on the labeling of the various thymidine nucleotide pools using thymine or thymidine as precursor. The results suggest that deoxyribonucleoside triphosphates may not be the precursors for DNA replication.

DNA Synthesis under Steady-State Conditions

Unless stated otherwise, the experiments reported in this paper were carried out with Escherichia coli strain 15 thy⁻, arg⁻, met⁻, tryp⁻ (E. coli TAMT) growing at 14°C in the presence of 5 μg/ml thymine. Under these conditions, E. coli TAMT has a generation time of about 400 minutes (8).

To study the kinetics of labeling of large and small DNA, bacteria were pulse-labeled with ^{3}H-thymine for periods ranging from 15 seconds to 10 minutes. The sedimentation analysis in alkaline sucrose gradients of the labeled DNA shows that thymine, like thymidine, is incorporated into both small and large DNA (Fig. 1). However, in contrast to thymidine, which is known to label Okazaki pieces preferentially, if not exclusively, less than 50% of the incorporated thymine appears in Okazaki pieces, even after the shortest labeling period. When pulse-labeling for periods between 15 seconds and one minute, the presence of cold thymidine has no effect on the distribution of label between large DNA and Okazaki pieces except that the total incorporation of ^{3}H-thymine into both kinds of DNA is increased by about 100%. (Thymine also labels a considerable amount of TCA-insoluble material that stays at the top of the gradient. This material is not DNA as judged from its buoyant density in CsCl; its possible relation to DNA synthesis is still unclear.

Since shortening the pulse appeared to reduce rather than increase the relative amount of label in Okazaki pieces, I labeled bacteria with tritiated thymine for only 10 seconds. Under these conditions only about 20% of the incorporated label is found in Okazaki pieces (Fig. 2A). Furthermore, when cold thymidine is given at the same time, the incorporation of ^{3}H-thymine into Okazaki pieces is completely suppressed (Fig. 2B). (The possibility that the presence of Okazaki pieces is being obscured by the very large amount of non-sedimenting material at the top of the gradient can be ruled out because, as shown in Fig. 2C, a 20-second pulse of ^{3}H-thymine given under the same conditions produced the familiar Okazaki peak.) Thymidine behaves very differently from thymine; two thirds of the label incorporated during a 10-second pulse appeared in small DNA pieces (Fig. 2D).

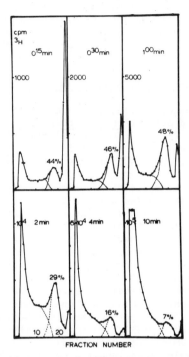

Fig. 1. Sedimentation analysis in alkaline sucrose gradients of DNA isolated from E. coli 15 TAMT labeled with ^{3}H-thymine under steady-state conditions. The bacteria were grown for several generations at 14°C in the following medium: 0.01M NaCl, 0.005M KCl, 0.02M NH$_4$Cl, 6x10^{-4}M KH$_2$PO$_4$, 2x10^{-4}M Na$_2$SO$_4$, 0.001M MgCl$_2$, 10^{-4}M CaCl$_2$, 10^{-5}M FeCl$_3$, 0.02M N-tris-(hydroxymethyl) methyl-glycine (pH 7.4) and 0.025M sodium lactate, supplemented with 5 μg/ml thymine and with 40 μg/ml of each arginine, methionine and tryptophan. The bacteria were chilled to 0°C under aeration, centrifuged, resuspended in fresh medium at a concentration of 2x10^{10} cells/ml and stirred in an open beaker at 14°C for 30 minutes. At time zero an equal volume of fresh medium was added containing 5 μg/ml ^{3}H-thymine (13C/mM). 0.5 ml samples were taken at the indicated times into 2 ml acetone of -30°C. The cells were centrifuged and resuspended in 1 ml 0.3M NaOH-0.1M EDTA. After adding 0.1 ml 20% sarkosyl-NL 97 the mixture was heated to 65°C for 5 minutes and then layered on an alkaline sucrose gradient (5-20% sucrose, 0.3M NaOH, 0.01M EDTA, 0.5M NaCl) and centrifuged in a Spinco SW-27 rotor at 26,000 for 16 hours. 24-fractions were collected and precipitated with 5% TCA in the presence of some bovine serum albumin.

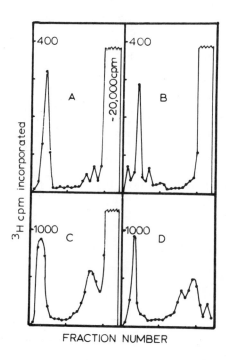

Fig. 2. Sedimentation analysis in alkaline sucrose gradients of DNA labeled under the following conditions: (A) 10 second pulse with ^{3}H-thymine; (B) 10 second pulse with ^{3}H-thymine in the presence of 1 μg/ml cold thymidine; (C) 20 second pulse with ^{3}H-thymine; (D) 10 second pulse with 0.1 μg/ml ^{3}H-thymidine. (Other conditions as in Fig. 1 except that centrifugation was at 27,000 rpm for 20 hours.)

These experiments show that if thymine is used as label, most of the label appears first in large DNA. This immediately suggests that large DNA is the precursor of Okazaki pieces (rather than the other way around), that DNA replication is not discontinuous and that the creation of Okazaki pieces serves some other function. The preferential incorporation of thymidine into Okazaki pieces suggests that thymidine is a rather specific precursor for the reaction that joins up the pieces. This interpretation would explain the fact that, when compared to thymine, the incorporation of thymidine into DNA accounts for only a small fraction of the total DNA synthesis. The preferential incorporation of thymine into the replication complex has also been observed by Fuchs and Hanawalt (9); after a 10-second pulse at 37°C, more than 95% of the incorporated thymine was found associated with the fork as compared with only 20% of the incorporated thymidine.

A New Model of DNA Replication

As a working hypothesis I propose the following model of DNA replication (Fig. 3): DNA is replicated continuously at the fork through simultaneous elongation of both new strands. Although the precise mechanism of this reaction is still obscure its preference for thymine rather than thymidine seems to indicate that the replication process is fundamentally different from repair synthesis and possibly involves a different precursor pool.

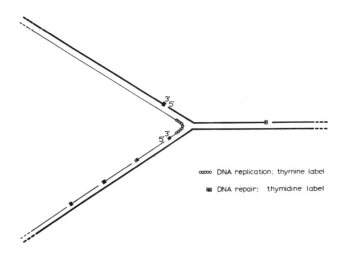

Fig. 3. Model of structure of replicating fork. (Explanations in the text.)

Okazaki pieces arise from single strand interruptions that are being introduced into one of the two new DNA strands by specific nucleases which recognize short nucleotide sequences spaced at about 3,000 nucleotide intervals along the bacterial chromosome. (These enzymes may not distinguish between parental and new strands and might therefore produce similar nicks in the parental strand of the other daughter duplex.) Thus the average

distance of the first chain interruption from the fork is about
1,500 nucleotides or half the size of an Okazaki piece (6).
At 14°C the growing point takes 10 seconds to move along this
length of DNA. Therefore, after pulse-labeling with ^{3}H-thymine
for periods longer than 10 seconds, the amount of label in
Okazaki pieces will approach 50%. The average lifetime of the
Okazaki pieces at the growing point must be about one minute
because the relative amount of label in large DNA starts to increase after pulse-labeling for longer than one minute (Fig. 1).
During that time the fork travels a distance of 9,000 nucleotides or three Okazaki pieces. The number of chain interruptions in the newly synthesized DNA should accordingly be four.

Between different bacterial strains and under different
growth conditions the number of strand interruptions in the
newly synthesized DNA seems to vary. When <u>E. coli</u> 15 TAMT is
grown in the presence of much higher thymine concentrations
(20-50 µg/ml) and pulse-labeled with either ^{3}H-thymine or
^{32}P-phosphate, little, if any, label is found in Okazaki pieces.
This suggests that the accumulation of strand interruptions in
the new DNA is a consequence of a relatively slow repair process caused by thymine deficiency and does not occur under normal conditions. (Under the same conditions thymidine is still
incorporated into Okazaki pieces, indicating that there may be
a second class of pieces which is not labeled by thymine.)

Little can be said about the mechanism of chain elongation at the growing point except that DNA synthesis is probably
not discontinuous. As thymine is found nearly exclusively in
large DNA after a very short pulse, it seems possible that the
two new DNA strands are covalently linked at the apex (as indicated in Fig. 3). Elongation of the new DNA strands would then
have to occur by insertion of nucleotide pairs at the apex.
Such an arrangement would have the advantage of making the
replication site inaccessible for repair enzymes. A similar
mechanism was recently proposed by Morgan (10). However, we
cannot exclude the possibility of two separate growing chains,
the small piece engaged in DNA replication at its 5'-end being
hidden under the large peak of nonsedimenting radioactive material.

The sealing of the single strand interruptions by polynucleotide ligase is presumably preceded by some repair synthesis. It seems reasonable to assume that it is this reaction
into which thymidine goes preferentially. If so, labeling with
^{3}H-thymidine should allow us to detect whether similar strand
interruptions occur in one of the two parental strands. The
following experiment shows that this is indeed the case:
<u>E. coli</u> B/r (autotroph for thymine) was grown for several gener-
ations in a "heavy" ^{13}C,^{15}N-containing medium, transferred to

"light" $^{12}C, ^{14}N$-medium for about a third of the generation time, and then pulse-labeled with ^3H-thymidine for 10 seconds at 14°C. Large and small DNA were separated by alkaline sucrose gradient sedimentation and banded separately in alkaline CsCl density gradients. Figure 4 shows that nearly all thymidine incorporated into both large and small DNA is covalently connected to heavy parental DNA. The slight shift to lighter density of the Okazaki pieces (Fig. 4A) suggests that a considerable region has been labeled with thymidine. It is possible that the sudden expansion of the TTP pool caused by the addition of thymidine leads to a more extensive replacement of nucleotides than the sealing reaction normally requires.

As mentioned before, we may have to distinguish between two classes of Okazaki pieces, those labeled exclusively with thymidine and those labeled with thymine or thymidine. From Fig. 4 it is clear that most of the Okazaki pieces that are labeled with thymidine come from parental strands. Since E. coli B/r, like strain 15 TAMT growing in the presence of 20-50 µg/ml thymine, probably has no Okazaki pieces at the growing point, most thymidine-labeled Okazaki pieces must arise from other regions of the chromosome. It is tempting to speculate that they represent cistrons that are engaged in transcription. The necessity of single strand interruptions for late mRNA synthesis in phage T4-infected E. coli was recently demonstrated by Riva et al. (11). It seems difficult to imagine how genes in a circular replicating E. coli chromosome (12) could be transcribed into RNA unless similar strand interruptions are introduced at both ends of the cistron thus allowing the DNA helix to rotate through the bulky RNA and protein synthesizing apparatus (rather than vice versa).

In contrast to the Okazaki pieces that are labeled only by thymidine, thymine-labeled pieces arise from new DNA. They can only be detected in thymine-requiring bacteria growing at low concentrations of thymine, and probably arise from the accumulation of specific interruptions in one of the two new DNA chains. It seems plausible that these strand interruptions function as swivel points for the rotation of the daughter helices during the replication of the circular chromosome. Similar nicks presumably occur in the parental strand of the other daughter helix. It is even possible that the interruptions in the parental strand are created ahead of the fork, thus providing the necessary swivel point for the unwinding of the parental double helix, and persist through the replication region into one daughter helix where they are eventually sealed.

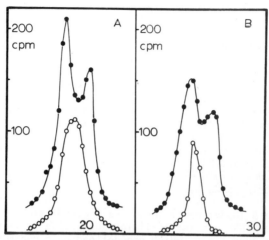

Fig. 4. Density analysis in CsCl of DNA pulse-labeled with ^{3}H-thymidine. <u>E. coli</u> B/r was grown for several generations in a "heavy" medium containing $^{15}NH_4Cl$ and ^{13}C-glucose. The cells were washed, resuspended in normal "light" medium (see Fig. 1), aerated for 15 min at 37°C, chilled to 0°C, centrifuged and resuspended in fresh "light" medium at a concentration of 1x10^{10} cells/ml. After another 30 min aeration at 14°C, 0.1 μg/ml of ^{3}H-thymidine was added. The reaction was stopped after 10 sec by the addition of an equal volume of cold acetone. After centrifugation and resuspension in tris-EDTA, the cells were lysed with lysozyme, pronase and sarkosyl, and the DNA sedimented through alkaline sucrose gradients. Fractions containing large and small DNA were pooled, dialyzed against 0.05M tris-0.01M EDTA (pH 7.4), concentrated and banded in CsCl. A, Okazaki pieces; B, large DNA. o-o-o, ^{3}H-label; •-•-•, ^{14}C-labeled "heavy" and "light" marker DNA.

Nucleotide Pool Studies

The ability of the cell to differentiate between thymine and thymidine, and to use them separately for two distinct reactions, DNA replication and repair synthesis, strongly suggests that the two reactions draw their immediate precursors from different pools.

From Kornberg's work on the DNA polymerase of E. coli we can assume that in vivo repair synthesis uses deoxyribonucleoside triphosphates as precursors. However, it has never been shown whether the same precursors are also being used for DNA replication. The only information on the nature of the replication precursor comes from the work of Price et al. (13), who have convincingly shown that the deoxyribonucleoside is already connected to its 5'-phosphate before it enters DNA; all DNA precursors must therefore contain deoxyribonucleoside 5'-monophosphate, presumably in some activated form. As it seemed somewhat unlikely that a bacterium contains two separate pools of deoxyribonucleoside 5'-triphosphates that would be labeled differently by thymine and thymidine, I studied the flow of thymine and thymidine through the various intracellular nucleotide pools into DNA, in the hope to find out whether DNA replication possibly uses precursors different from deoxyribonucleoside triphosphates.

Figure 5 (upper part) shows the incorporation of ^{3}H-thymine into both DNA and intracellular nucleotide pools under steady-state conditions, which are defined as a dynamic equilibrium between all precursor pools and are indicated by the linear uptake of thymine and the parallel final rate of incorporation into DNA. The difference between the two curves, uptake and incorporation, represents the size of the combined intracellular thymidine nucleotide pools.

With all pool sizes remaining constant during the incorporation of ^{3}H-thymine we expect the immediate precursor of DNA to reach its maximum radioactivity at the same time as, or slightly before, the rate of incorporation of thymine into DNA becomes maximal. As can be seen from Fig. 5, TMP is maximal within less than one minute, much too fast to be the immediate DNA precursor. However, both TDP and TTP reach their maximal levels after about 10 min, which is the same time the rate of incorporation into DNA becomes maximal. Both nucleotides must, therefore, be considered potential DNA precursors.

Under non-steady-state conditions the situation becomes somewhat more complicated. The presence of unlabeled thymidine (Fig. 5, lower part) causes a dramatic expansion of all thymine-derived nucleotide pools. TMP increases more than 20-fold over the steady-state pool size, and both TDP and TTP pools expand about 4-fold. Furthermore, the maximal rate of incorporation of thymine into DNA is reached much sooner than under steady-state conditions. (The reduction of the incorporation rate after 10 min is probably caused by the dilution of ^{3}H-thymine with cold thymine produced from thymidine by thymidine phosphorylase.) (14).

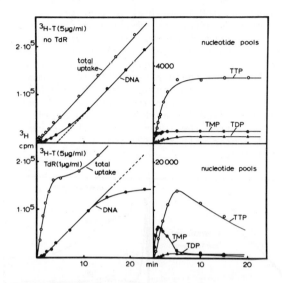

Fig. 5. Incorporation of ^{3}H-thymine into DNA and nucleotide pools under steady-state conditions (upper half) and in the presence of cold thymidine (lower half). Conditions of labeling were the same as described in Fig. 1. Uptake into the precursor pools (-o-o-) (left side) was determined by taking samples into a 200-fold excess of medium containing cold thymine and then adding an equal volume of 10% TCA 30 min later. Incorporation into DNA (-•-•-) was measured by taking samples directly into 5% TCA. Incorporation of counts into nucleotide pools (right side) was determined by taking samples into TCA (final concentration 5%). After extracting the TCA with ether the nucleotides were separated by two dimensional thin layer chromatography on PEI sheets (Brinkman, Westbury, N.Y.). 1st dimension: 1.6M LiCl; 2nd dimension: 4M ammonium formate, pH 3.4.

To follow the fate of thymidine in these experiments, two identical cultures of bacteria were exposed to a mixture of thymine and thymidine, only one of which was radioactively labeled in each culture. The specific activities of both labels were identical so that quantitative comparisons between the thymine- and thymidine-derived nucleotide pools were made possible. The results of this experiment are shown in Fig. 6.

The initial rate of thymidine incorporation into DNA accounts for only about 10% of the total DNA synthesis as measured by the thymine incorporation, and this rate levels off rather quickly. Looking at the thymidine-derived nucleotide pools we notice that TMP is maximal in the very first sample taken only 10 seconds after the addition of ^{3}H-thymidine. With regard to its preferred uptake by the cell it seems possible that thymidine is phosphorylated to TMP at the bacterial membrane, perhaps concomitantly with its transport into the cell. Furthermore, the rather rapid rise of TTP to its maximal level within less than one minute, sooner than TDP reaches its maximum, seems to suggest that thymidine-derived TMP does not mix with the thymine-derived TMP-pool but is converted to TTP directly without a free TDP intermediate; the so expanded TTP pool then equilibrates with the TDP pool which thus receives thymidine-label last.

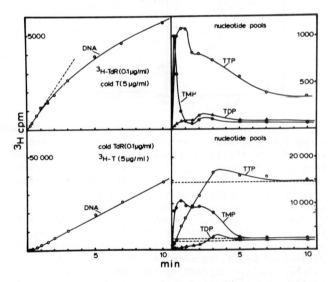

Fig. 6. Incorporation of thymine and thymidine into DNA and nucleotide pools both precursors being present at the same time. Thymine-derived steady-state nucleotide pools, measured in a parallel culture growing only in ^{3}H-thymine, are indicated by dotted lines. The specific activity of both thymine and thymidine was 7.15C/mM. (Labeling conditions as in Fig. 1.)

Although the maximal rate of thymine incorporation is reached within 20 to 30 seconds, the steady-state level of the TTP pool is not reached until about 2.5 min. (In this experiment a lower concentration of thymidine was used than in the previous experiment; the thymine-derived nucleotide pools are consequently less expanded than shown in Fig. 5.) It seems possible therefore, that under these non-steady-state conditions thymine bypasses at least part of the preexisting unlabeled TTP pool on its way into DNA.

To quantitatively test this possibility I did the following experiment which allowed me to measure specific activities of all thymidine nucleotide pools at various times during labeling with ^{3}H-thymine in the presence of unlabeled deoxyadenosine, and to compare these specific activities to the rate of thymine incorporation into DNA. (Deoxyadenosine having the same effect as thymidine in pushing thymine into DNA (15) shows the additional advantage of not breaking down to thymine and thus not diluting the tritiated thymine.) A culture of 15 TAMT was divided into two parts. Part A was labeled with ^{3}H-thymine, part B was growing in cold thymine of the same concentration. After 16 min (enough time to completely label all nucleotide pools with ^{3}H-thymine) part A was diluted fourfold with fresh medium containing cold thymine and deoxyadenosine, and part B received ^{3}H-thymine and deoxyadenosine. Incorporation of label into both DNA and nucleotide pools was monitored.

Figure 7 shows the incorporation of label into DNA for both cultures. Whereas under steady-state conditions it takes 10 min to achieve full incorporation rate, the presence of deoxyadenosine reduces this interval to about 20 seconds. Similarly, after the shift from ^{3}H-thymine to cold thymine and deoxyadenosine the new rate (one fourth of the old rate) is established within seconds.

The nucleotide pools are shown in Fig. 8. The decrease of the existing nucleotide pools after the shift to cold medium as obtained from culture A (after correcting for the additional incorporation of ^{3}H-thymine at one fourth the original specific activity, which can be calculated from the values obtained from culture B)(Fig. 8, upper part) shows, when combined with the data on the newly labeled pools from culture B (Fig. 8, lower part), the proportion of old unlabeled to new ^{3}H-labeled nucleotides in each pool at various times during the labeling with ^{3}H-thymine.

Assuming that the pools are well mixed, the specific activity of the nucleotide that is on the main pathway into DNA should approach that of the medium within 20 sec, the time when the rate of incorporation of label into DNA becomes maximal. From Fig. 9 it is clear that the specific activities of both TDP and TTP do not reach that of the medium for a long time; at 20 seconds they have attained only about 30% of the maximal value, while incorporation of label into DNA is already maximal. We must conclude, therefore, that a large part, if not all, of the intracellular TDP and TTP is not in the main pathway to DNA. In contrast, TMP is at full specific activity within a few seconds. Its activity dips briefly to around 70% but regains the 100% level soon afterwards. Despite this dip in specific activity, which is apparently caused by some breakdown of preexisting TTP, most of the TMP appears to be on the main pathway to DNA. (It is conceivable that a similar dip also occurred in the incorporation rate but was not detectable by our methods, or that the new TMP pool was not mixed with the breakdown TMP.)

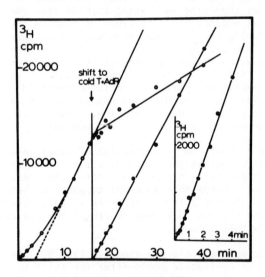

Fig. 7. Incorporation of ^{3}H-thymine into DNA under various conditions. A concentrated culture of E. coli 15 TAMT (2×10^{10} cells/ml) was split into two parts. At time zero an equal volume of fresh medium containing 5 μg/ml ^{3}H-thymine (15C/mM) was added to part A while part B received the same amount of medium containing cold thymine. After 16 min, part A was diluted 4-fold with fresh medium containing 5 μg/ml cold thymine and 0.2 μg/ml deoxyadenosine, part B was diluted 2-fold with medium containing 5 μg/ml ^{3}H-thymine (7.5C/mM) and 0.2 μg/ml deoxyadenosine. 200λ-samples were taken into TCA (final concentration: 5%) and 10 λ subsamples filtered for determining DNA counts. A,-o-o-o-; B,-●-●-●- (cpm of samples taken from 0 to 16 min were divided by four to correct for difference in cell concentration.)

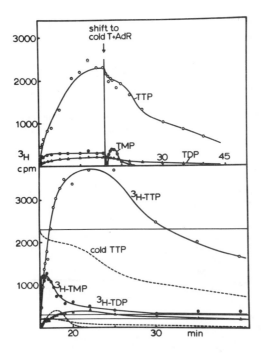

Fig. 8. Change in nucleotide pool composition during labeling with ^{3}H-thymine. The TCA samples from the experiment described in Fig. 7 were extracted with ether and chromatographed as described in Fig. 5. The upper graph shows the incorporation of ^{3}H-thymine into the nucleotide pools of culture A under steady-state conditions; after 16 min it shows the reduction of the labeled nucleotides in these pools caused by the close with cold thymine. (The values after 16 min were corrected for the additional incorporation of ^{3}H-thymine at one fourth the original specific activity. This was done by subtracting one fourth of the pool values obtained from culture B.) The lower graph shows the incorporation of ^{3}H-thymine into the nucleotide pools of culture B. The presence of unlabeled nucleotides, as obtained from the upper graph, is indicated by dotted lines. The steady-state pool sizes are marked by solid lines.

To summarize all results, I have shown that under steady-state conditions both TDP and TTP appear to be more immediate precursors of DNA than TMP, which is labeled much too quickly. However, in the presence of deoxyadenosine (or thymidine) only TMP seems to be used for DNA synthesis while both TDP and TTP appear to be bypassed. The very rapid establishment of maximal incorporation rate of thymine into DNA in the presence of deoxyadenosine further suggests that the real DNA precursor pool is very small, perhaps only one tenth the size of the deoxyribonucleoside triphosphate pool. Since TMP cannot be the immediate precursor of DNA (because of its behavior under steady-state conditions) we are left with two possibilities as to the nature of the real DNA precursor: Either there exists a small separate deoxyribonucleoside triphosphate pool of only about one tenth the size of the total triphosphate pool, or the real DNA precursor is another, yet unknown, form of activated deoxyribonucleoside 5'-monophosphate.

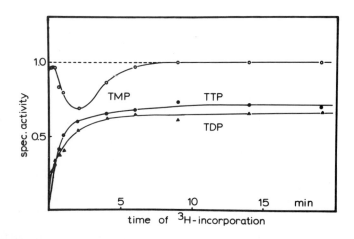

Fig. 9. Relative specific activities of the nucleotide pools during the incorporation of ^{3}H-thymine and deoxyadenosine, expressed as fractions of the specific activity of the ^{3}H-thymine. The specific activities were obtained from Fig. 8 (lower part), dividing the amount of ^{3}H-labeled nucleotide by the sum of both labeled and unlabeled nucleotides.

With regard to the distinct labeling patterns observed with thymine and thymidine (see Fig. 6) and for reasons to be discussed later, I consider a compartmentation of the deoxyribonucleoside triphosphate pool rather unlikely and strongly favor the idea of a different DNA precursor. A possible scheme as to how this precursor might be labeled by thymine or thymidine is presented in Fig. 10.

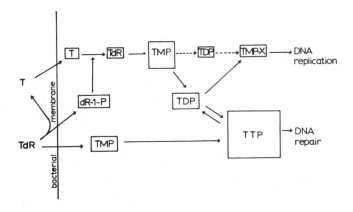

Fig. 10. Tentative scheme of thymine and thymidine metabolism (explanations in the text).

Under steady-state conditions thymine is converted sequentially to thymidine, TMP, TDP and TTP. The postulated precursor for DNA replication, TMP-X, is synthesized from TDP, which is in complete equilibrium with TTP, the precursor for repair synthesis, thus leading to the observed labeling pattern of both TDP and TTP. In the presence of a deoxyribose donor like thymidine or deoxyadenosine, the increased concentration of intracellular deoxyribose-1-phosphate (16) causes a large expansion of the TMP pool. It is conceivable that the large excess of TMP now competes successfully with TDP for sites on the enzyme that activates TMP to TMP-X (perhaps intermediary phosphorylation to TDP can occur on this enzyme complex) thus temporarily cutting off both TDP and TTP from participation in DNA replication.

Although most of the thymidine is presumably broken down by thymidine phosphorylase to thymine and deoxyribose-1-phosphate, some of the thymidine seems to escape this breakdown (at least as long as the activity of thymidine phosphorylase is low) and is converted directly to TTP via a separate, possibly membrane-bound TMP pool. This TTP can only participate in DNA replication via TDP. However, this pathway is blocked because of the simultaneous increase in the thymine-derived TMP pool caused by the simultaneously occurring breakdown of thymidine to deoxyribose-1-phosphate. Consequently, during a short pulse, ^{3}H-thymidine will only go into repair synthesis, while thymine goes preferentially into replication. Of course, during longer pulses this exclusive handling of thymidine will disappear.

The very small size of the DNA precursor pool may provide some clue as to its nature. The pool of all four deoxyribonucleoside triphosphates in <u>E. coli</u> amounts to only about 1.2×10^5 molecules per cell, which is about 1% of its total DNA content (17). Considering the relative concentrations of both ribo- and deoxyribonucleotides as well as the rates of RNA and DNA synthesis, Maniloff (18) has recently calculated that the average lifetime of a deoxyribonucleoside triphosphate molecule at the active site of the polymerase is about 2 orders of magnitude shorter than that of a ribonucleoside triphosphate molecule. Assuming that the time required for proper recognition of the monomer by the polymerase is approximately the same for RNA and DNA synthesis, one wonders whether the intracellular concentration of deoxyribonucleoside triphosphates is sufficient to support the observed rate of DNA replication by means of a random diffusion and collision process. With the discovery that the DNA precursor pool is even smaller, possibly by as much as an order of magnitude, we are compelled to consider processes of monomer selection other than random collision between precursors and polymerase.

It has been known for some time that gene 32 of the bacteriophage T4 is essential for DNA replication and that the gene product is required in stoichiometric amounts. Alberts has recently found that the gene-32-protein has the remarkable property of binding cooperatively to single stranded DNA, and proposed that it is involved in the unwinding of the double helix (19). As each T4-infected cell contains enough gene-32-protein to provide about 160 molecules per growing point, one could imagine that a considerable region of the parental double helix is separated ahead of the replication site (about 800 nucleotides on each side) and that the exposed single strands function as template for the prealignment of deoxyribonucleotides in some activated form. The DNA polymerase would then only have to connect these nucleotides and may itself have no function in recognizing and selecting the proper nucleotides. As this alignment process is occurring simultaneously over an extended region, the collision lifetime of an individual nucleotide with the template is no longer limiting to the rate of DNA replication. No doubt proteins analogous to the gene-32 product will be found in <u>E. coli</u> and other organisms. A model of this rather speculative mechanism is shown in Fig. 11.

As to the nature of the DNA precursor we can only speculate. Since no other forms of thymidine-nucleotides other than mono-, di-, and triphosphates showing the right kinetics of labeling can be extracted from the cell it is possible that the precursor is either rapidly destroyed in the usual extraction procedure or that it is a deoxyribonucleoside monophosphate linked to a carrier molecule which is too big to be extracted from the cell. Carrier molecules linked to complementary nucleotide pairs might actually help to stabilize the template-precursor complex which single nucleotides would not be able to form. The TCA-insoluble, non-sedimenting material that is labeled with thymine but not with thymidine (see Fig. 1) might prove to be such a charged carrier molecule or a derivative thereof.

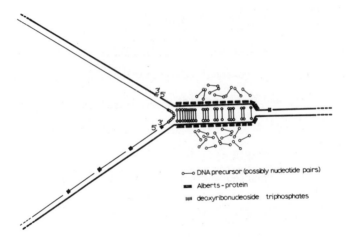

Fig. 11. A possible mechanism for DNA replication.

Whatever the exact mechanism of DNA replication may be, I think I have presented some evidence that deoxyribonucleoside triphosphates may not be the immediate precursors of this reaction. This conclusion agrees with results of Cairns and Denhardt who found that carbonmonoxide or cyanide stops DNA replication without depleting the intracellular deoxyribonucleoside triphosphate pools (8), and might explain why the recently developed in vitro systems of DNA replication depend on the presence of ATP (20,21).

References

(1) P. DeLucia and J. Cairns. Nature 224 (1969) 1164.

(2) S. Mitra, P. Reichard, R.B. Inman, L.L. Bertsch and A. Kornberg. J. Mol. Biol. 24 (1967) 429.

(3) R. Okazaki, T. Okazaki, K. Sakabe, K. Sugimoto and A. Sugino. Proc. Natl. Acad. Sci. U.S. 59 (1968) 598.

(4) R. Okazaki, T. Okazaki, K. Sakabe, K. Sugimoto, R. Kainuma, A. Sugino and N. Iwatsuki. Cold Spring Harbor Symp. Quant. Biol. 33 (1968) 129.

(5) R. Okazaki, K. Sugimoto, T. Okazaki, Y. Imae and A. Sugino. Nature 228 (1970) 223.

(6) R.E. Bird and K.G. Lark. J. Mol. Biol. 49 (1970) 343.

(7) M. Deutscher. Personal communication.

(8) J. Cairns and D.T. Denhardt. J. Mol. Biol. 36 (1968) 335.

(9) E. Fuchs and P. Hanawalt. J. Mol. Biol. 52 (1970) 301.

(10) R. Morgan. Nature 227 (1970) 1310.

(11) S. Riva, A. Cascino and E.P. Geiduschek. J. Mol. Biol. 54 (1970) 103.

(12) J. Cairns. J. Mol. Biol. 6 (1963) 208.

(13) T.D. Price, R.A. Darmstadt, H.A. Hinds and S. Zamenhof. J. Biol. Chem. 242 (1967) 140.

(14) W.E. Razzell and P. Casshyap. J. Biol. Chem. 239 (1964) 1789.

(15) A. Munch-Petersen. Biochim. Biophys. Acta 142 (1967) 228.

(16) A. Munch-Petersen. Biochim. Biophys. Acta 161 (1968) 279.

(17) J. Neuhard. Biochim. Biophys. Acta 129 (1966) 104.

(18) J. Maniloff. J. Theoret. Biol. 25 (1969) 339.

(19) B. Alberts and L. Frey. Nature 227 (1970) 1313.

(20) D.W. Smith, H.E. Schaller and F.J. Bonhoeffer. Nature

226 (1970) 711.

(21) R.E. Moses and C.C. Richardson. Proc. Natl. Acad. Sci. U.S. 67 (1970) 674.

Acknowledgements

    I am very grateful to John Cairns whose encouragement and criticism during the preparation of this paper were of invaluable help.

    This work was done during the tenure of an Established Investigatorship of the American Heart Association and was supported by research grants from the American Cancer Society and the National Science Foundation.

DISCUSSION

U.Z. LITTAUER: I just wanted to comment on the work of Manor and Deutscher that you referred to in the beginning of your talk. They measured the step-time of nucleotide synthesis and compared it to the overall rate of DNA synthesis. However, this was done with a thymidine-requiring mutant. Dr. Manor has now repeated this experiment with $E.\ coli$ B and found that the nucleotide step-time is shorter than observed with the mutant; so we can no longer comment on the presence or absence of Okazaki pieces.

A. TOMASZ: Did you look at the incorporation of thymine versus thymidine in repair negative cells, for instance in the Pol A1 mutant?

R. WERNER: No, I have not used any system other than $E.\ coli$ 15 TAMT and B/r.

J. CAIRNS: There has been one report about Okazaki pieces in Pol A by Kuempel at Colorado. He concludes that the Okazaki pieces, or maybe pieces that are somewhat bigger, persist longer in Pol A than they do in Pol A$^+$ strain, as if the joining up of these pieces proceeds slower in Pol A than in Pol A$^+$.

B. ALBERTS: I would like to mention that in the T4 phage system there is very good evidence that T4 DNA polymerase is at least one of the real polymerase acting in the replicating apparatus. I wonder how you reconcile the fact that this enzyme uses

5'deoxyribonucleoside triphosphates efficiently with the idea that there is quite a different precursor for DNA replication?

R. WERNER: You might have more of these strand interruptions in replicating phage T4 DNA, possibly caused by the high frequency of recombination, for the repair of which the T4 DNA polymerase is required. It is conceivable that the Kornberg polymerase, or the T4 polymerase in this case, is part of the whole replicating machinery, but I feel rather sure that DNA replication uses different precursors.

W. FIRSHEIN: Didn't Okazaki in 1968, in the Cold Spring Harbor Symposium, report on two to six-second pulses where he found that only small 10-11S fragments were produced which then increased in size? Also, in your thymidine-thymine enzymes, have you taken into account, for example, that other enzymes involved in thymidine metabolism could contribute to the interactions in ways other than the manner in which you predict?

R. WERNER: Apparently there are different pathways into thymine-derived nucleotides. For example, it was reported recently that there is a pathway from dCTP to TTP directly. In a thymine-requiring mutant we must, therefore, have at least two mutations, one in the thymidylate synthetase and another one blocking the conversion of dCTP to TTP; a third mutation is required to allow the bacterium to grow on low thymine concentrations. As I have used for my studies a thymine-requiring strain, none of these enzymes could be responsible for the observed incorporation pattern. As to your first question concerning Okazaki's experiments with six-second pulses, there is something I haven't mentioned in this talk: one of the main difficulties in working with DNA is that it is difficult to stop the reaction. Thymidine incorporation continues in the presence of KCN, and this is what occurred in most of the experiments Okazaki has done. In one experiment, Okazaki labeled cells with ^{3}H-thymidine for 6 seconds at 8°C. John Cairns and I repeated this experiment last winter at Cold Spring Harbor. We took samples into either TCA or KCN and ice. In the TCA samples, we found nearly no counts; the KCN samples, however, had an appreciable amount of tritium incorporated. In my experiments I stopped incorporation by adding cold acetone. This treatment inhibits DNA synthesis immediately. Another possibility is, as John Cairns found, the addition of 10% pyridine. I have not observed an increase in the S-value of Okazaki pieces that are labeled with thymine.

S.B. GREER: My comment refers not to your kind of Okazaki pieces, but Okazaki's kind of short pieces that occur first; although there is a low deoxyribose pool in the cell, it is possible that perhaps in one out of five hundred or fifteen hundred nucleotides, there occurs the incorporation of a deoxyribose without a base. When this DNA is placed in an alkaline sucrose gradient, these sites will serve as sites of degradation. Thus the low molecular weight fragments could be a consequence of sensitivity to alkali. One must further assume that the sites void of purine or pyrimidine bases are repaired after a short period. This would distinguish the lability of new DNA vs. old DNA to alkali.

R. WERNER: It is probably difficult to test whether some deoxyribose molecules are incorporated into DNA directly or whether some bases are removed later during the isolation of the DNA.

S.B. GREER: Perhaps Dr. Kornberg would comment on the extent of incorporation of deoxyribose or deoxyribose triphosphates in the *in vitro* system that he studied.

A. KORNBERG: The *in vitro* system I can cite is DNA polymerase, the one reputed to have so little to do with replication; DNA polymerase does not incorporate deoxyribose triphosphate to a measurable degree.

J. CAIRNS: It is obvious from Dr. Werner's work that people in the future must be very careful when they label cells with thymidine. I remember that Rollin D. Hotchkiss, in the summary of a symposium we had at Cold Spring Harbor in 1968, speculated on the chaos that would be produced if somebody wrote a paper entitled "Three new thymidine-containing proteins associated with DNA".

ENZYMATIC MECHANISMS IN DNA REPLICATION

ROBB E. MOSES, JUDITH L. CAMPBELL,
ROGER A. FLEISCHMAN, and CHARLES C. RICHARDSON
Department of Biological Chemistry
Harvard Medical School

Abstract: Escherichia coli cells treated with toluene are capable of carrying out DNA replication when provided with all four deoxyribonucleoside 5'-triphosphates, ATP, Mg^{++}, and K^+. Replication in vitro occurs at a rate comparable to that observed in vivo and continues for up to 0.3-fold replication; it is semiconservative in nature; and it is abolished at restrictive temperatures in temperature-sensitive mutants. In pulse labeling experiments toluene-treated E. coli cells incorporate radioactively labeled dTTP into DNA fragments which sediment slowly (10S) in alkaline sucrose gradients. These fragments can be chased into DNA of high molecular weight in the presence of DPN, but not in the presence of NMN. Formation of high molecular weight DNA in toluene-treated cells is inhibited by the presence of N-ethylmaleimide, nalidixic acid, or deoxyhydroxymethylcytosine 5'-triphosphate. Toluene-treated E. coli polAl, lacking DNA polymerase I, shows normal levels of ATP-dependent synthesis. A new DNA polymerase (polymerase II) has been purified to homogeneity from E. coli polAl as well as from wild-type cells. Polymerase II differs from polymerase I in chromatographic properties, sensitivity to sulfhydryl-blocking agents, response to antibody, and template requirements. In two DNA_{ts} mutants, CRT43-26 and CRT266-26, neither polymerase activity is temperature sensitive in vitro.

INTRODUCTION

The precise mechanism of bacterial DNA replication remains to be elucidated. The bacterial chromosome is replicated semiconservatively, and a number of studies indicate a linear, sequential replication of the chromosome starting from a fixed point on the helix. This model of unidirectional replication implies that both new strands are laid down in the same direction at a single fork. Since the two polynucleotide chains of the template are of opposite polarity, the elongation of one of the new strands must be toward the 5'-end of its template, while the elongation of the other must proceed toward the 3'-end of the template. A model of discontinuous synthesis of DNA has been proposed in an attempt to reconcile DNA synthesis at a single fork with known enzymatic mechanisms. Indeed, many studies now suggest that fragments of low molecular weight serve as precursors of the bacterial chromosome. Models of DNA replication must further account for observations that the chromosome is associated with the cell membrane and that replicating DNA appears in membrane fractions. It is also clear that the control and initiation of replication are closely associated with cell growth and division. For reviews of these topics see References 1-4.

Although a variety of enzymes of DNA metabolism have been identified and characterized, very little is known about their specific roles in vivo (3). For example, the DNA polymerases isolated from several sources incorporate nucleotides only at the 3'-termini of DNA strands, and thus can account for the replication of only one of the two parental strands. The isolation of a mutant of E. coli polA1 (5), which has greatly reduced levels of DNA polymerase activity in extracts, but which nevertheless replicates DNA normally, has raised questions about the role of DNA polymerase in vivo. Similarly, DNA ligases are thought to be involved in the covalent joining of precursor fragments of DNA. However, the isolation of mutants with reduced levels of DNA ligase which can grow normally (3,6) has emphasized the uncertainty in assigning so specific a role to this class of enzymes.

RESULTS

A DNA Replication System in vitro

Since DNA replication appears to involve multiple enzymes and structural components as well as a fragile chromosome, one approach to understanding this process has been to isolate

replicating complexes and to develop replicating systems that closely resemble viable cells.
Buttin and Kornberg (7) have described a technique for measuring intracellular DNA synthesis in cells treated with EDTA-Tris, but the synthesis observed is mainly dependent on endonuclease activity (8). Smith, Schaller, and Bonhoeffer (9) have gently lysed cells imbedded in agar to obtain a system which can carry out semiconservative DNA replication. Several laboratories (10-15) have used more extensive lysis procedures to obtain membrane fractions capable of carrying out limited DNA replication in vitro. However, the usefulness of these systems is often limited by the brief period of replication and by the requirement for certain bacterial strains or cell fractions having low levels of DNA polymerase activity.

Toluene-treated cells: Escherichia coli cells treated with toluene become permeable to compounds of low molecular weight including deoxynucleoside triphosphates (16). Although such cells are no longer viable, they maintain their structure (Fig. 1) and many of their physiological functions.

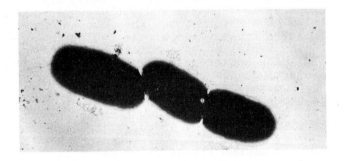

Fig. 1. Electron micrograph of E. coli cells incubated with 1% toluene for 10 min at 37°

When provided with the appropriate substrates (17), they are capable of carrying out extended semi-conservative DNA replication. Furthermore, DNA replication can be selectively measured and distinguished from repair synthesis (17).

Hoffmann-Berling (18) has used a similar system of ether-treated cells to study intracellular φX174 DNA replication.

After treatment with toluene for approximately 10 min, the cells display an optimal rate of DNA synthesis in the presence of the four deoxynucleoside 5'-triphosphates, ATP, Mg^{++}, and K^+ (Fig. 2). The maximum rate of synthesis is 1.5×10^3 nucleotides per cell per sec at 35°, a rate comparable to the rate observed during replication in vivo.

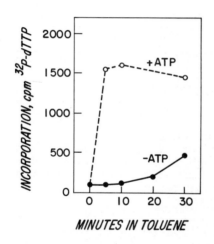

Fig. 2. Time course of toluene treatment. E. coli W3110 cells were treated with 1% toluene for the times indicated and then assayed for DNA synthesis with or without ATP as previously described (17).

DNA replication: The ATP-dependent DNA synthesis has been shown to be DNA replication by several methods (17): 1) The rate of synthesis is comparable to that observed in vivo and continues up to 0.3 replications. 2) The DNA synthesized is identical in size to that of DNA isolated from growing cells. 3) Synthesis is semiconservative as determined by pycnographic analysis. 4) The synthesis can be abolished at the restrictive temperature in DNA temperature-sensitive mutants (TABLE I), an observation also made by Mordoh, Hirota, and Jacob (19).

TABLE I

DNA synthesis in toluene-treated DNA_{ts} mutants

Strain	DNA synthesis (+ ATP)	
	$33°$	$43°$
	%	%
Wild-type	100	51
CRT266-26	100	12
CRT43-26	100	8

Cells were grown at $28°$, treated with toluene, and assayed for ATP-dependent DNA synthesis at $33°$ and at $43°$ (17). E. coli CRT266-26 and CRT43-26, obtained from Dr. G. Buttin, rapidly cease DNA replication at $41°$ in vivo (20).

Characteristics of the reaction: The availability of a replication system in vitro allows the modification of the pool sizes of low molecular weight compounds and the study of their effect on DNA replication (TABLE II). All four deoxynucleoside 5'-triphosphates are required for DNA replication; if one or more are omitted from the reaction mixture, there is no appreciable replication. The deoxynucleoside 5'-monophosphates cannot replace the corresponding 5'-triphosphates. However, the deoxynucleoside 5'-diphosphates are equally as effective as the 5'-triphosphates. As shown in Fig. 2 and TABLE II, there is a 10 to 20-fold stimulation in replication upon the addition of ATP. A major portion of the synthesis observed in the absence of ATP is associated with endonuclease activity (17). DNA replication in vitro is sensitive to sulfhydryl-blocking agents such as N-ethylmaleimide, and to nalidixic acid.

TABLE II

Properties of DNA replication in toluene-treated E. coli

Conditions	DNA synthesis
	%
Control	100
− dATP	<1
− 4dXTP + 4dXMP	<1
− 4dXTP + 4dXDP	95
− ATP	10
+ NEM	3
+ NAL	35

Toluene-treated E. coli ER22 cells were prepared and assayed for DNA synthesis under the conditions described. The complete system contained the four deoxynucleoside 5'-triphosphates, ATP, Mg^{++}, and K^+ (17).

Inhibition by dHMCTP: The DNA of the T-even phages contains hydroxymethylcytosine (HMC) in place of cytosine, and glucose linked to the HMC groups in ratios characteristic for each type of phage (21). Failure of glucosylation of HMC residues exposes the phage DNA to specific restriction by certain hosts (22). It was therefore of interest to examine the effect of introducing HMC into the bacterial DNA. Such an experiment is feasible with toluene-treated cells.

E. coli W3110 does not support the growth of non-glucosylated T-even phages. When deoxyhydroxymethylcytosine 5'-triphosphate (dHMCTP) is substituted for dCTP in the reaction mixture containing toluene-treated E. coli W3110, incorporation of radioactivity into acid-precipitable DNA is reduced to less than 5% of that obtained with dCTP alone (TABLE III). In fact, the same reduction is observed even when dCTP and dHMCTP are present

together in equal amounts. In contrast,
E. coli K12r$_2$,$_4^-$r$_6^-$, permissive to phages containing non-glucosylated DNA (22), replicates DNA at only a slightly reduced rate in the presence of dHMCTP. Furthermore, this mutant incorporates HMC into its DNA as measured by the incorporation of ^{14}C-dHMCTP. These results demonstrate that the restriction mechanism for non-glucosylated T-even phage is present in the host cell before phage infection, and can recognize HMC residues in its own DNA as well as that in the T-even phages. Glucosylation thus appears to protect the HMC-containing phage DNA from restriction.

TABLE III

Effect of dHMCTP on DNA replication

Strain	Phenotype	DNA synthesis	
		dCTP	dHMCTP
		%	%
W3110	Non-permissive to T*-even	100	<5
K12r$_2$,$_4^-$r$_6^-$	Permissive to T*-even	100	70

Toluene-treated cells were assayed for ATP-dependent DNA synthesis in the standard assay (17) containing dCTP, or with dHMCTP replacing dCTP. T*-even phages contain HMC, but little or no glucose (22).

Joining of pulse-labeled fragments of DNA: Sakabe and Okazaki (23) were the first to observe that, after a very short pulse of radioactively labeled thymidine, the labeled DNA sedimented considerably more slowly (10S) than did the bulk of the E. coli DNA. Evidence that the pulse-labeled DNA was an intermediate in DNA replication came from pulse-chase experiments in which it was shown that the fragments could be converted to DNA of high molecular weight. As shown in Fig. 3, similar results can be obtained with toluene-treated cells of E. coli. When toluene-treated cells are incubated for brief periods at reduced temperature in

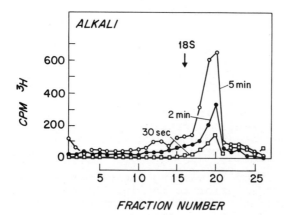

Fig. 3. Alkaline sucrose gradient of pulse-labeled toluene-treated E. coli. E. coli D110 (polA1 end⁻) cells were treated with toluene and incubated at 25° in the DNA synthesizing reaction mixture (17) containing a 10-fold excess of cells and ^{3}H-dTTP (12 mC/ μmole) for the times indicated. Under these reaction conditions the rate of DNA synthesis is reduced 30-fold. The reaction was stopped and the cells were lysed in 0.1 N NaOH - 0.001 M EDTA - 1% Sarkosyl. The lysate was sedimented through a 5 to 20% sucrose gradient containing 0.7 M NaOH - 0.3 M NaCl - 0.001 M EDTA in a Spinco type SW50.1 rotor for 3 hrs at 49,000 rpm. Fractions were collected, and the acid-insoluble radioactivity was measured. φX174 DNA was present as a marker (18S).

the standard DNA synthesizing assay containing ^{3}H-dTTP of high specific radioactivity, all of the radioactively labeled DNA sediments in alkali as 10S fragments. Furthermore, there is an accumulation of these fragments during the 5'-min period of incubation (Fig. 3). Further incubation results in the formation of DNA of higher molecular weight without an appreciable change in the amount of 10S fragments. Although the results shown were obtained with E. coli polAl, similar results have been obtained with wild-type E. coli (E. coli endI$^-$).

If the toluene-treated cells are incubated with an excess of cold dTTP following the pulse of ^{3}H-dTTP, all of the radioactivity is chased into material of high molecular weight. If the joining of these fragments is catalyzed by the E. coli DNA ligase, then it should be possible to prevent joining by eliminating DPN, the cofactor for the ligase (24,25), from the reaction mixture. This was accomplished by washing the toluene-treated cells in buffer prior to the incubation with the deoxynucleoside triphosphates. In order to inhibit the reaction further and to discharge any ligase-AMP present in the toluene-treated cells, nicotinamide mononucleotide (NMN), a product of the ligase reaction, was also included in the reaction mixture. Not only did the washed toluene-treated cells accumulate 11S fragments in the presence of NMN, but furthermore a major portion of the fragments could not be chased into DNA of high molecular weight in the presence of NMN (Fig. 4). However, if the chase experiment was carried out in the presence of DPN, all of the radioactivity in the 10S material was transferred to faster sedimenting material. These results implicate the E. coli DNA ligase in the covalent joining of the Okazaki Fragments.

A New DNA Polymerase in E. coli

The predominant deoxynucleotide polymerizing activity (polymerase I) found in extracts of wild-type E. coli has been purified to homogeneity and studied extensively (26). Although earlier studies (27-29) have suggested the presence of additional DNA polymerase activities in E. coli, there has not been sufficient data to establish their identity. The isolation by DeLucia and Cairns (5) of a mutant with markedly reduced

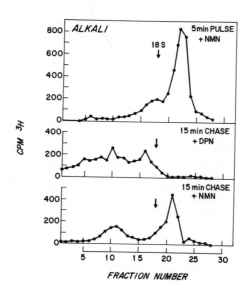

Fig. 4. Effect of DPN on conversion of DNA fragments to DNA of high molecular weight. E. coli D110 (polA1 end⁻) cells were treated with toluene, collected by centrifugation, re-suspended in 20 mM NMN - 50 mM potassium phosphate buffer (pH 7.4), and then incubated at 25° for 5 min in the standard DNA synthesizing reaction mixture (17) containing ^{3}H-dTTP (12 mC/µmole) and 5 mM NMN (a). Incubation was then continued for an additional 15 min at 25° in the presence of a 100-fold excess of unlabeled dTTP and either 0.3 mM DPN (b) or 5 mM NMN (c). The cells were lysed and analyzed by sedimentation in alkaline sucrose gradients as described in Fig. 3.

levels of polymerase activity in cell extracts greatly facilitated a search for additional deoxynucleotide polymerizing activities in E. coli. The mutation in this strain, designated polA1, has been mapped and shown to be an amber mutation (30). Five additional mutants have been isolated that map at the same location as polA1; all of the pol⁻ mutants are more sensitive than wild-type E. coli to ultraviolet light and to methylmethanesulphonate (30; DeLucia and Cairns, pers. comm.; Peacey and Gross, pers. comm.). Direct evidence that the pol gene is the structural gene for polymerase I has been obtained by Kelley and Whitfield (pers. comm.). They have purified what appears to be DNA polymerase I from an E. coli strain with a pol mutation and shown that it differs from the wild-type enzyme in several characteristics.

As shown in TABLE IV, extracts of E. coli polA1, prepared by conventional methods, do not contain DNA polymerase activity when assayed with a variety of templates. DNA replication, however, occurs

TABLE IV

Polymerase activity in extracts of E. coli W3110 and E. coli polA1

Template	W3110	polA1
	% Activity	
E. coli DNA	100	<0.2
Activated DNA	50	<0.4
Heat-denatured DNA	60	<0.4
M13 DNA	62	<0.3
Poly dAT	55	<0.4
Poly dA	79	<0.3
Poly dT	3	---
Poly dA + Poly dT	25	<0.1

Extracts of E. coli W3110 (wild-type) and E. coli polA1 were prepared by sonic irradiation and assayed for DNA polymerase activity in the standard assay (31) containing the templates indicated.

normally in E. coli polA1 cells treated with toluene (17). The pattern of ATP-dependent DNA synthesis during toluene treatment of polA1 closely resembles that obtained with wild-type E. coli (Fig. 5). However, there is no synthesis in the absence of ATP, even after prolonged treatment with toluene, indicating that the mutant is deficient in repair synthesis. In vitro DNA synthesis in polA1 has also been observed in spheroplasts (9), in membrane fractions (12,13), and in cells treated with ether (Vorberg and Hoffmann-Berling, pers. comm.).

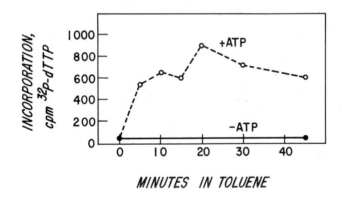

Fig. 5. Time course of toluene treatment with E. coli polA1. E. coli polA1 was treated with 1% toluene for the times indicated and then assayed for DNA synthesis with or without ATP as previously described (17).

Purification of polymerase II: A new DNA polymerase, polymerase II, has been purified to homogeneity from toluene-treated E. coli polA1 (31).

Since the chromatographic properties of polymerases I and II differ significantly, it has also been possible to purify polymerase II from wild-type cells (32). The last step of purification of polymerase II, chromatography on phosphocellulose, is shown in Fig. 6.

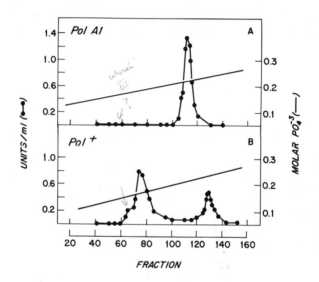

Fig. 6. Phosphocellulose chromatography of DNA polymerase II from E. coli polAl (a) and E. coli W3110 (b). The purification and assay of activity have been previously described (31,32).

Comparison of polymerases I and II: DNA polymerases I and II both require a DNA primer, the four deoxynucleoside 5'-triphosphates, and Mg^{++}. However, the two enzymes can be distinguished by several criteria. In addition to their chromatographic properties, DNA polymerases I and II can be distinguished by their sensitivity to sulfhydryl-blocking agents, their template requirements, and their inhibition by antibody (TABLE V).

TABLE V

Comparison of polymerase I and II activities

Condition	Polymerase I	Polymerase II
	%	%
Control	100	100
+ 2 mM DTT	99	198
+ 3 mM NEM	98	3
+ poly d(A-T)	240	5
+ antibody to polI	10	85

The phosphocellulose fractions of polymerase I and II purified from E. coli W3110 were assayed in the standard assay (17) containing "activated" salmon sperm DNA. The rate of synthesis for each enzyme is recorded as 100% (Control). Dithiothreitol (DTT) or N-ethylmaleimide (NEM) were present in the reaction mixtures as indicated. When poly d(A-T) was present, the "activated" DNA was omitted.

Polymerase I activity is not stimulated by sulfhydryl compounds nor inhibited by N-ethylmaleimide, whereas polymerase II activity shows a 2-fold stimulation with dithiothreitol and is completely inhibited by N-ethylmaleimide. Polymerase I prefers poly d(A-T) to partially degraded DNA, but polymerase II does not use poly d(A-T) as template. Antibody to polymerase I has no significant effect on polymerase II activity.

DNA_{ts} mutants: The thermosensitivity of DNA replication in the two "quick" stop mutants discussed above could be the result of a temperature-sensitive DNA polymerase. Since it has previously been shown that polymerase I is not heat labile in these mutants (20,33), it was of interest to examine polymerase II. Polymerases I and II were purified from E. coli CRT43-26 and CRT266-26 (32). Polymerase I and polymerase II, purified from both mutants grown at 33°, have increased activity at the elevated temperature (TABLE VI), suggesting that neither of these enzymes is responsible for the thermosensitivity in these mutants.

TABLE VI

Activity of DNA polymerases I and II from DNA_{ts} strains

	CRT43-26		CRT266-26	
	33°	43°	33°	43°
	%	%	%	%
Polymerase I	100	205	100	210
Polymerase II	100	210	100	193

The phosphocellulose fractions of polymerases I and II purified from E. coli CRT43-26 and E. coli CRT266-26 were assayed in the standard assay (31) at 33° and 43°.

DISCUSSION

Toluene-treated cells provide a system useful in the study of DNA replication in vitro, since DNA synthesis in this system closely parallels that occurring in vivo. The major alteration with regard to DNA replication is the permeability of toluene-treated cells to small molecules. As a result, it has been possible to study the effects of compounds of low molecular weight on DNA replication. For example, deoxynucleoside 5'-triphosphates appear to be the direct precursors for DNA replication, and DPN has been implicated as a cofactor in the covalent joining of DNA fragments.

Similarly, it has been possible to study the effect of a base analogue, hydroxymethylcytosine, on DNA replication. The lack of DNA synthesis with dHMCTP could be due to rapid degradation of newly synthesized DNA containing HMC, or to inhibition of further DNA replication by the incorporation of HMC, or to inhibition by dHMCTP of some earlier step in DNA replication.

The observation that toluene-treated E. coli, permissive to non-glucosylated phages, can incorporate HMC and replicate DNA normally implies that the effect is related to modification and restriction in vivo. Two novel findings are that DNA replication in toluene-treated cells is stimulated by ATP and inhibited by sulfhydryl-blocking agents, suggesting that a number of enzymes or structural components are involved in DNA replication.

Toluene treatment of E. coli polA1, a mutant lacking DNA polymerase I, permitted an in vitro assay of DNA replication in this mutant. Using such a preparation a new DNA polymerase, polymerase II, has been purified to homogeneity. Kornberg and Gefter (34), as well as Knippers (35), have also purified polymerase II from polA1 cells. Although the purified polymerase II has properties which distinguish it from polymerase I, we felt that it was necessary to isolate polymerase II from wild-type strains in order to demonstrate that the new activity was not a result of the mutation in polA1. Following the method developed for purification of polymerase II from E. coli polA1, the enzyme has been purified from wild-type E. coli. It is clearly separable from polymerase I by chromatography on DEAE and phosphocellulose, and it is indistinguishable from polymerase II purified from polA1 cells. Assuming equal recovery of polymerases I and II during purification, polymerase II represents 5 to 10% of the polymerizing activity found in wild-type cells. Loeb, Slater, Ewald, and Agawal (pers. comm.) used antiserum specific to polymerase I to obtain evidence that wild-type E. coli, as well as polA1 cells, contain an additional DNA polymerizing enzyme. Since our purification procedure is suitable for the purification of both polymerases I and II, the absence of polymerase I at all stages of purification of polymerase II from polA1 indicates a complete absence of polymerase I in extracts of this mutant.

Polymerase II has been purified using a partially degraded DNA as primer, and thus far there is no evidence that the catalytic activity of the enzyme differs from that of other polymerases. The purified enzyme is not stimulated by ATP as is replication in toluene-treated cells, but it is inhibited by N-ethylmaleimide.

At present the roles of polymerases I and II in DNA replication are unknown. It is possible that both polymerases play a role in replication, but that either enzyme can suffice alone. For example, Kingsburg and Helinski (pers. comm.) have shown that the product of the pol gene is essential for the maintenance of the col E$_1$ factor. Only detailed genetic analysis of the function of polymerases I and II, and further enzymatic studies on these two enzymes, can answer these questions. Attempts to attribute the thermosensitivity to DNA$_{ts}$ mutants to polymerases I and II have thus far been unsuccessful. Gross, Grunstein, and Witkin (pers. comm.) have found that recA$^-$ derivatives of polA1 are non-viable. An understanding of the role of the recA function might therefore be helpful in elucidating the roles of the two DNA polymerases in vivo.

REFERENCES

1. N. Sueoka, in: Molecular Genetics, Vol. 2, ed. J. H. Taylor (Academic Press, New York, 1967) p. 1.

2. A. Ryter. Bacteriol. Rev. 32 (1968) 39.

3. C. C. Richardson. Ann. Rev. Biochem. 38 (1969) 795.

4. K. G. Lark. Ann. Rev. Biochem. 38 (1969) 569.

5. P. DeLucia and J. Cairns. Nature 224 (1970) 1164.

6. M. Gellert and M. L. Bullock. Proc. Nat. Acad. Sci., U.S.A. 67 (1970) 1580.

7. G. Buttin and A. Kornberg. J. Biol. Chem. 241 (1966) 5419.

8. G. Buttin and M. Wright. Cold Spring Harbor Symp. Quant. Biol. 33 (1968) 259.

9. D. W. Smith, H. Schaller, and F. J. Bonhoeffer. Nature 226 (1970) 711.

10. F. R. Frankel, C. Majumdar, S. Weintraub, and D. M. Frankel. Cold Spring Harbor Symp. Quant. Biol. 33 (1968) 495.

11. A. T. Ganesan. Cold Spring Harbor Symp. Quant. Biol. 33 (1968) 45.

12. R. Knippers and W. Strätling. Nature 266 (1970) 713.

13. R. Okazaki, K. Sugimoto, T. Okazaki, Y. Imae, and A. Sugino. Nature 228 (1970) 223.

14. E. A. Linney and M. Hayashi. Biochem. Biophys. Res. Commun. 41 (1970) 669.

15. D. T. Denhardt and A. B. Burgess. Cold Spring Harbor Symp. Quant. Biol. 33 (1968) 449.

16. R. W. Jackson and J. A. DeMoss. J. Bacteriol. 90 (1965) 1420.

17. R. E. Moses and C. C. Richardson. Proc. Nat. Acad. Sci., U.S.A. 67 (1970) 674.

18. H. Hoffmann-Berling, in: Molecular Genetics, eds. H. C. Wittmann and M. Schuster (Springer, Berlin, 1968) p. 38.

19. J. Mordoh, Y. Hirota, and F. Jacob. Proc. Nat. Acad. Sci., U.S.A. 67 (1970) 773.

20. G. Buttin and M. Wright. Cold Spring Harbor Symp. Quant. Biol. 33 (1968) 259.

21. A. Kornberg, Enzymatic Synthesis of DNA (John Wiley and Sons, New York, 1961).

22. H. R. Revel and S. E. Luria. Ann. Rev. Genetics 4 (1970) 177.

23. K. Sakabe and R. Okazaki. Biochim. Biophys. Acta 129 (1966) 651.

24. S. B. Zimmerman, J. W. Little, C. K. Oshinsky, and M. Gellert. Proc. Nat. Acad. Sci., U.S.A. 57 (1967) 1841.

25. B. M. Olivera and I. R. Lehman. *Proc. Nat. Acad. Sci.*, *U.S.A.* 57 (1967) 1700.

26. A. Kornberg. *Science* 163 (1969) 1410.

27. K. Hori, H. Fujiki, and Y. Takagi. *Nature* 210 (1966) 604.

28. A. G. Lezius, S. B. Hennig, C. Mengel, and E. Metz. *Eur. J. Biochem.* 2 (1967) 90.

29. L. F. Cavalieri and E. Carrol. *Proc. Nat. Acad. Sci.*, *U.S.A.* 59 (1968) 951.

30. J. Gross and M. Gross. *Nature* 224 (1969).

31. R. E. Moses and C. C. Richardson. *Biochem. Biophys. Res. Commun.* 41 (1970) 1557.

32. R. E. Moses and C. C. Richardson. *Biochem. Biophys. Res. Commun.* 41 (1970) 1565.

33. F. Bonhoeffer. *Z. Vererbungslehre* 98 (1966) 141.

34. T. Kornberg and M. L. Gefter. *Biochem. Biophys. Res. Commun.* 40 (1970) 1348.

35. R. Knippers. *Nature* 228 (1970) 1050.

The work described in this paper was sponsored by grants from the National Institutes of Health (AI-06045) and the American Cancer Society (P-486). R. E. Moses was supported by a Public Health Service Fellowship (5-FO3-GM42,968). C. C. Richardson is the recipient of a Public Health Service Research Career Program Award (GM-13,634).

DISCUSSION

S.H. CHANG: You talk about the ATP requirement for DNA replication in the toluene treated $E.\ coli$ cells. Is this due to the ATP requirement for ligase reaction?

C.C. RICHARDSON: No. These studies were all carried out with wild-type $E.\ coli$ whose cofactor is DPN. It is thus unlikely that ATP is having any effect on ligase action. I should also point out that the stimulation by ATP is not absolutely specific for ATP. ATP can be replaced by the other deoxynucleoside triphosphates, the best of which is dATP, although it is only about 45% as effective.

A. TOMASZ: Is there any evidence that the toluenized cells can carry out more than one round of DNA replication per cell? That is, can there be re-initiation?

C.C. RICHARDSON: As I mentioned, if synthesis is allowed to continue at 37° with optimal ATP levels, then it is possible to obtain 0.3 replication, i.e. 30% of the chromosome is replicated. Furthermore, as cells approach stationary phase, the amount of synthesis they can carry out after toluene treatment is reduced progressively; toluene-treated stationary phase cells cannot carry out ATP-dependent DNA synthesis. Therefore, I do not think that toluene-treated cells can initiate a new round of replication.

W. HSIANG: From all you have said about polymerase II, I interpret it to be the same enzyme as the one Knippers and Stratling describe as membrane-bound polymerase. They had shown that the membrane-bound polymerase showed very short-lived activity, whereas in their latest paper the activity after the enzyme was purified was extended from 2 min. to about 90 min. Do you think that the short-lived activity is reflected in your slides showing 95% activity for the polymerase I compared to 5% for the polymerase II?

C.C. RICHARDSON: I am not sure I understand the question.

W. HSIANG: In studies that Smith and Knippers did with membrane-bound polymerase, they found that their enzyme was active only up to 2 or 5 min., whereas after they purified this enzyme, the enzyme was active for about 90 min.

C.C. RICHARDSON: The membrane system they describe would appear to be an $in\ vitro$ replicating system in which, presumably, all the components necessary to carry out DNA replication are present. This system was unstable, as reflected in the limited amount of DNA replication. When polymerase II is purified free from the complex, one is no longer measuring DNA replication, but instead a repair-type reaction. For example, the ATP requirement is lost.

W. HSIANG: How long does polymerization last with the purified enzyme?

C.C. RICHARDSON: Synthesis can continue for prolonged periods if the enzyme is supplied with sufficient amounts of template and precursors. I should stress, however, that with the purified enzyme we are most likely measuring only a repair type synthesis. Once the single-stranded regions in the template have been repaired, DNA synthesis stops.

R. WERNER: Have you tested some of the recombination-negative mutants for the polymerase, especially *rec A*?

C.C. RICHARDSON: We were certainly very much interested in examining *rec A*. We have purified polymerase II from a temperature-sensitive mutant, a missense mutant, and an *amber* mutant of *rec A*. Polymerase II is present in normal amounts in all three mutants. However, it is possible that, like polymerase I, polymerase II has multiple enzyme functions. As a result it is possible that an unidentified activity of polymerase II could be defective in *rec A* mutants.

J. CAIRNS: I think that Gefter has found the same thing.

R. NOVAK: Have you tried to recombine the flow-through from your phosphocellulose column with your purified polymerase II to see if you can regain the ATP dependency?

C.C. RICHARDSON: Unfortunately, we lose ATP requirement as soon as the toluene cells are lysed.

J. WILLIS: Have you tried short deoxyribonucleotide oligomers as primers with polymerase II to see if there is activity similar to that of Bollum's terminal addition enzyme from calf thymus gland?

C.C. RICHARDSON: We tried a number of oligonucleotides, and these do not promote any DNA synthesis in the absence of a complementary template. I should point out that the base composition of the DNA that is synthesized with polymerase II using "activated" DNA templates reflects the base composition of the templates.

W. FIRSHEIN: If your cells (the toluenized cells) are really just more permeable, it is hard to see why the monodeoxynucleotides do not also act as good precursors in the presence of ATP, particularly as the diphosphates work. Perhaps the enzymes are too far removed to make the triphosphates. Could you, for example, add something like creatine phosphate and the phosphokinase to boost the activity of kinases? Alternatively, if there is something basic missing, could you for example use ribonucleoside diphosphates as precursors to show the activity of reductase, because I think that you have a great system to test which pathways are used here.

C.C. RICHARDSON: I really cannot explain why the monophosphates are not effective. We have not carried out the experiments that you mention.

S.B. WEISS: Have you looked at the direction of chain growth or what would happen if both polymerases I and II were put together with a given template?

C.C. RICHARDSON: We are in the process of carrying out these studies. Preliminary evidence indicates that it adds nucleotides in the 3'- direction.

M. RABINOVITZ: You indicated that it was not possible to tell whether the deoxynucleoside diphosphates in your toluene-treated cells were made into triphosphates via the diphosphokinase and ATP. Is it possible to use the methylene analogs of the deoxynucleoside triphosphates to prevent terminal phosphate cleavage but still permit pyrophosphorylytic cleavage of the methylene diphosphonate. This should eliminate the ATP requirement if it served only for deoxynucleoside triphosphate regeneration.

C.C. RICHARDSON: We have considered this experiment, but have not yet tested these analogues.

STUDIES ON TERMINATION DURING
IN VITRO DNA-DEPENDENT RNA SYNTHESIS

Allan R. Goldberg and Jerard Hurwitz
Division of Biology, Department of Developmental Biology and Cancer
Albert Einstein College of Medicine, Bronx, N.Y.

Abstract: A number of laboratories have demonstrated that during DNA-dependent RNA synthesis catalyzed by RNA polymerase at high ionic strength, RNA synthesis is stimulated and release of RNA chains occurs. This stimulation is due to reinitiation of RNA chains. Roberts discovered a protein called rho which causes release of RNA chains. These chains are considerably smaller in length than RNA synthesized in its absence, and for this reason rho is considered to be a chain terminator.

The mechanism of RNA chain release by both high ionic strength and rho has been studied in detail. At 0.2 M KCl, *in vitro* termination of RNA synthesis occurs in the absence of added rho factor. Under these conditions, RNA polymerase is released from the DNA template and can act catalytically to reinitiate new RNA chains. In contrast, rho cannot act at salt concentrations which favor salt-mediated termination. Only at lower salt concentrations does rho cause release of RNA chains. However, under these conditions there is no reinitiation of new RNA chains, presumably because RNA polymerase is not released. Thus, rho appears to be a protein which stops the translocation activity of RNA polymerase.

INTRODUCTION

The DNA-dependent synthesis of RNA is a complex process governed by a variety of factors and conditions. The synthesis of RNA can be divided into at least 5 discrete reactions which can be studied. These include: (1) the binding of RNA polymerase to DNA; (2) the binding of nucleoside triphosphates to the enzyme-DNA complex; (3) the initiation of RNA chains; (4) the elongation of RNA chains by the addition of nucleotides to the 3'-hydroxyl end of the growing chain and (5) the termination of RNA growth with the release of newly formed RNA chains from the ternary complex. Included in the latter step is the release of enzyme which can then restart the entire cycle.

A complete review of all of these reactions is beyond the scope of this article. The present communication will focus on the termination of RNA chains, specifically comparing two factors which control this process: namely, high ionic environment and a protein factor called rho.

EXPERIMENTAL AND RESULTS

Purification and Properties of ρ-Factor

ρ-Factor was isolated from *E. coli* and purified as described by Roberts (1). The factor has a sedimentation constant of 8-10S and is relatively homogeneous, showing one visible band on cellulose acetate electrophoresis under non-denaturing conditions. Polyacrylamide gel electrophoresis in SDS indicates our preparation of ρ-factor to be about 70-80% pure. The predominant band which is observed (Fig. 1) possesses the same electrophoretic mobility ascribed to ρ-factor by Roberts (1). In addition, this band becomes intensified during the purification procedure, while the minor bands diminish.

RNA polymerase preparations used in the experiments described below appear to be completely free of detectable ρ-factor (Fig. 1). In confirmation of Roberts' findings, ρ-factor is free of detectable RNase activity since it has no effect on the sedimentation of either Qβ-RNA or an *in vitro* RNA product (transcribed from T4 DNA) in SDS-formaldehyde-sucrose density gradients. ρ-Factor preparations are devoid of DNA-endonuclease activity since λ DNA, after incubation with ρ-factor (2 μg) showed the same sedimentation profile as untreated λ DNA in an alkaline sucrose gradient.

A number of laboratories have demonstrated that during RNA synthesis, the release of RNA chains can occur. This release is dependent on the ionic strength of the reaction mixtures (2, 3, 4, 5). These workers have demonstrated that concomitant with the release of RNA chains, RNA polymerase is released allowing reinitiation of new chains. Thus in the presence of 0.2 M KCl, RNA polymerase can act catalytically forming more moles of RNA chains than moles of enzyme present.

RNA chains synthesized in the presence of ρ are considerably smaller in length than RNA found in its absence and for this reason ρ has been called a chain terminator. The work described below will present evidence suggesting that release of RNA chains mediated by ρ-factor and salt may be by different mechanisms.

It might be reasoned that release of RNA chains in a high salt environment is necessary for the release of RNA polymerase and for subsequent reinitiation. Interestingly, ρ-factor does cause the release of RNA from λ DNA (1) but as can be seen in Table 1 such release does not allow either increased reinitiation or prolonged RNA synthesis at high rates. In confirmation of

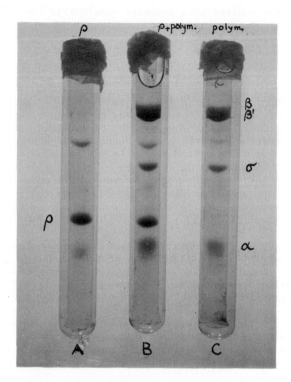

Fig. 1. SDS-polyacrylamide gel electrophoresis of ρ-factor and RNA polymerase. Solutions of ρ-factor and RNA polymerase (kindly supplied by Dr. Umadas Maitra of this department) were dialyzed overnight against 0.01 M sodium phosphate, pH 7.0 and were then adjusted to 0.1 M 2-mercaptoethanol and 1.0% SDS. They were heated to 100°C. for 1 minute and then subjected to electrophoresis according to the procedure of Shapiro *et al.*, (6) on 5% polyacrylamide gels containing 0.1% SDS. The gels then were stained for 2 hours with 0.25% Coomassie Brilliant Blue in methanol-acetic acid-water (5:1:5) and destained with 5% methanol in 7.5% acetic acid. A, 8 µg ρ-factor and 8 µg RNA polymerase: C 8 µg RNA polymerase.

Roberts' data ρ-factor causes a depression of nucleotide incorporation which can be used as a measure of this factor.

Because high ionic strength affects the kinetics of transcription it seems relevant to examine the effect of [KCl] on the

Table 1

Kinetics of Initiation of Transcription from λ DNA

Time (min)	Total Nucleotide Incorporation (mµmoles)		γ-32P Nucleotide Incorporation (µµmoles)		Average Chain Length (nucleotides)	
	no ρ	+ ρ	no ρ	+ ρ	no ρ	+ ρ
5	1.32	0.59	0.77	0.54	1720	1090
15	3.46	1.71	1.18	1.39	2930	1230
40	6.08	2.53	1.75	2.19	3470	1165
80	8.69	3.88	2.36	3.03	3680	1280
160	10.90	4.46	3.08	3.55	3540	1235

Table 1. Reaction mixtures of 0.5 ml contained: 0.05 M Tris (pH 7.9); 0.004 M 2-mercaptoethanol; 0.01 M magnesium acetate; 0.10 M KCl; 15 µg λS7 DNA; 200 µg/ml bovine serum albumin; 3 µg RNA polymerase; 1.5 µg ρ-factor where added; 0.15 mM ATP, GTP, and UTP. Reaction mixtures contained either γ-^{32}P-ATP (2.7 x 10^3 cpm/µµmoles) or γ-^{32}P-GTP (2.7 x 10^3 cpm/µµmole). Incubation was at 37° and 100 µl aliquots were removed at the specified times. Synthesis was stopped by chilling and the addition of 0.4 ml H$_2$O, 0.1 ml albumin (10 mg/ml), 0.1 ml 0.1 M pyrophosphate, 0.1 ml ATP or GTP (30 µmoles/ml), and 0.3 ml 7% perchloric acid (PCA). After 5 minutes at 0°C the precipitate was collected by centrifugation and was redissolved by adding 0.1 ml pyrophosphate, 0.1 ml ATP or GTP, and 0.2 ml ice cold 0.2 N NaOH. After dissolution of the pellet, 5% TCA and pyrophosphate were added and the resulting pellet was collected by centrifugation after 5 more minutes at 0°C. This washing procedure was repeated three more times and the pellet was resuspended in 1.0 ml 0.5 N NH$_4$OH and counted in Bray's scintillation fluid. Total nucleotide incorporation was calculated by multiplying the number of mµmoles of ^{3}H-UTP incorporated by four. γ-^{32}P nucleotide incorporation represents the sum of ATP and GTP incorporation.

ability of ρ-factor to depress nucleotide incorporation with λ T4, or T7 DNA as templates. ρ-Factor has no effect on λ DNA transcription at [KCl] > 0.1 (Fig. 2) but shows its maximum activity at 0.1 M KCl. In contrast, transcription from T4 or T7 DNA is affected most by ρ-factor at salt concentrations which do not favor maximal RNA synthesis or release of RNA chains by RNA polymerase (Figs. 3 and 4).

Kinetics of Initiation

The rate and extent of RNA synthesis on λ DNA were studied in the presence of 0.1 M KCl, and as observed by Roberts (1), nucleotide incorporation is decreased about two to three-fold in the presence of saturating amounts of ρ-factor (Table 1). Although RNA synthesis is depressed by ρ-factor, initiation of RNA chains, measured by incorporation of γ-^{32}P nucleoside triphosphates (GTP and ATP) is unaffected by ρ-factor. The average chain length of RNA synthesized in the absence of ρ-factor is about 3500 nucleotides, while its presence reduced this value to 1200 nucleotides. The average size of the RNA chains calculated from initiation studies was approximately the same as those observed in sucrose density gradients. The

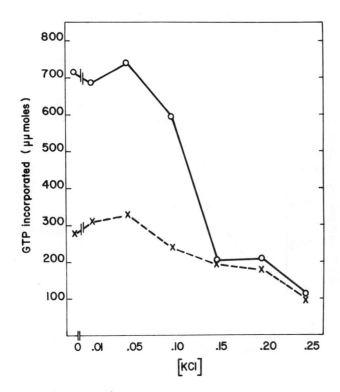

Fig. 2. Effect of [KCl] on the ability of ρ-factor to inhibit chain propagation on λ DNA. Reaction mixtures of 0.25 ml contained: 0.05 M Tris (pH 7.9); 0.004 M 2-mercaptoethanol; 0.01 M magnesium acetate; KCl as indicated; 1.5 μg λ S7 DNA (T); 0.15 mM UTP; 0.15 mM CTP; 0.15 mM ATP; 0.1 mM ^{3}H-GTP (3.6 x 10^3 cpm/mμmole); 50 μg bovine serum albumin; 2 μg RNA polymerase; and 1.0 μg ρ-factor when added. Reaction mixtures were incubated at 37°C for 20 minutes. Synthesis was determined as in Fig. 2 and RNA was counted as described. O———O no factor; X-------X with factor.

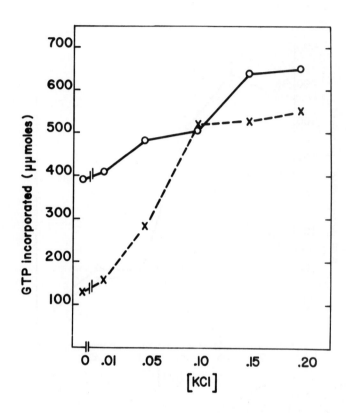

Fig. 3. Effect of [KCl] on the ability of ρ-factor to inhibit chain propagation on T4 DNA. The conditions of synthesis were as in Fig. 2, except with 1 μg T4 DNA. Treatment of the RNA product after 20 minutes of incubation at 37°C, was as described in Fig. 2 O———O no factor; X-----X with factor.

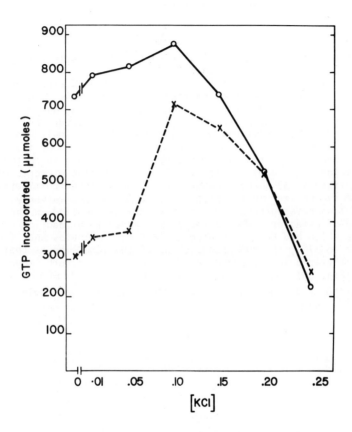

Fig. 4. Effect of [KCl] on the ability of ρ-factor to inhibit chain propagation on T7 DNA. The conditions of synthesis were as in Fig. 2, except with 1 μg of T7 DNA. Treatment of the RNA product after 20 minutes of incubation at 37°C, were as described in Fig. 2. O————O no factor; X-------X with factor.

majority of the RNA transcribed from λ DNA in the presence of ρ-factor sediments with a value between 7-12S.

ρ-Factor can Recognize a Pre-initiated Complex

Experiments on the kinetics of initiation clearly illustrate that ρ-factor does not influence the extent or rate of initiation of RNA chains. However, it is not clear from these experiments whether ρ-factor can produce shorter RNA chains if it is not part of the initiation complex. To answer this question, RNA synthesis was initiated (in the presence and absence of ρ-factor) and terminated under conditions which would yield very small RNA, i.e., incubation at 15°C for 2 minutes. Figure 5a shows that the RNA produced in that time interval was less than 4S in the presence or absence of ρ-factor. If these reaction mixtures were then shifted to 37°C and incubated for an additional 20 minutes, the RNA synthesized in the presence of ρ-factor was markedly smaller and less heterogeneous in size than RNA synthesized in the absence of ρ-factor. When ρ-factor was added to a reaction mixture which had been incubated at 15°C for 2 minutes and then shifted to 37°C, the RNA produced was of the same size as that RNA synthesized in the presence of ρ-factor during both incubations (Fig. 5b). Thus ρ-factor exerts its effect after initiation of RNA chains and need not be part of the initiation complex.

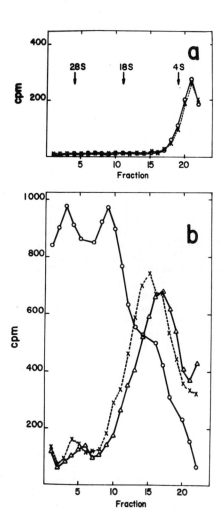

Figure 5
(Legend see next page)

Fig. 5. Effect of ρ-factor on the size of RNA when initiation of synthesis is begun in the presence or absence of ρ-factor. Reaction mixtures of 0.25 ml contained: 0.05 M Tris (pH 7.9); 0.004 M 2-mercaptoethanol; 0.01 M magnesium acetate; 0.01 M KCl; 2 μg λC_1 b2 DNA; 0.15 mM GTP; 0.15 CTP; 0.15 mM ATP; 0.1 mM α-^{32}P-UTP (2 × 10^4 cpm/mμmole); 50 μg bovine serum albumin; 2 μg RNA polymerase; 1 μg ρ-factor where indicated. The reaction mixtures of 6a were incubated at 15°C for 2 minutes, with and without ρ-factor. Two of the reaction mixtures of 6b were incubated at 15°C for 2 minutes without factor, and a third was incubated with factor. ρ-Factor was then added to one of the reaction mixtures which had been incubated previously without factor. All three reaction mixtures of 6b were then incubated at 37° for 20 minutes. In all cases, synthesis was stopped by chilling and the addition of SDS to a final concentration of 0.1%. Precipitated SDS was removed by centrifugation and the supernatants were heated to 65°C for 5 minutes. The reaction mixtures were layered on 5.2 ml 5-20% sucrose gradients in 0.1 M NaCl, 0.05 M Tris (pH 7.5), 0.001 M EDTA followed by centrifugation for 3.0 hours at 55,000 rpm and 5°C in the Spinco SW65 rotor. Fractions were collected and precipitated with 5% TCA in the presence of 0.01 M pyrophosphate and filtered on Millipore filters. ^{14}C-labeled HeLa cytoplasmic RNA was centrifuged in one of the tubes in 6a and 6b as a marker at 4S, 18S, and 28S. Double label counting was done in toluene-liquofluor on a liquid scintillation spectrometer. O———O without factor; X-------X factor added at 0 minutes; Δ———Δ factor added at 2 minutes.

DISCUSSION

Possible Mechanisms of Action of ρ-Factor

It is clear that ρ-factor affects chain propagation rather than initiation. The work of Roberts (1) indicated that some *in vitro* transcription on λ DNA is initiated at known λ promoters. Mutants defective in either r-strand or l-strand transcription originating outwards from the immunity region (specifically the λ sex and X_{13} mutants) can influence the amount of transcription *in vitro* in the presence of ρ-factor. Thus the size of the RNA synthesized from these regions in the presence of ρ roughly corresponds to predicted values. However, at present, it is not clear as yet whether ρ-factor can affect termination *in vitro* precisely at genetically determined sites.

The data presented above indicate that ρ causes the release of RNA chains by inhibiting their growth; there is no evidence for the release of RNA polymerase in the process. Thus ρ affects chain propagation rather than initiation; how ρ acts in this process is unclear. Evidence has been obtained which indicates that ρ binds to DNA (Table 2). By using a PEG-dextran phase

Table 2

Binding of ρ to DNA

DNA ADDED	INHIBITION OF RNA BY PEG PHASE
mμmole	%
0	54
0.05	49
0.14	47
0.28	27
0.55	15
0.83	9
2.75	<5

Table 2. The binding reaction was carried out in a total volume of 0.1 ml containing 0.01 M potassium phosphate buffer (pH 6.5), 2×10^{-4} M DTT, 5.0% polyethylene glycol (PEG), 4.5% Dextran, ρ-factor, and varying amounts of λ DNA. All reactants were mixed at 0° in a 3 ml test tube and were incubated for 2 min, at 37°. The phases were separated by brief centrifugation in 100 μl hematocrit tubes. An aliquot of the upper PEG phase was removed and added to a reaction mixture containing 0.1 M KCl, as described in the legend to Figure 2, in order to measure the extent of inhibition of RNA synthesis by the ρ-factor which had partitioned into the PEG phase. The above method for studying the binding of macromolecules to each other was suggested to us by Dr. Philip Silverman of this institution.

partition, it can be shown by direct measurement that ρ distributes primarily into the PEG-phase. In the presence of DNA, ρ which is complexed to DNA partitions within the dextran phase. Thus the loss of inhibition due to the addition of DNA in this simple partition reflects binding to DNA. Under these conditions, it is found that the concentration of ρ bound to DNA at saturation level corresponds to 1 molecule (MW 200,000) per 350 nucleotides of λ DNA. If, indeed, the effects of ρ are solely due to its binding ability to DNA, it is clear that there are too many ρ molecules bound to explain the biological specificity observed with this protein. Further work is essential before we will understand the significance of this interaction.

The following possible modes of action of ρ-factor do not appear to be ruled out by existing information:

I. ρ-Factor may function by recognizing RNA polymerase as part of the dynamic ternary complex of DNA-enzyme-RNA and cause release of RNA chains by altering the conformation of the complex. If ρ-factor interacts with the subunits of the core enzyme (i.e., similar to the manner of σ), it is clear that it need not do so at the initiation step. As shown above, ρ-factor can recognize a DNA-enzyme-RNA complex and terminate RNA synthesis. The signal for termination may lie in either the DNA strand which is being transcribed or in the synthesized complementary RNA strand. When RNA polymerase with ρ-factor attached (perhaps at the same site which σ occupied before initiation occurred) reaches a termination site, a conformational change is induced in the enzyme by the factor thereby causing the release of RNA.

II. Since ρ-factor can terminate RNA synthesis even after the initiation step it is possible that it functions by recognizing sites on DNA ahead of the synthesizing complex thereby preventing the movement of the enzyme through specific genetically prescribed locations. Steric blockage of RNA polymerase by ρ-factor may effect the release of RNA.

III. The possibility that ρ-factor is a specific nuclease that recognizes a termination site has not been ruled out by the data presented above. This nuclease action can either be specific for an RNA-DNA complex or a particular complex involving polymerase. Further studies are in progess to elucidate the mechanism of ρ-factor action.

FIGURE 6

Models of Salt Action and rho Action in Termination

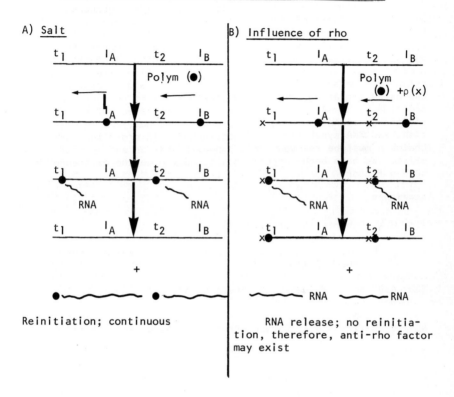

In this model, l_A and l_B refer to initiation sites for RNA synthesis while t_1 and t_2 refer to termination sites.

The studies presented above can be summarized by Figure 6. The models presented can be described as follows: (a) In salt-mediated termination, RNA polymerase binds at promoter-initiation sites, and RNA chains grow until termination sites are reached. The termination sites must be unique nucleotide sequences which signal the release of RNA chains and polymerase molecules. Released polymerase molecules can then reinitiate and repeat the cycle. (b) In the case of ρ action, evidence suggests that ρ blocks elongation. Our present working hypothesis consists of a model in which the binding of ρ to termination sites is independent of polymerase. The polymerase initiates and generates RNA chains by migrating down the DNA chain until it reaches the site occupied by ρ. At this point the RNA is released but the polymerase is not. Thus, this process does not lead to reinitiation, and RNA synthesis is discontinuous. In order to start new chains ρ must be removed; this suggests that an anti-ρ factor may exist. We have begun experiments to determine the existence of such a factor.

REFERENCES

1. J.W. Roberts, Nature, 224 (1969) 1168.

2. E. Fuchs, R.L. Miliette, W. Zillig and G. Walter, Eur. J. Biochem. 3 (1967) 183.

3. A.G. So, E.W. Davie, R. Epstein and A. Tissieres, Proc. Nat. Acad. Sci. 58 (1967) 1739.

4. U. Maitra and F. Barash, Proc. Nat. Acad. Sci. 64 (1969) 779.

5. J.P. Richardson, Nature, 225 (1970) 1109.

6. A. Shapiro, E. Vinuela and J.V. Maizel, Biochem. and Biophys. Res. Comm. 28 (1967) 815.

7. A.R. Goldberg and M. Howe, Virology, 38 (1969) 200.

*This work was supported by grants from the U.S. National Institutes of Health, the National Science Foundation and the American Cancer Society.

DISCUSSION

S.S. COHEN: Can you tell us whether rho is only present in uninfected or λ-infected *coli*?

J. HURWITZ: Yes, rho is present in uninfected *coli* as well as in λ-infected *coli*.

S.S. COHEN: And is it destroyed in T-even phage infection?

J. HURWITZ: That has not been looked at yet. As far as we know, rho is capable of acting on T_4, but we have not examined rho in T_4-infected cells in terms of its quantitative aspects.

F. HUIJING: The fact that rho is present at the end of such a sequence, together with the enzyme, still prevents initiation at the start; so if you were doing experiments where you had excess enzyme as compared to the DNA, would you not have any further reaction or would you accumulate a second enzyme at the end of the sequence, maybe with the RNA still attached?

J. HURWITZ: The important thing in running these experiments is that one uses limited amounts of DNA. If you have an excess of rho, it doesn't matter how much enzyme you add, you still find that the RNA which is produced is blocked in terms of the size. You don't go past the rho termination.

R.L. NOVAK: I'd like to ask a question about the chemical modifications of the polymerase. Were all of Isihama's modification studies performed on complete enzyme. That is, on core enzyme puls sigma?

J. HURWITZ: As far as we can tell enzyme is more than 80-85% loaded with sigma.

R.L. NOVAK: Some of our modifications of the core enzyme differ from those found by Isihama for the complete enzyme. For example, sulfydryl modifying reagents do not inhibit DNA binding or initiation, but inhibit polymerization. The same studies also indicate that the core enzyme may be a heterogeneous population of at least two kinds of molecules - one type of polymerase being inactivated by modifiers such as Koshland's reagent and the other being totally resistant.

J. HURWITZ: So far we have not seen heterogeneity with regard to α, β and β' but I must say that the separation is very very difficult. Certainly the β and β' subunits can be looked at by virtue of their large size.

P.K. QASBA: Do you think that the high salt-mediated termination found *in vitro* is also operating *in vivo*?

J. HURWITZ: I don't know. The only thing which I can say is that the *in vitro* RNA made off of T_4DNA (this is the work done by Maitra and his colleagues) in the presence of high salt can be specifically competed for by very early *in vivo* T_4 messenger RNA. This doesn't answer the question specifically in terms of termination sites. The salt concentration *in vivo*, based on various studies carried out with *E. coli* mutants which are blocked in potassium transport, is apparently very close to 0.2M. But the important point to us is that if you are going to invoke the salt-mediated termination with *E. coli* DNA, the effects of rho are going to be blocked under these conditions. We are in a dilemma: when one acts, the other is not going to act; when there is high salt, rho isn't going to act; when there is low salt, only rho is going to act.

P.K. QASBA: High salt-mediated termination *in vitro* may be just mimicking the termination process at non-biological termination sites. *In vivo* termination may be directed by rho-like factors specific for each phage or bacteria.

J. HURWITZ: These are possibilities.

REPAIR OF DAMAGES DNA IN HUMAN AND OTHER EUKARYOTIC CELLS

J. E. Cleaver
Laboratory of Radiobiology
University of California, San Francisco

> Words strain,
> Crack and sometimes break, under the burden,
> Under the tension, slip, slide, perish,
> Decay with imprecision, will not stay in place,
> Will not stay still.
>
> The Four Quartets, T. S. Eliot, 1934

At one time radiobiologists interpreted most of their observations in terms of target theory and obtained some very useful hints as regards gene size, virus size and structure, and chromosome structure (1). In recent years there has been an apostasy from target theory and an emergence of a new cult based on repair theory. But the word "repair" is unfortunately used all too often as a magic word to confer relevance and significance to many uninterpretable radiation experiments. I am sure that many things we observe in irradiated cells, even those things associated with what we think to be error correcting processes, have little to do with processes which are genuinely responsible for recovery of damaged cells. Many novel and perturbed reactions occur in a radiation-damaged cell population. Some may be repair which culminates in the recovery of some cells. Others may be ineffective attempts at repair or merely pathological reactions. All of these reactions may occur simultaneously in a cell population, but the most interesting for us are those which represent repair which culminates in recovery. Considerable advances have been made in the study of repair processes in bacteria due in large measure to the versatility with which mutants can be made and selected. In addition, in vitro enzymological studies have begun to elucidate the DNA-protein interactions which underly the enzymatic recognition and correction of damage to DNA. In vitro experiments with mammalian repair enzymes are yet to be begun, but we can now infer a great deal about the nature of one repair process, excision repair. Other repair processes undoubtedly occur, but little can be said with any certainty about their nature and importance in mammalian cells; the following discussion is therefore restricted to excision repair.

Xeroderma Pigmentosum — Symptoms and Cell Cultures

The strongest evidence for the importance of excision repair in mammalian cells comes from the human herediatry skin disease xeroderma pigmentosum (XP) (2-9). This disease is inherited as an autosomal recessive gene (11), the karyotype is normal, and the outstanding clinical symptom is the induction by sunlight of all types of cancers of the skin (e.g., squamous carcinomas, melanomas, sarcomas, keratoacanthomas (12), and occasionally mental retardation (3,12)).

Most experimental work has been done, of necessity, on fibroblasts which are the only cell type from skin biopsies that can be cultured for long periods. XP biopsies are rather more difficult than biopsies from normal or heterozygous XP donors to develop into fibroblast cultures. This may be due more to the general diseased state of the skin than to any specific feature of homozygous XP fibroblasts. Once in culture they are similar to normal fibroblasts in most characteristics of growth (Table I) although results vary between different XP donors and different experimenters. It is reassuring that our results with fibroblasts are thus far consistent with results obtained from epidermal cells in vivo (13) and peripheral lymphocytes (14).

Table I. Tissue culture characteristics of fibroblasts from normal and XP adult donors.[a]

	Normal	XP
Doubling time	21 hr[b], 42-48 hr	26 hr[b], 39 hr
Plating efficiency[c]	10-16.5%	6.3-12%
S phase	8-9 hr	8-8.5 hr
G_2	8-10 hr	7-10 hr
DNA chain growth[d]	0.9-1.8 μ/min[d]	1.1 μ/min
Lifetime in culture[e]	50-70 generations	50-70 generations
SV40 transformation frequency [f]	0.025%	0.03%, 0.023%

[a]My own unpublished data except where specified. All measurements made within 20 weeks of taking biopsies. [b]Reference (5). [c]The plating efficiency may vary by a factor of 2 with culture age (16). [d]Based on isopycnic gradient method of Painter and Schaefer (17). [e]S. Goldstein, unpublished data. Normal donors (16). [f]Reference (18).

Tests for Excision Repair

If we ask whether xeroderma pigmentosum is a human analogue of UV-sensitive bacteria defective in some stage of excision repair, then the following experimental tests can be made of the model for excision repair (Fig. 1):

(a) Are individual cells of xeroderma pigmentosum genotype sensitive to UV light?

(b) How well do XP fibroblasts support the growth of UV-damaged viruses?

(c) Are thymine dimers formed and removed from DNA?

(d) Are single strand gaps made and joined during excision?

(e) Are short regions of new bases inserted into DNA to replace excised dimers?

(f) Are short regions of new bases inserted to repair single strand breaks?

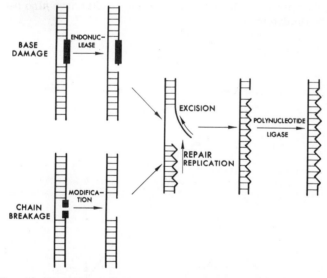

Fig. 1. Heuristic scheme for the operational steps in excision repair of damaged bases (e.g., pyrimidine dimers) and broken strands (e.g., ionizing radiation damage) by a common pathway. Initial step for each kind of damage has some unique features but the excision, replication and ligase steps can be common.

Definite answers can be given to all of these questions except for (d). An answer to (d) has to await refinement of present alkaline sucrose gradient techniques for studying a small number of breaks in large DNA molecules. Despite contrary claims, attempts to demonstrate single strand breaks during excision have been equivocal due to extreme variability in gradient profiles (5,15).

Sensitivity of Xeroderma Pigmentosum Cells

The clinical symptom of xeroderma pigmentosum patients is a high incidence of UV-induced skin cancers, but this could be due to a variety of systemic factors other than the inherent properties of individual cells (12). It is important to demonstrate that the XP genotype has some phenotypic expression in individual cells. The only method at our disposal is colony formation by single cells, but unfortunately primary fibroblasts have a low plating efficiency (a maximum of 10 to 20% (9,16)). Use of this method presupposes that the cells which form colonies are a random sample of the population and are typical of the whole population in terms of their UV sensitivity. Granted these presuppositions, XP fibroblasts are much more sensitive than normal cells (Fig. 2); similar results have also been obtained by Goldstein (19).

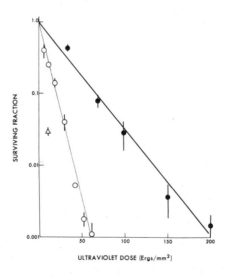

Figure 2

Fig. 2. Single cell survival curve for normal and XP fibroblasts irradiated with UV light at 14 ergs/mm^2 (mainly 254 nm). Plating efficiencies of unirradiated cells was between 9 and 18%. Normal fibroblasts ●, XP6 ○, XP7△. Bars denote 95% confidence limits and lines are drawn by regression analysis using all data points, except those for XP7. Normal cells n = 1.0, D_0 = 29 ergs/mm^2. XP6 n = 0.9, D_0 = 9 ergs/mm^2.

Host Cell Reactivation of UV-damaged Viruses

Some classes of UV-sensitive bacteria have a reduced ability to support the reproduction of UV-damaged phage, presumably because the same enzyme system repairs host DNA and infecting phage DNA (19). Experiments which illustrate this phenomenon in human cells have been done both with herpes virus (6) and SV40 virus (18) (Fig. 3). XP fibroblasts appear in these experiments to be analogues of the bacteria which are defective in host cell reactivation (20).

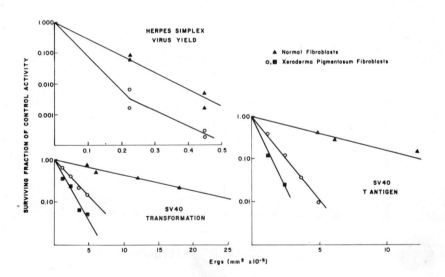

Fig. 3. Percentage remaining virus activity of irradiated viruses grown in normal or XP cells. Top, survival of herpes virus (6); bottom, T antigen induction by SV40 (18); right, transformation frequency by SV40 (18). Data redrawn from cited publications.

Excision of Pyrimidine Dimers from Human Cells

The number of dimers formed in the DNA of human cells increases linearly with UV dose if the cells are fixed immediately after irradiation (Fig. 4).

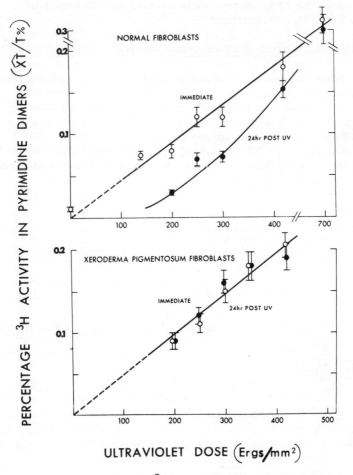

Fig. 4. Percentage of ^{3}H in pyrimidine dimers ($\widehat{TT}/T$) as a function of UV (254 nm) dose. O, value immediately after irradiation; ●, value 24 hr after irradiation. Top, normal fibroblasts; bottom, XP fibroblasts. Bars denote standard errors for 3 or more determinations or a standard deviation of 10% for a single determination (8).

When cells are allowed to grow for a time after irradiation before fixation and measurement, normal cells excise some of the dimers from their DNA, whereas xeroderma pigmentosum cells apparently do not (Fig. 5). However, only about 50 to 75% of the dimers are excised from normal human cells, even at low UV doses (5,8,21). Bacteria are much more efficient in excision and some can excise nearly 100% of the dimers from their DNA. Although the difference between normal and XP fibroblasts looks distinctive, it is less so if one calculates the number of dimers which remain in DNA. At a dose of 100 ergs/mm^2 approximately 10^6 dimers per cell are formed in DNA. When excision of 75% is complete in normal cells, they still have 2.5×10^5 dimers as compared to 10^6 in XP cells. Why the difference between 10^6 and 2.5×10^5 dimers can make such a large difference in relative survival of the two cell types is not easy to understand. Finally, with respect to dimer studies, the assay is not sufficiently precise to distinguish between a low and a completely negligible excision level. Other studies make it likely that low levels of excision occur in xeroderma pigmentosum cells (2,7,9,22).

Insertion of New Bases to Repair DNA

Replacement of excised dimer-containing regions requires synthesis of very small amounts of DNA in short regions, "repair replication" (2,7,9,23-25). This has to be detected in the presence of the relatively large amounts of semiconservative DNA replication that occur during the cell cycle.

Two methods can be used to distinguish repair replication from semiconservative replication. One method consists of detecting by autoradiography small amounts of radioactive thymidine (^{3}HTdR) incorporated into cells during repair replication when the cells are incapable of semiconservative replication (2,4,7,13,26-29). DNA repair detected in this way is called "unscheduled synthesis" (29) to distinguish it from the normal "scheduled" DNA synthesis that gives rise to heavily labeled cells. A particularly elegant demonstration (13) of such repair in normal skin, and its absence in XP skin, can be seen if ^{3}HTdR is injected intradermally soon after exposure of the skin to UV light (Fig. 5).

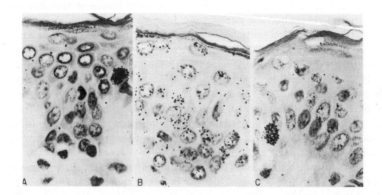

Fig. 5. (A) Normal human epidermis not subjected to UV irradiation. There is dense labeling of one basal cell in normal S phase DNA synthesis. (B) Normal human epidermis irradiated with 13.6×10^4 ergs/mm^2 UV light. There is sparse labeling of cells throughout the epidermis. (C) Epidermis of an XP patient irradiated with 13.6×10^4 ergs/mm^2 UV light. There is one densely labeled basal cell but no sparse labeling. All skin injected with ^{3}HTdR (10 μCi/ml, 11 Ci/mmole) intradermally and biopsies taken 1 hr later. Photographs kindly supplied by J. H. Epstein (13).

Repair Replication

Another method to detect repair replication was devised by Pettijohn and Hanawalt (23). This takes advantage of the difference between the short pieces synthesized during repair and the long strands synthesized by semiconservative replication. Tritiated bromodeoxyuridine (^{3}HBrUdR, molecular weight 308) is used as a heavy analogue of ^{3}HTdR (molecular weight 242) and is incorporated into DNA in its place. The long chains formed by semiconservative replication are denser than normal and ^{3}H-labeled. The pieces synthesized by repair replication are such small patches in long molecules that the density of the molecules is unchanged, but they too will be ^{3}H-labeled. Thus, the presence of ^{3}H-labeled molecules of normal density is an indication that repair replication has occurred. DNA can be fractionated on the basis of its density by means of centrifugation through cesium chloride gradients (Fig. 6) which show that XP fibroblasts perform much less repair replication than normal cells. Most XP fibroblasts from donors in North America have residual repair capacity that is below 25% of normal (Table II (9)) but higher levels have been found in some Dutch cases (22).

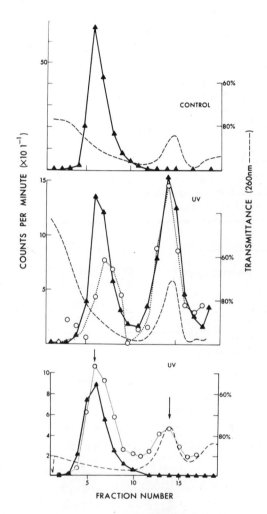

Fig. 6. Equilibrium density gradients from normal and xeroderma pigmentosum cells irradiated with 200 ergs/mm^2 ultraviolet light and labeled for 4 hours with ^{3}HBrUdR. Arrows mark positions of normal density (1.698 gm/cc) and bromouracil substituted density (1.751 gm/cc). Dashed line indicates the total amount of DNA in the gradients (high values in the first few fractions are an artefact). Top, unirradiated normal cells. Center, normal (▲) and heterozygous xeroderma pigmentosum (O) fibroblasts irradiated with 200 ergs/mm^2. Note the large peak of radioactivity at a density of 1.698 gm/cc which indicates

repair replication. Bottom, two different fibroblast cultures, from unrelated homozygous xeroderma pigmentosum patients, irradiated with 200 ergs/mm^2 UV. Note the low repair replication peak at 1.698 gm/cc in one case (O) and the absence in the other case (▲).

Table II. Relative levels of excision repair (measured during 4 hr after 220 ergs/mm^2 UV light, 254 nm).*

			Percent
Normal skin fibroblasts			100
Normal skin fibroblasts (SV40 transformed)			121
Fibroblasts (isolated by amniocentesis)			103
HeLa (cervical carcinoma)			94
Xeroderma pigmentosum			
Heterozygote	♀	XPH1	67, 84, 87
	♀	XPHK §	100
	♂	XPH11M	44, 48
	♀	XPH11F	47, 42
Homozygote		XPJ	6
		XP6	5.1, 5.3
		XP7	5
		XP11	<4
	(SV40 transformed)	XP7	<16
Homozygote		XP1 §†	9
		XP2 §†	25
		XPIW §	0
		XPKM §†	0
		XPKF §†	0

*Standard deviation is 20 percent of quoted value. (Upper limit cited when no distinct normal density ^{3}H peak was detectable in the isopycnic gradient.) (9)

§ de Sanctis Cacchione syndrome. XP1, 2 diagnosed by E. Klein, Roswell Park Institute. XPIW, KM, KF diagnosed by W.B. Reed, UCLA. This form of XP has mental abnormalities in addition to skin symptoms.

† Siblings.

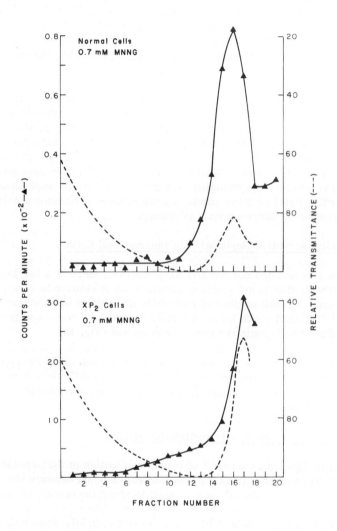

Fig. 7. Isopycnic gradients of normal and XP cells grown for 1 hr in BrUdR, 3 µg/ml, for 45 min in BrUdR plus 0.7 mM nitro nitrosoguanidine, and labeled for 4 hr in ^{3}HBrUdR (20 µCi/ml, 3 µg/ml plus 10^{-3} M hydroxyurea). FUdR at 10^{-6} M added throughout the incubation. ^{3}H ▲——, transmittance at 260 nm -----.

Damage from ionizing radiations and chemical mutagens is repaired to normal extents by XP fibroblasts (4, 30, 31) (Fig. 7). These agents appear to cause non-enzymatic breakage of the DNA strands (32, 33) and repair of the breaks presumably bypasses the initial enzymatic steps required for repair of UV damage (pyrimidine dimers) (Fig. 1).

It should be emphasized that XP and normal cells constitute the only example in which there is a correlation between the amount of repair replication and the level of survival after irradiation (9, 30). In general, the amount of repair replication increases with both the amount of damage to DNA bases and the level of cell killing (4, 24, 25, 30). Consequently, much of what we call repair replication does not result in recovery from radiation damage. Either repair replication is ineffective itself or other competing degradative processes dominate in the presence of large amounts of damage.

Non-Semiconservative Replication in Unirradiated Cells

In unirradiated normal and XP fibroblasts there is a low level of DNA synthesis that is not semiconservative but is similar to repair replication. A small number of molecules are able to incorporate low levels of ^{3}HBrUdR which increases their density by a much smaller amount than would semiconservative replication (Fig. 8).
The level attained by XP cells is below the maximum they can perform and the XP mutation has therefore no influence on this form of repair replication. Such replication may be repair of naturally occurring damage to DNA bases or of breaks that could arise, for example, during RNA synthesis.

Size of Patch Made by Repair Replication after UV

DNA labeled as a result of repair replication in the presence of ^{3}HBrUdR has a density indistinguishable from normal because the ^{3}HBrU in the repair patches is too small to affect the density of molecules that are large relative to the patch size. After sonication to low molecular weights, repaired DNA appears to be slightly denser than normal DNA (Fig. 9) and the increase in density can be used to determine the percentage replacement of T by BrU in the repaired DNA. At present, however, the density increase is just on the limit of resolution in these gradients at a molecular weight of slightly less than 10^6, and smaller sizes are difficult to obtain by sonication. At a molecular weight of 10^6 and a dose of 200 ergs/mm^2 there is less than 1 pyrimidine dimer per molecule on average. The degree of replace-

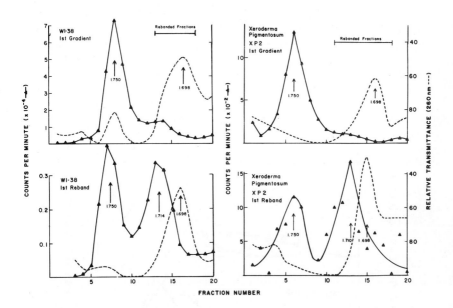

Fig. 8. Isopycnic gradient profiles from normal and XP cells grown for 30 min in 3 µg/ml BrUdR, for 4 hr in ^{3}HBrUdR (20 µCi/ml, 3 µg/ml plus 10^{-3}M hydroxyurea) and a further 30 min in 3 µg/ml BrUdR. FUdR at 10^{-6}M was added throughout the experiment. Gradient calibrated by the positions of normal DNA and M. luteus DNA (1.731 g/cc), latter not shown in figure. Top, initial gradients, bottom, first reband of DNA from normal density regions of initial gradients. ^{3}H ▲, transmittance at 260 nm -----.

ment represented by the density increase in Fig. 9 can then be used to estimate an upper limit for the average patch size, assuming there is just one patch per molecule. The size is less than 200 nucleotides.

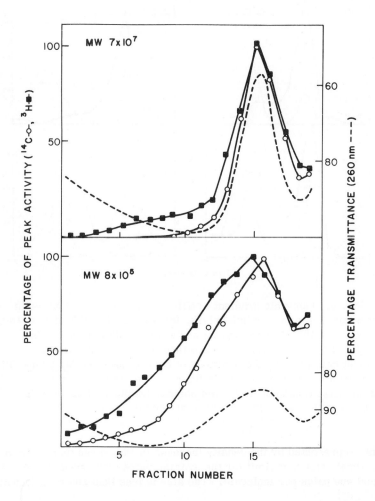

Fig. 9. First rebands in isopycnic gradients of normal density DNA from HeLa cells labeled with ^{3}HBrUdR (see Fig. 6 for protocol) by repair replication mixed with DNA from HeLa cells labeled with ^{14}CTdR. Top, DNA sheared gently to a uniform high molecular weight. Bottom, DNA sonicated to a uniform low molecular weight. Counts normalized to a percentage of peak activity. ^{3}H ―■―, ^{14}C ―○―, transmittance at 260 nm -----.

Site of Enzymatic Defect in Xeroderma Pigmentosum

Since xeroderma pigmentosum cells seem to be unable to excise dimers from their DNA but do repair breaks to normal extents (4, 31), it is reasonable to presume that an enzyme involved in excision is affected. The disease is genetically recessive (11, 12), so the affected enzyme is either present at a much reduced level or in a structure which impairs its function. The enzymes involved in repair have only been isolated from some bacteria, notably from M. luteus in recent work by L. Grossman (34-36). In this organism three enzymes are involved in excising dimers (Fig. 10): (1) an endonuclease which cleaves the chain near a dimer to leave 3'phosphoryl and 5'hydroxyl termini, the latter adjacent to the dimer, (2) a 3'phosphomonoesterase to remove the terminal phosphate, (3) an exonuclease to degrade the dimer-containing sequence. At least two more enzymes are required for completion of repair — a polymerase to lay down a new strand to replace the excised region, and a ligase to make the final link to complete the repair patch. In E. coli the exonuclease and polymerase functions seem to be on a single enzyme (37). A similar sequence of operations seems to occur in human cells and the enzyme affected in xeroderma pigmentosum is probably the endonuclease (4).

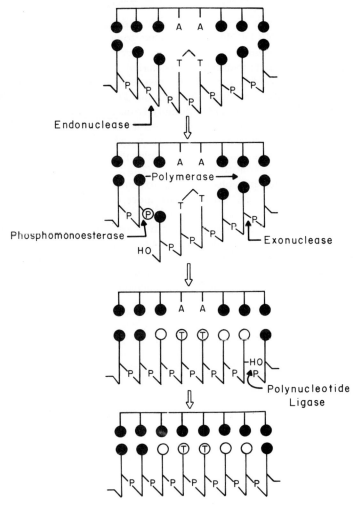

Fig. 10. Scheme for the biochemical steps in repair of pyrimidine dimers in DNA (redrawn from Grossman et al. (36)).

Repair Levels in Xeroderma Pigmentosum Heterozygotes

Heterozygotes of XP do not show any unambiguous symptoms, but we might expect the enzyme activity in heterozygotes to be half that in normal cells (38, 39). In our initial observations on cultured fibroblasts, however, the repair levels of heterozygous and normal cells were identical (4) (Fig. 11, top), and similar results have also been obtained by Bootsma et al. (22). Recently, we have studied a family in which both parents appear to have reduced repair levels (Fig. 11, bottom) although there is neither known consanguinity nor clinical symptoms of sensitivity (9).

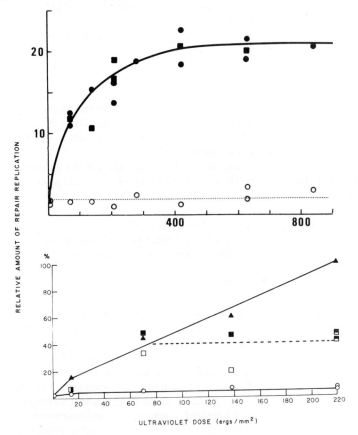

Fig. 11. Relative repair levels in normal, heterozygous and homozygous XP fibroblasts from various donors. Top, normal ●, heterozygous XP (de Sanctis Cacchione) ■, homozygous XP (de Sanctis Cacchione) O. Levels based on grain numbers in autoradiographs of unscheduled synthesis (4). Bottom, normal ▲, heterozygous XP □ (mother), ■ (father), homozygous XP O. Levels based on specific activities in isopycnic gradients (see Fig. 6 and Table II).

The dose response of repair in these two heterozygotes suggests that the repair level saturates at a lower level than in normal cells, and below 70 ergs/mm^2 heterozygous and normal cells are similar. In most of the homozygous XP cases studied the reduced repair level is found throughout the whole dose range (22). Repair replication levels are not, however, direct measurements of activity of repair enzymes. Until we have some direct measurements similar to those available for the UV specific endonuclease of M. luteus (34-36), these observations are only suggestive of differences amongst XP heterozygotes.

DNA Repair in Other Diseases

XP is a disease associated with a defect in the initial step of excision repair. At present, we do not know of any other human disease associated with defects in other steps, although several have been investigated (2, 7). No animal analogues are available either (40); DNA repair is normal in the hairless mouse (41) and the cancer cye syndrome in Hereford cattle (42), both of which are light sensitive conditions with some similarity to XP.

It would be interesting also if there were human analogues of other UV-sensitive bacterial mutants such as rec⁻ (43) or pol⁻ (44, 45), but such human mutants might not survive embryogenesis.

Repair Replication in Non-Human Mammalian Cells

We might conclude from the failure of dimer excision in XP that cells which do not excise should all be UV sensitive mutants. Such is not the case. There is no doubt that repair replication seems to be widespread in mammalian cells and has been observed in human, Chinese hamster, mouse, rabbit, marsupial, and cow cells (25-29, 42). Excision of pyrimidine dimers into acid soluble fragments has only been reported for human and primate cells (5, 8, 20, 46); attempts to detect excision from Chinese hamster, mouse, pig and marsupial cells have been uniformly unsuccessful (47-50). The isolated report (50) of excision in Ehrlich ascites cells has remained unconfirmed and the cells used had unusually high UV resistance. Since repair replication occurs even in species where no excision of dimers has been detected (25-28), some form of excision must occur and one must assume that either non-dimer regions are excised or that methods for detection of dimer excision are currently inadequate. The former alternative is unsatisfactory, because dimers are clearly important repaired lesions in microorganisms (51) and man (5, 7, 8, 21). A simple resolution of the problem is to assume that dimers are excised in acid-insoluble fragments in non-primate cells. Such fragments need not be larger than 10 to 15 nucleotides long to be acid-insoluble. If this proves correct, then the exonuclease step of excision repair has distinct differences amongst various mammalian species. In non-primate cells the exonuclease may degrade DNA less extensively than in human cells or microorganisms.

Conclusions — Excision Repair in Mammalian Cells

It appears that mammalian cells have a repair pathway similar to excision repair in bacteria but only in human cells is the pathway similar in all respects. In other animal cells the excision step has yet to be detected and the role of pyrimidine dimers is still a mystery, though much evidence suggests that excision in some form should occur. In human XP cells the defect in excision repair is similar to that in uvr⁻ hcr⁻ bacterial mutants (2-9) and the discovery of the basis of this disease has been invaluable in proving the existence and importance of excision repair in higher organisms. Similar mutations in animals, if they became available, would be invaluable in studying all aspects of excision repair and its relationship to radiation carcinogenesis. Little work has been done on the isolation of any enzymes associated with excision repair in mammalian cells and this may be the next step to anticipate. We hope that such studies will be as instructive as those of the research groups of Grossman on M. luteus (34-36) and Kornberg on E. coli (37). Whether the human DNA polymerase will prove as versatile as its bacterial counterpart remains to be seen.

Speculations — Excision Repair and Carcinogenesis

What are the lessons to be learned from the association of defective UV repair with enhanced UV carcinogenesis in xeroderma pigmentosum? The first is that the disease is a special case and most forms of carcinogenesis probably have nothing to do with defective repair (7) and cancer cells have perfectly competent repair processes (2,7). We must beware of making a dramatic example into a general rule. Instead, xeroderma pigmentosum must be viewed in the context of what we currently believe about carcinogenesis. It would be easy, but naive, to extend bacterial results to man and say that since uvr⁻ hcr⁻ bacteria show an elevated UV-induced mutation rate, so should XP cells. Hence, here is evidence for the somatic mutation theory of cancer (52). Any of the following possibilities are consistent with current knowledge and ignorance concerning causes of cancer and can be applied to XP.

(a) Increased somatic mutation rate from sunlight.

(b) Increased incidence of chromosome aberrations.

(c) Increased DNA alterations leading to altered differentiation.

(d) Increased transformation by oncogenic viruses.

(e) Increased induction of latent viruses.

(f) Secondary consequence of accelerated cell death or aging.

Some of these possibilities can be excluded on an experimental basis. Accelerated aging does not appear to occur in cultured XP cells (Table I) so this possibility can be excluded. Unlike the situation in some malignant diseases (53, 54), XP cells are transformed at normal rates by SV40 virus (Table I) (18). Irradiation of recipient cells before infection, however, increases transformation by SV40 (55, 56) and therefore XP cells might be more readily transformed by oncogenic viruses in the presence of sunlight.

Of the available possibilities, unrepaired alterations in DNA which lead to increased mutations or increased oncogenic viral transformation may be the most readily testable hypotheses for actinic carcinogenesis in XP. There is a possibility (Fig. 12) that the recombinational repair mechanism (43, 57, 58) could be a common means for both mutation production and oncogenic viral transformation, and the mutational and viral theories may not necessarily be mutually exclusive. Which theories prove correct in subsequent research, and whether there is any lesson to be learned from XP for carcinogenesis in general, we must wait to see.

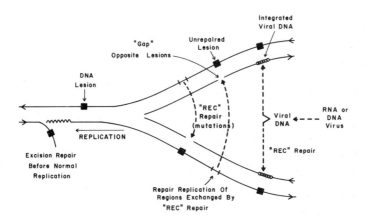

Fig. 12. Heuristic scheme containing features of both mutation and viral oncogenesis. Unrepaired lesions in DNA cause gaps to be left in the newly synthesized daughter strands; the gaps are filled by a recombinational process from the parental strand of the other DNA molecule (57). Since the "rec" process is error prone (59), it may also incorporate viral DNA that may be present in the cell from oncogenic DNA or RNA viruses. The combination of mutations and integrated viral DNA produces the altered malignant cell.

ACKNOWLEDGMENTS

Three groups of people deserve acknowledgment for their invaluable assistance to my own part in this work. First and foremost, the XP patients and their parents who have cooperated with little hope of immediate benefit; second, the many dermatologists who have aided me in my trespass into their field; and third, the research staff and assistants of the Laboratory of Radiobiology, whose contributions in ideas and hard work have been invaluable. Among the latter, Mr. G. H. Thomas deserves special mention for performing the major portion of our own technical work.

REFERENCES

1. D. E. Lea, Actions of Radiations on Living Cells (Cambridge University Press, Cambridge, 1946).
2. J. E. Cleaver, Nature 218 (1968) 652.
3. W. B. Reed, B. Landing, G. Sugarman, J. E. Cleaver, and J. Melnyk, J. Amer. Med. Assn. 207 (1969) 2073.
4. J. E. Cleaver, Proc. Natl. Acad. Sci. 63 (1969) 428.
5. R. B. Setlow, J. D. Regan, J. German, and W. L. Carrier, Proc. Natl. Acad. Sci. 64 (1969) 1035.
6. A. S. Rabson, S. A. Tyrell and F. Y. Legallais, Proc. Soc. Exp. Biol. Med. 132 (1969) 802.
7. J. E. Cleaver, J. Invest. Derm. 54 (1970) 8.
8. J. E. Cleaver and J. E. Trosko, Photochem. Photobiol. 11 (1970) 547.
9. J. E. Cleaver, Int. J. Rad. Biol. (1970) in press.
10. D. W. Smith and P. Hanawalt, Biophys. J. 7 (1967) 78.
11. H. El-Hefnawi, S. M. Smith and L. S. Penrose, Ann. Human Genet. 28 (1965) 273.
12. A. Rook, D. S. Wilkinson and F. J. Ebling, Textbook of Dermatology, Vol. I (Blackwell, Oxford and Edinburgh, 1968) p. 62.
13. J. H. Epstein, K. Fukuyama, W. B. Reed and W. L. Epstein, Science 168 (1970) 1477.
14. P. G. Burk, M. A. Lutzner and J. H. Robbins, Clin. Res. 17 (1969) 614.

15. E. Ben-Hur and R. Ben-Ishai, Photochem. Photobiol. (1971) in press.
16. S. Goldstein, J. W. Littlefield and J. S. Soeldner, Proc. Natl. Acad. Sci. 64 (1969) 155.
17. R. B. Painter and A. W. Schaefer, J. Mol. Biol. 45 (1969) 467.
18. S. A. Aaronson and C. D. Lytle, Nature 228 (1970) 359.
19. S. Goldstein, Proc. IVth Int. Congress Radiation Research, Evian, France (1970) in press.
20. C. S. Rupert and W. Harm, in: Advances in Radiation Biology, Vol. 2, eds. L. Augenstein, R. Mason and M. R. Zelle (Academic Press, New York, 1966) p. 1.
21. J. D. Regan, J. E. Trosko and W. L. Carrier, Biophys. J. 8 (1968) 319.
22. D. Bootsma, M. P. Mulder, F. Pot and J. A. Cohen, Mutation Res. 9 (1970) 507.
23. D. Pettijohn and P. C. Hanawalt, J. Mol. Biol. 9 (1964) 395.
24. J. E. Cleaver, Biophys. J. 8 (1968) 775.
25. J. E. Cleaver, Photochem. Photobiol. 12 (1970) 17.
26. R. E. Rasmussen and R. B. Painter, Nature 203 (1964) 1360.
27. R. E. Rasmussen and R. B. Painter, J. Cell Biol. 29 (1966) 11.
28. R. B. Painter and J. E. Cleaver, Radiation Res. 37 (1969) 451.
29. B. Djordjevic and L. J. Tolmach, Radiation Res. 32 (1967) 327.
30. J. E. Cleaver, Proc. IVth Int. Congress Radiation Research, Evian, France (1970) in press.
31. W. J. Kleijer, P. H. M. Lohman, M. P. Mulder and D. Bootsma, Mutation Res. 9 (1970) 517.
32. C. J. Dean, M. G. Ormerod, R. W. Serianni and P. Alexander, Nature 222 (1969) 1042.
33. B. B. Strauss and T. Hill, Biochem. Biophys. Acta 213 (1970) 14.
34. J. C. Kaplan, S. R. Kushner and L. Grossman, Proc. Natl. Acad. Sci. 63 (1969) 144.
35. L. Grossman, Proc. IVth Int. Congress Radiation Research, Evian, France (1970) in press.

36. L. Grossman, J. Kaplan, S. Kushner and I. Mahler, Ann. Inst. Super. Sanita 5 (1969) 318.
37. R. B. Kelley, R. B. Atkinson, J. A. Huberman and A. Kornberg, Nature 224 (1969) 495.
38. T. T. Puck and H. Z. Hill, Proc. Natl. Acad. Sci. 57 (1967) 1676.
39. D. Y. Hsia, Med. Clinics North Amer. 53 (1969) 857.
40. J. E. Cleaver, Comp. Path. Bull. 2 (1970) 4.
41. J. H. Epstein, K. Fukuyama and W. L. Epstein, J. Invest Derm. 51 (1968) 445.
42. J. E. Cleaver, J. Lett, M. Zelle and R. Kainer (in preparation).
43. A. Rorsch, P. van de Putte, J. E. Mattern and H. Zwenk, in: Genetical Aspects of Radiosensitivity: Mechanisms of Repair (IAEA, Vienna, 1966) p. 105.
44. J. M. Boyle, M. C. Patterson and R. B. Setlow, Nature 226 (1970) 708.
45. L. Kanner and P. C. Hanawalt, Biochem. Biophys. Res. Comm. 39 (1970) 149.
46. J. D. Regan, R. B. Setlow, W. L. Carrier and W. H. Lee, Proc. IVth Int. Congress Radiation Research, Evian, France (1970) in press.
47. J. E. Trosko, E. H. Y. Chu and W. L. Carrier, Radiation Res. 24 (1965) 667.
48. J. E. Trosko and M. R. Kasschau, Photochem. Photobiol. 6 (1967) 215.
49. M. Klimek, Photochem. Photobiol. 5 (1966) 603.
50. M. Horikawa, O. Nikaido and T. Sugahara, Nature 218 (1968) 489.
51. R. B. Setlow, Science 153 (1966) 379.
52. L. Szilard, Proc. Natl. Acad. Sci. 45 (1959) 30.
53. G. J. Todaro, H. Green and M. R. Swift, Science 153 (1966) 1252.
54. S. A. Aaronson and G. J. Todaro, Virology 36 (1968) 254.
55. E. J. Pollock and G. J. Todaro, Nature 219 (1968) 520.
56. C. D. Lytle, K. B. Hellman and N. C. Telles, Int. J. Rad. Biol. 18 (1970) 297.

57. W. D. Rupp and P. Howard-Flanders, J. Mol. Biol. 31 (1967) 291.
58. J. E. Cleaver and G. H. Thomas, Biochem. Biophys. Res. Comm. 36 (1969) 203.
59. E. Witkin, Brookhaven Symposium in Biology 20 (1967) 17.

DISCUSSION

R. WARREN: Have you tried, or do you know if anyone else has tried, cell hybridization experiments to see if you can correct the defect in xeroderma? You might possibly see if the deSanctis-Caccioni variety is different from the more common type of xeroderma as an approach to such hybridization experiments.

J.E. CLEAVER: At the moment such experiments haven't been attempted with xeroderma cells since it has been our main concern to define the condition first. This is the type of experiment which is to be done in the future.

R. WARREN: Also, I am wondering if there is a decrease in the propensity for inducing chromosome breaks in xeroderma pigmentosum as compared to the normal. If so, you could infer that this nuclease is somehow involved in chromosome breakage.

J.E. CLEAVER: Nothing is known concerning chromosome breakage in this cell system; I said the karyotype is normal in the disease but experiments on UV-induced chromosome aberrations haven't been done, at least haven't been published yet.

J. LEVI: Can you tell me if, in these patients with xeroderma pigmentosum who have induced cell carcinomas, these basal cell cancers also have the defect in repair; can they not also excise these dimers?

J.E. CLEAVER: We have not measured, strangely enough, any of the tumor cells; this somehow constitutes a logistical problem in negotiating with clinical departments. Tumors tend to be a more infected source of tissue than normal skin. We have measured one keratotic lesion; this has the same characteristic as the unaffected skin on the patients. But one would assume that the tumor does not represent a back mutation.

J. LEVI: In that you would assume the tumor would also lack the ability to excise dimers, this would be an interesting tumor which could not only be induced by UV but also could be cured by UV.

N. PENNEYS: Are the fibroblasts from xeroderma pigmentosum more sensitive to mustard compounds than normal? You would expect this, I think, and it might have some clinical inportance in treating these patients.

J.E. CLEAVER: I have not measured sensitivity but I have measured repair replication levels. The levels which are induced by nitrogen mustard in xeroderma cells and normal cells is quite low compared with the amount induced by ultraviolet light, and in our experiments doesn't exceed the low level which the xeroderma pigmentosum cells were capable of performing, so that extra sensitivity was not exhibited by the xeroderma cells in this case. Very little of the nitrogen mustard damage requires excision repair.

S. GREER: You did not mention the fact that UV lesions in cells in which BUDR has been incorporated can be repaired, but I wondered if, as a clinical probe, the patient with xeroderma pigmentosum might receive an infusion of BUDR. The toxic effects of BUDR may have been overemphasized in the past, so that it could reasonably be used as a tool for further characterization of the lesion. I do not propose that it be used with radiation but as a method of stimulating repair in those individuals.

J.E. CLEAVER: It wouldn't help because of the point which I was trying to make that the repair replication that we measured is not always an expression of functional repair. If you put in nonradioactive BrUDR and irradiate with UV light you form a special type of photoproduct or spectrum of BrU photoproducts which result in greatly increased lethality. These photo products result in breaks which are repairable by xeroderma pigmentosum. However, this does not result in increased survival, it results in increased death.

M. MICHAELIS: Are there any observations on ocular tissues?

J.E. CLEAVER: I am sorry, could you repeat the question?

M. MICHAELIS: Are there any observations on ocular tissue, e.g. on viral infections of the cornea? Do you have any experimental data on this?

J.E. CLEAVER: No.

INTERACTIONS OF THERMAL PROTEINOIDS WITH POLYNUCLEOTIDES

S. W. FOX, J. C. LACEY, JR.,
and T. NAKASHIMA

Department of Biochemistry
and
Institute of Molecular Evolution
University of Miami

Abstract: Requirements for the formation of microparticles from thermal proteinoids and polynucleotides have been explored. In some of the experiments, selective interactions are observed. A number of the properties of the particles formed are reported. One of the properties of these particles is a codon-related selective incorporation of individual amino acids from the adenylate.

INTRODUCTION

In addition to the many specific enzyme-nucleic acid interactions in contemporary biological systems, three localizations of special relationships between proteins and nucleic acids are known. These are: 1) the association of DNA with chromosomal proteins to yield chromatin, 2) the association of rRNA with ribosomal proteins to yield ribosomes and 3) the synthesis of specific proteins on an mRNA template. The latter system involves the relationship between proteins and nucleic acids that we know as the genetic code. While we know what the genetic code is, we do not know the molecular basis for its origin. Furthermore, we cannot state whether the relationships between proteins and nucleic acids in chromatin and in ribosomes are the same as those of the genetic code.

The experiments in this paper were designed for the primary objective of gaining information relative to a molecular basis for the origin of the genetic code. In order that the experiments fit into an evolutionary context, thermal proteinoid was used as a model for prebiotic protein (1). The experiments include those that model to some extent each of the three natural systems noted above.

EXPERIMENTAL

Preparation and analysis of proteinoids

Thermal proteinoids can be simply prepared by heating to 160-200° C for several hours a dry mixture of amino acids having the appropriate composition (2). The proteinoids used in this study were dialyzed 3 days against water, filtered and lyophilized to obtain the polymer, which was usually of light amber color. Proteinoids so produced and properly purified have a molecular weight of several thousand, are each limited in heterogeneity (3), possess a variety of enzymic activities, and readily form particulate structures (1).

Partial racemization occurs during the heating, so that the product contains a mixture of D and L optical isomers. The basic proteinoids are cross-linked, presumably due to amide bond formation between the ε-amino groups of some of the lysine residues and ω-carboxyls of aspartic and glutamic acids.

Analyses of the composition of the proteinoids were carried out by: 1) hydrolysis in 6 N HCl for 48 hr. in an evacuated sealed tube at 110° C, and 2) chromatography of the hydrolyzates on a Phoenix analyzer (K-5000).

Polynucleotides

sRNA and calf thymus DNA were obtained from Calbiochem. Enzymically synthesized homopolyribonucleotides were obtained from Miles Laboratories, Inc.

Formation and assay of particles

Solutions of thermal proteinoid and polynucleotide, as indicated for each experiment, were prepared separately and then mixed. The interaction was studied by either measuring the turbidity as absorbance at 600 nm or by centrifuging and analyzing the precipitate. The precipitate was completely dissolved at pH 12 and the absorbances of the proteinoid and polynucleotide were found to be additive at each wavelength, 235 and 260 nm. Consequently, through measuring the absorbance at 235 and 260 nm, and knowing the extinction coefficients of each component, the amount of proteinoid and polynucleotide in the precipitate was calculated.

Incorporation of amino acids into microparticles from homopolyribonucleotides and thermal proteinoids

In order to study the effect of the composition of the nucleoproteinoid microparticles on the incorporation of amino acids into those particles, the procedure shown in Fig. 1 was used. The microparticles which were centrifuged were assayed on a scintillation counter for the amount of radioactive amino acid incorporated from the radioactive amino acyl adenylate. The adenylate preparation and purification was according to Berg (4).

RESULTS AND DISCUSSION

Earlier studies were carried out by Waehneldt (5,6), using a series of thermal proteinoids having varying lysine content from 7 to 47 mole percent and basic to acidic amino acid ratios (B/A) of 0.3 to 5.0. These studies indicated that complexes of the proteinoids with DNA tended to be fibrous in nature but that those with sRNA were spherical (Fig. 2). The data also showed that the proteinoid must have a net positive charge (B/A > 1.0) in order to form complexes with sRNA at low ionic strengths. The composition of the particles tended to be constant at about 45 wt % RNA. Waehneldt also showed that the particles dissolved at NaCl concentrations of 0.2-0.3 M. The particles dissociated at high pH, a fact which permitted relevant analyses.

These data are interpreted to signify that the thermal proteinoid-sRNA particles are formed, at least in part, by electrostatic interaction of the positively charged residues in the proteinoid with the negatively charged ribose phosphate backbone of the polynucleotides.

The experiments described thus far were not designed to reveal selectivities of interaction. Yuki (7) prepared thermal proteinoids having widely different compositions. One, for example, was lysine-rich and arginine-free and another arginine-rich and lysine-free. Yuki then observed turbidity formation by mixing solutions of these proteinoids with homopolyribonucleotides. The data in Fig. 3 show that the lysine-rich (arg-free) proteinoid formed particles preferentially with polypyrimidines whereas the arginine-rich (lys-free) proteinoid formed particles preferentially with polypurines.

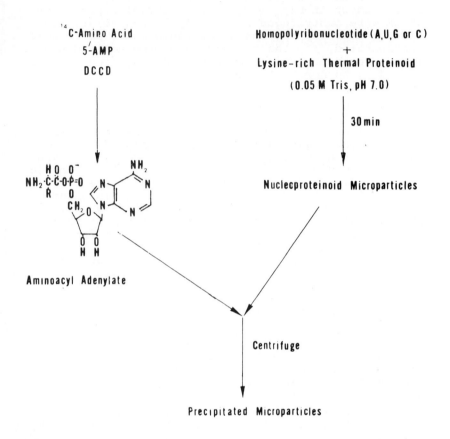

Fig. 1. Schematic diagram for studying the incorporation of radioactive amino acids into preformed nucleoproteinoid particles by exposing them to aminoacyl adenylates (Nakashima and Fox, 1971).

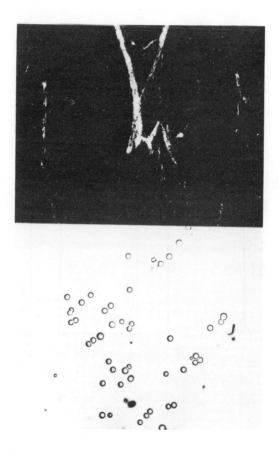

Fig. 2. Structures resulting from the interaction of lysine-rich thermal proteinoid and DNA (above) and sRNA (below) at low ionic strength.

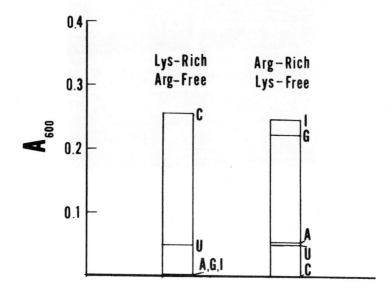

Fig. 3. Formation of microparticles from solutions of lysine-rich (arginine-free) or arginine-rich (lysine-free) proteinoids (1.0 mg/ml) with homopolyribonucleotides (0.1 μmole/ml). Buffer was 0.05 M tris at pH 7.0 and room temperature. Turbidity was measured as A_{600} 10 minutes after mixing (Yuki and Fox, 1969).

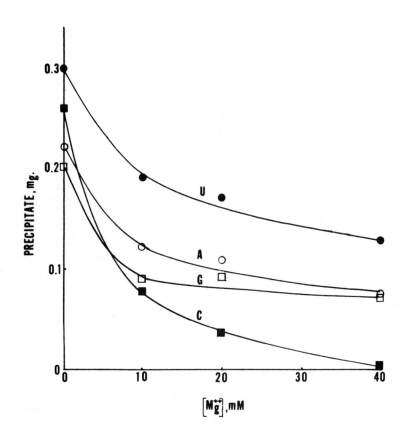

Fig. 4. Effect of [Mg^{++}] on the total wt. of precipitate from a mixture of thermal proteinoid 20Y (24.5 mole % lys) at 1.0 mg/ml concentration with polyribonucleotide (A, ○ ; U, ● ; G, □ ; C, ■) at 0.1 μmole/ml on a residue basis in tris buffer, 0.05 M and pH 7.0 at room temperature (Lacey and Fox, 1971).

Since these results are based on turbidity, their significance is qualitative rather than quantitative. However, they do display differences in polynucleotide preference depending on the composition of the proteinoid.

Table 1 shows the variation in total amount of precipitate (proteinoid plus polynucleotide) that resulted from a set of proteinoids having different compositions but all containing 20-25 mole percent lysine. The essential point of the data in this table is that differences in proteinoid composition yield different patterns of interaction with polynucleotides. The data do not show a consistency with respect to the genetic code.

TABLE 1

Effect of varying composition of thermal proteinoid on the total precipitate with polyribonucleotides (8)

Composition of proteinoid	Hierarchy of interaction with polynucleotides (total ppt., μg)
Eleven amino acids	U(204) > A(172) > C(108) = G(116)
Ditto minus aspartic acid	A(243) > G(149) = U(143) > C(79)
Ditto minus glycine	U(265) = C(260) = A(256) > G(168)
Ditto minus leucine	G(141) = A(135) > U(86) > C(45)
Ditto minus proline	G(136) > A(112) > U(79) > C(41)

A series of lysine-rich proteinoids was next used. These comprised six proteinoids in which the lysine content varied from 16 mole percent to 55 mole percent lysine. In order to predict which polynucleotide might interact with these proteinoids preferentially, we calculated the composition of the mRNA which would be equivalent to each proteinoid. The calculations were made by using all codon assignments and assuming that all codons occur with equal frequency. The results are shown in Table 2. All of these proteinoids have calculated mRNA compositions rich in adenylic acid. If a codonic interaction were to prevail, the proteinoids would interact preferentially with poly A, or if an anticodonic relationship prevailed, poly U would be selected.

By using one of the proteinoids (24.5 mole percent lysine), the effect of [Mg^{++}] on the total

TABLE 2

Calculated compositions of mRNAs equivalent to thermal proteinoids

Proteinoid number	Mole % lysine	Mole % nucleotide in mRNA			
		A	G	C	U
10Y	15.9	31.9	30.6	19.6	17.9
20Y	24.5	36.6	29.0	17.2	17.2
30Y	34.8	43.2	27.4	15.5	13.9
40Y	47.3	48.8	22.9	14.5	13.8
50Y	51.8	51.8	23.1	13.2	11.9
60Y	54.7	57.2	25.8	9.7	7.4

precipitate was explored. Fig. 4 shows that the amount of total precipitate is diminished with each of the polynucleotides with increasing [Mg^{++}]. Polyuridylic acid gives the most total precipitate at all [Mg^{++}] studied. These results indicate a competition between Mg^{++} and basic proteinoid for interaction with the polynucleotides; the importance of electrostatic interactions is again emphasized.

Figure 5 shows the effect of varying the lysine content of the proteinoids in experiments without Mg^{++} and with 20 mM Mg^{++}. The general effect of Mg^{++} is to reduce the total precipitate, although this effect is less noticeable at high and low lysine contents. In the experiments without Mg^{++}, the minimal requirement for plus charges on the proteinoid is seen between 16 mole percent lysine and 24.5 mole percent lysine. The ratios of basic to acidic (B/A) amino acids in these proteinoids are approximately 1.7 and 2.3 respectively. The requirement for a B/A ratio > 1.0, established by Waehneldt for proteinoid-nucleic acid interactions, was for low ionic strengths. The present experiments were conducted in 0.05 M tris buffer; the presence of the buffer apparently causes a higher B/A ratio requirement in the proteinoid.

Figure 5 also exhibits a maximal amount of total precipitate in the range of 25 to 35 mole percent lysine where maximal differences between the polynucleotides are noted. The fact that poly U yields the most precipitate with proteinoids having 25 to 35 mole percent lysine either with or without Mg^{++} is significant at the 95% confidence level. These

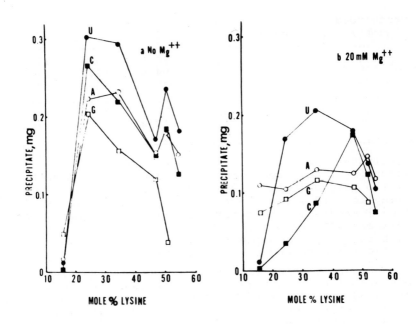

Fig. 5. Effect of mole % lysine in the thermal proteinoid on the amount of total precipitate from a mixture of thermal proteinoids (1.0 mg/ml) and polyribonucleotides (A, ○ ; U, ● ; G, □ ; C, ■) in tris buffer (0.05 M) at pH 7.0 and room temperature, a. no Mg^{++}; b. 20 mM Mg^{++}. Initial polynucleotide concentration 0.1 μmole/ml on a residue basis (Lacey and Fox, 1971).

results are consistent with an anticodonic effect.
However, the fact that the largest amount of precipitate is observed with poly U may be explained on the basis of its having the most open structure of all the polynucleotides. This openness may permit its negatively charged backbone to be more available for interacting with proteinoid. We must infer that some of the differences in the amount of precipitate of thermal proteinoids and polynucleotides are due to differences in the openness of structure of the polynucleotide.

In spite of the above inference, under some conditions selectivities in the interaction of proteinoids and polynucleotides were observed. However, these selectivities did not exhibit a regular pattern consistent with codonicity or anticodonicity. We believe that, in any event, the use of polymer-polymer interactions permits the manifestation and consequently the study of selectivities in such systems. In the past, attempts to study amino acid-polynucleotide interactions have been largely unfruitful.

The experiments we have discussed up to this point have concerned the formation of particulates from the interaction of thermal proteinoids with polynucleotides. These nucleoproteinoid microparticles have been of interest as models for evolutionary precursors of nucleoproteinoid organelles, such as ribosomes. We wondered if such particles might display preferential incorporation of amino acids.

Studies of this sort have been performed with the compounds that are involved in contemporary protein biosynthesis, the amino acid adenylates (9). Since they can be prepared chemically, and since they react to produce peptides in aqueous solution (10), they have been studied in the search for the molecular interactions which might be the essence of an organelle-directed coded synthesis of polyamino acids. A variety of such experiments has been performed for some time in a constructionistic mode (11). When the condensation of radioactive adenylates is allowed to take place in the presence of nucleoproteinoid microparticles, radioactive amino acids are incorporated into the microparticles. Although such incorporation of amino acids proves to be a selective process, many experiments have demonstrated that the selectivities are highly sensitive to variations in conditions. For example, as shown in Fig. 6, if

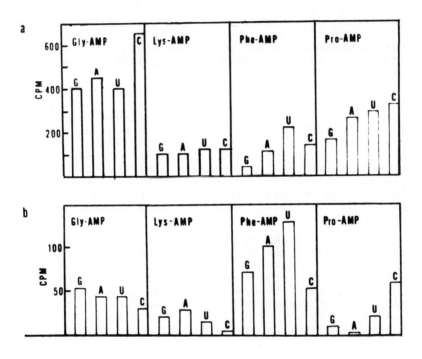

Fig. 6. Incorporation of radioactive amino acids into nucleoproteinoid microparticles preformed from lysine-rich proteinoid and different polynucleotides as shown. Amino acids introduced as the adenylates (see Fig. 1). Initial polynucleotide concentrations: a) 0.75 mg/ml; b) 0.25 mg/ml (Nakashima and Fox, 1971).

0.75 mg/ml polynucleotide is used, phe and pro reveal a codonic relationship in their incorporation into the particles, gly is anticodonic, and lys is not selective. However, if 0.25 mg/ml polynucleotide is used in preparing the microparticles, a codonic incorporation of amino acids into the particles is observed. These experiments have been carried out with four different proteinoids, with the codonic relationship being observed in all cases. Further data are now being collected to allow statistical analyses of the results (11).

The results with selective incorporation of amino acids into microparticles under a particular set of conditions gave a codonic relationship. Under different conditions, different results were obtained. These results indicate that these recognition processes are very sensitive to conditions and indeed might be switchable from anticodonic to codonic. Although their results do not involve a codonic-anticodonic switch, the experiments of Leng and Felsenfeld (12) and of Shapiro et al. (13) show that poly-L-lysine has a preference for A-T rich regions of DNA. This preference can be switched to G-C simply by adding tetramethylammonium chloride to the system. Similarly, protein synthesizing systems are very sensitive to changes in conditions, e.g. variation in [Mg^{++}] from an optimum value causes misreading of the messenger and incorporation of the incorrect amino acid (14).

SUMMARY

The experiments reported have led to the following conclusions:
1. Thermal proteinoids form microparticles when mixed with polynucleotides in aqueous solution.
2. Particle formation as in (1) requires a net positive charge on the proteinoid. This requirement varies with the ionic strength.
3. Nucleoproteinoid particles are dissociated by high ionic strength and high pH.
4. Mg^{++} competes with thermal basic proteinoid for interaction with

polynucleotides. The interactions are seen to be, at least in part, electrostatic in nature.

5. Selectivities are displayed in the interaction of thermal proteinoid with polynucleotides. Under certain conditions, these selectivities at first appeared to be consistent with an anticodonic effect. Further exploration of this point revealed that the rigidity of structure of the polynucleotide is one important factor in these interactions.

6. Under empirically selected conditions, exposure of preformed nucleoproteinoid microparticles to radioactive aminoacyl adenylates causes a codon-related selectivity of incorporation of amino acids into the microparticles (11).

Interactions of basic proteinoids and polynucleotides are being studied further in the attempt to classify selectivities more precisely. The experiments with preformed microparticles and adenylates are being extended to define the nature of the recognition and of the incorporation.

REFERENCES

1. S. W. Fox, K. Harada, G. Krampitz, and G. Mueller, Chem. Eng. News 48 (1970) 80.

2. S. W. Fox and K. Harada, in: A Laboratory Manual of Protein Chemistry, Vol. 4, eds. P. Alexander and H. P. Lundgren (Pergamon Press, 1966), p. 127.

3. S. W. Fox and T. Nakashima, Biochim. Biophys. Acta 140 (1967) 155; T. Nakashima, S. W. Fox, and C. Wang, Arbeitstagung über extraterrestriche Biophysik und Biologie, Marburg (Red. H. Bücker, Frankfurt, Juni 1968), p. 223.

4. P. Berg, J. Biol. Chem. 233 (1958) 608.

5. S. W. Fox and T. V. Waehneldt, Biochim. Biophys. Acta 160 (1968) 246.

6. T. V. Waehneldt and S. W. Fox, Biochim. Biophys. Acta 160 (1968) 239.

7. A. Yuki and S. W. Fox, Biochem. Biophys. Res. Comm. 36 (1969) 657.

8. J. C. Lacey and S. W. Fox, in preparation.

9. G. D. Novelli, Ann. Rev. Biochem. 36 (1967) 449.

10. G. Krampitz and S. W. Fox, Proc. Natl. Acad. Sci. U.S. 62 (1969) 399.

11. T. Nakashima and S. W. Fox, in preparation.

12. M. Leng and G. Felsenfeld, Proc. Natl. Acad. Sci. U.S. 56 (1966) 1325.

13. J. T. Shapiro, M. Leng, and G. Felsenfeld, Biochemistry 8 (1969) 3219.

14. A. G. So, J. W. Bodley, and E. W. Davie, Biochemistry 3 (1964) 1977.

We wish to acknowledge the support of Grant no. NGR 10-007-008 of the National Aeronautics and Space Administration. Contribution no. 185 of the Institute of Molecular Evolution. We thank Mr. C. R. Windsor and Mrs. Ania Mejido for technical assistance.

DISCUSSION

G.D. NOVELLI: Dr. Lacey, were any attempts made to ascertain whether the incorporated amino acids were linked together with one another or linked to the protein in the particles.

J.C. LACEY: Well, of course, this is a very interesting question and of paramount importance. Dr. Nakashima is working on it, but he hasn't answered it yet. We can't say what the form of the incorporated amino acid is.

ON THE STRUCTURE OF THE REPLICATION APPARATUS

BRUCE ALBERTS
Dept. of Biochemical Sciences
Princeton University, Princeton, N.J. 08540

Abstract: T4 bacteriophage gene 32 codes for a DNA-binding protein which is essential for DNA replication. The properties of "32-protein", together with the genetics of T4 DNA replication, suggest that the DNA, T4 DNA polymerase, 32-protein and at least 2 to 4 additional proteins combine to produce a convoluted structure through which unwound template DNA strands pass as replication proceeds.

INTRODUCTION

Although the first DNA polymerase was discovered over 15 years ago (1), and a large amount of outstanding work has followed, the detailed enzymology of DNA replication remains one of the major unsolved problems in molecular biology today. One reason why progress in this field has been limited is that only within the last year has an _in vitro_ system which replicates double-stranded DNA at _in vivo rates_ been developed; even in this crude system, replication continues for only about one minute at $37°$ (2). This appears to reflect a basic instability of the replication apparatus which is not shared by either the ribosome (and its associated factors) or by RNA polymerase, since purified _in vitro_ systems for both protein synthesis and RNA transcription which operate at close to _in vivo_ rates can be obtained with relative ease.

Recently, a new type of DNA-binding protein has been isolated which seems to be the main structural protein of the replication apparatus in the T4 bacteriophage system (3, 4, 5). This protein is the product of T4 gene 32. The properties and mode of action of "32-protein", together with the genetics of T4 DNA replication, appear to allow the general nature of the T4 replication apparatus to be delineated. The model which results helps to explain why the system that replicates DNA is so unstable _in vitro_ and, if correct, may point the way to an eventual solution of the replication problem.

PROPERTIES OF PURIFIED 32-PROTEIN AND ITS BINDING TO DNA

Through the use of appropriate mutant bacteriophages, one of the major DNA-binding proteins synthesized after T4 bacteriophage infection of E. coli can be identified as the product of T4 gene 32 (3, 4). Elution from a single-stranded DNA-cellulose column with 2M NaCl followed by DEAE-cellulose chromatography yields 32-protein which is electrophoretically homogeneous (5). As detailed in previous publications (3, 4, 5), the native protein consists of a single polypeptide chain, with a molecular weight of about 35,000 daltons. The 32-protein carries a net negative charge at pH 7, yet shows tight, salt-sensitive binding to single-stranded DNA.

The stoichiometry of the complex which 32-protein forms with single-stranded DNA at low salt concentrations has been examined by sucrose gradient sedimentation of a fixed quantity of ^{3}H-labelled 32-protein in the presence of varying amounts of the circular, single-stranded DNA from bacteriophage fd. The results are shown in Fig. 1. At lower concentrations of DNA, two distinct peaks of radioactive protein are seen; one sediments rapidly with the DNA, the second at the slow rate characteristic of the free protein. The free protein peak is absent above a weight ratio of DNA to protein of 1 : 12. The complex therefore contains about one protein molecule of 35,000 daltons for every 10 single-stranded DNA nucleotides.

Although 32-protein binds tightly to all single-stranded DNA's tested, including the synthetic polynucleotides poly dA, poly dI, and poly dT, no binding of the purified protein to double-stranded DNA or to R17 RNA was detected by sucrose gradient sedimentation at 4°.

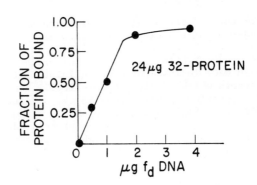
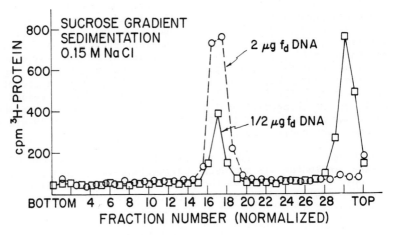

Fig. 1. Stoichiometry of 32-protein interaction with single-stranded DNA. The indicated quantity of purified fd DNA was mixed with 24 μg of ^{3}H-32-protein in 0.2 ml of 0.02 M Tris-HCl, pH 8.1, 0.5 mM Na$_3$EDTA, 0.15 M NaCl, 100 μg/ml bovine serum albumin (BSA), 10% glycerol, 1 mM β-mercaptoethanol at 4°. After 20 min., the mixture was layered at 4° onto a 5 ml, 5 - 30% sucrose gradient containing the same buffer. Following centrifugation, fractions were collected and monitored for radioactivity by standard techniques. Recoveries of ^{3}H-protein added averaged about 75%. The stoichiometry of the complex appears to be unchanged with 0.01 M MgCl$_2$ added, or with single-stranded T4 DNA substituted for fd DNA under the above conditions.

The binding of 32-protein to single-stranded DNA is highly cooperative. As illustrated by the model in Fig. 2, this has been explained by proposing that protein monomers line up in close juxtaposition along a DNA strand. Due to favorable protein-protein interactions, an incoming 32-protein monomer strongly prefers to bind to the 10 nucleotides adjacent to a previously bound molecule of 32-protein rather than to an isolated stretch of 10 nucleotides. At least three different types of observations support the validity of this model (for details see Ref. 5):

1) The 32-protein monomers self-aggregate in the absence of DNA at a concentration of 0.5 mg/ml or higher (see, for example, Fig. 6, below), suggesting that the cooperative binding to DNA is the result of direct stabilizing interactions between adjacent protein molecules.
2) As predicted (Fig. 2), large clusters of bound 32-protein molecules are observed to be interspersed with long stretches of bare DNA under conditions of DNA excess.
3) As judged either by translational frictional coefficients or by electron microscopy, the complexing of 32-protein with a single-stranded DNA molecule greatly extends its otherwise highly-folded conformation.

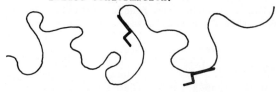

WEAK BINDING AT LOW 32-PROTEIN CONCENTRATION

STRONG BINDING AT HIGH 32-PROTEIN CONCENTRATION

Fig. 2. Model for the cooperative binding of 32-protein to single-stranded DNA.

DENATURATION AND RENATURATION OF DNA WITH 32-PROTEIN

Histones and polyamines bind to double-helical DNA more tightly than to single strands and thereby raise the temperature for DNA denaturation. Conversely, the strong selective affinity of 32-protein for single-stranded DNA drastically lowers the thermal denaturation temperature of double-stranded DNA. At $37°$, with concentrations of 32-protein near the physiological level (200 μg/ml), poly dAT is rapidly denatured, while T4 DNA retains its native state due to the added stability conferred by G-C base-pairs (5). Nevertheless, the cooperative binding of 32-protein can be shown by electron microscopy to generate transient regions of local denaturation in the T4 double helix under similar conditions (H. Delius and B.A., ms in preparation).

At elevated ionic strengths, complementary strands of denatured DNA can pair rapidly with each other in vitro to reform a double helix. This renaturation process is optimal at about $65°$, and sharply decreases in rate at lower temperatures (6). This decline in rate is attributed to the tight intrastrand folding of denatured DNA, which begins when the temperature is lowered more than $25°$ below the Tm of the double helix (7). These folds of hair-pin base-paired loops make the DNA bases relatively inaccessible, and thereby prevent complementary single strands from finding satisfactory interstrand pairings. This is purely a kinetic effect, since the equilibrium stability of the double helix relative to single strands is greater at the lower temperature.

In addition to destabilizing the double helix, the binding of 32-protein to DNA single strands forces them to adopt a much more open conformation, and thereby increases their rate of renaturation under physiological conditions by as much as 1000-fold (5). A plot of the effect of 32-protein on renaturation rates is given in Fig. 3, in order to emphasize that a decrease in helix stability can accelerate the renaturation, as well as the denaturation, of DNA.

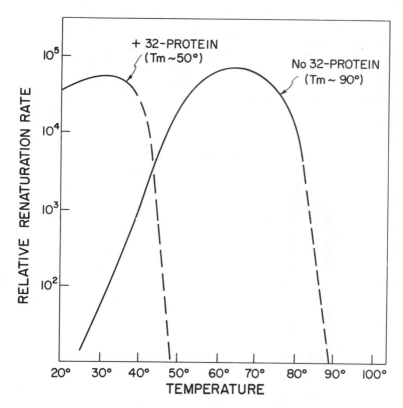

Fig. 3. Approximate effect of 32-protein on the temperature dependence of DNA renaturation rates. The "no 32-protein" curve has been estimated from the results of Studier (7).

THE PROTEINS ESSENTIAL FOR T4 DNA REPLICATION

As first shown by Tomizawa and coworkers (8), 32-protein is required for an early step in genetic recombination in the T4 bacteriophage system. This can be explained by the dual action of 32-protein in transiently opening up local regions of native DNA while at the same time facilitating helix formation between matching single strands, both of which may be essential for the recombination process (see Ref. 5).

Of concern in the present context is that 32-protein is also one of several gene products required for T4 DNA replication (9, 10). That the role of this protein in replication is a direct one, independent of its role in genetic recombination, is clearly demonstrated by studies with T4 mutant ts P7, which makes a temperature-sensitive 32-protein. As first shown by Riva, Cascino and Geiduschek (11), cells infected with this mutant bacteriophage synthesize T4 DNA at normal rates at $25°$, but stop replication completely within 1 minute after a shift to $42°$ (Fig. 4, M. Curtis, Senior Thesis, Princeton University, 1970).

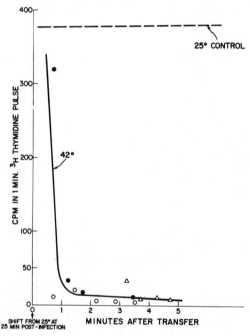

Fig. 4. Dependence of the rate of T4 DNA synthesis on the continued function of gene 32-protein. E. coli B3 at 3×10^8 cells/ml in M_Θ minimal medium containing 0.3% casein hydrolysate and 20 μg/ml thymidine were infected at $25°$ with a multiplicity of 5 T4 ts P7 phage per bacterium. After 25 min., the infected cells were diluted fivefold into prewarmed media (without additional thymidine). At various times, thymidine-^{3}H was added for 1 min. to an aliquot. At the indicated time, incorporation was stopped with an equal volume of cold 10% trichloroacetic acid, and the precipitate collected and counted.

Similar experiments indicate a <u>direct</u> role in T4 DNA replication for at least three other gene products: genes 43, 41, and 45 (11). Amber mutants in any of these genes or in genes 62 and 44 synthesize very little or no DNA, even though all deoxyribonucleotide triphosphate precursors are present (9, 12). Therefore, in addition to 32-protein, it appears that at least five T4-induced proteins play a central role in building an active T4 replication apparatus. The function of only one of these proteins is known: gene 43-protein is the T4 DNA polymerase (13). This polymerase has an absolute requirement for a single-stranded DNA template, and needs a pre-existing 3' hydroxyl-terminated primer chain onto which it polymerizes 5' deoxyribonucleotide triphosphate precursors in the 5' to 3' direction (13, 14).

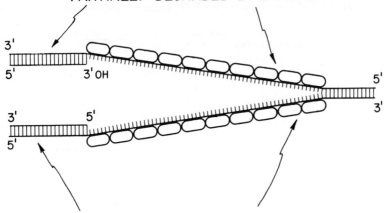

Fig. 5. Schematic view of a simple replication fork containing 32-protein.

THE INTERACTION OF T4 DNA POLYMERASE AND 32-PROTEIN

How might the products of genes 43 and 32 function together in the replication apparatus? The requirement of T4 DNA polymerase for a single-stranded DNA template and the ability of 32-protein to unwind the double helix suggest that 32-protein might precede the polymerase, exposing the template and aligning the bases in a conformation optimal for polymerase action.

In this case, the "repair" of DNA, partially degraded by exonuclease-III, by T4 DNA polymerase in the presence of 32-protein should provide a system for study of the synthetic events on at least one side of the replication fork (Fig. 5). A detailed study of this system (J. Huberman, A. Kornberg, and B.A., ms submitted for publication) has revealed that:

1) 32-protein stimulates the *in vitro* rate of polymerization of T4 DNA polymerase on an exonuclease-III-degraded DNA template 5 - 10 fold. The rate of stimulated synthesis is about 10% of that measured for growing point movement *in vivo*, and is greatest with a level of 32-protein sufficient to bind most or all of the template strand present.

2) The 32-protein from a temperature-sensitive mutant in gene 32 (*ts* P7) is also temperature-sensitive in the stimulation of T4 DNA polymerase *in vitro*, suggesting that this stimulation is related to the intracellular function of 32-protein in DNA replication.

3) The activity of DNA polymerase from *E. coli* on an exonuclease-III-degraded DNA template is not significantly affected by 32-protein. In addition, in the absence of DNA, a weak complex is formed between 32-protein and T4 DNA polymerase, while no such complex is formed with *E. coli* DNA polymerase (Fig. 6). These observations could be explained if some specific fit between polymerase and template strands complexed with 32-protein is essential for the stimulation of DNA synthesis observed.

STRUCTURE OF REPLICATION APPARATUS

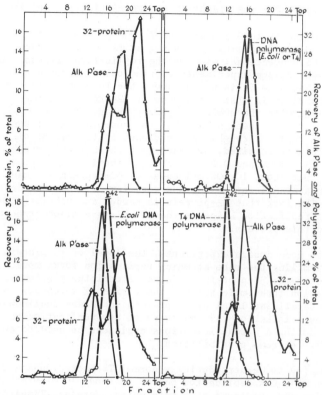

Fig. 6. Sucrose gradient sedimentation of DNA polymerases in the presence of a high concentration of 32-protein. Sedimentation was through 5 ml, 5 - 30% sucrose gradients containing 10 mM MOPS, pH 7.4, 50 mM KCl, 0.5 mM Na_3 EDTA, 10 mM $MgCl_2$, 1 mM β-mercaptoethanol, 100 μg/ml bovine serum albumin, and 10% glycerol. The following proteins, in 0.1 ml of sucrose-gradient buffer containing 10 μg of E. coli alkaline phosphatase, were layered onto the respective gradients: a) 200 μg of ^3H-32-protein; b) either 1.5 μg of T4 DNA polymerase, or 1 μg of E. coli DNA polymerase; c) 200 μg of ^3H-32-protein plus 1 μg of E. coli DNA polymerase; d) 200 μg of ^3H-32-protein plus 1.5 μg of T4 polymerase. If, in an experiment similar to d), only 30 μg of 32-protein is used, the T4 DNA polymerase peak sediments behind the alkaline phosphatase marker, as in b). Thus, the complex of 32-protein and T4 polymerase is a loose one. Note that, at the concentration used, an uniquely stable aggregate of 32-protein appears. This species is believed to arise when end to end aggregates form which are long enough to circularize.

IMPLICATIONS FOR THE STRUCTURE OF THE REPLICATION APPARATUS

The notion that 32-protein and T4 DNA polymerase function together in DNA replication does not at first sight drastically alter the conventional view of the replication process. However, gene dosage experiments carried out by Snustad have revealed that gene 32 is unique among the T4 genes affecting DNA metabolism in that the amount of 32-protein synthesized directly limits the number of progeny phage produced (15). More recently, it has been shown that the total amount of DNA made in these cells is roughly proportional to the amount of 32-protein synthesized, at least over the six-fold range explored (N. Sinha and D.P. Snustad, ms submitted for publication). Yet about 10,000 molecules of 32-protein are made per cell in a normal infection, and very few, if any, are destroyed (4). The fact that 32-protein is needed in such large amounts for DNA replication suggests that it plays a major structural role in this process.

In the gene dosage experiments, the level of 32-protein could be controlling the rate at which replication forks move. At the other extreme, 32-protein could be determining the number of replication forks made but not their rate of movement. Since the intracellular level of 32-protein increases continuously throughout infection (4), analysis of how replication rates change during a normal infectious cycle can be used to evaluate these two extreme possibilities.

According to Werner, during a normal T4 bacteriophage infection at $25°$, DNA replication begins at about 10 minutes post-infection and continues at a rate which increases linearly until 30 minutes post-infection, when the rate stabilizes at a constant maximum (16). This increase in the rate of replication is caused by a parallel increase in the number of replication forks from zero to 60 per infected cell. The crucial point is that each replication fork moves at a constant rate which is independent of the time after infection, and therefore independent of the amount of 32-protein present (16). Moreover, if the protein synthesis inhibitor chloramphenicol is added to T4-infected cells during the period when the rate of DNA synthesis (and therefore the number of replication forks) is increasing, DNA synthesis continues, but at a rate which gradually decreases from that established just prior to the moment of chloramphenicol addition (17). The simplest interpretation of this result is that generation of each new replication fork requires at least one protein present in limiting amounts; it is therefore the rate of synthesis of this protein which determines how rapidly new replication forks are made. Among known genes, this protein could only be the product of gene 32, since the products of the other T4 genes which affect

DNA replication are clearly synthesized in large excess (15).

Thus, we arrive at the following general model for the structure of the T4 replication apparatus:

> A replication unit capable of synthesizing DNA contains a <u>fixed</u> number, "X", of 32-protein molecules. (As 10,000 molecules of 32-protein are enough to make at least 60 replication forks, X cannot be more than 170.) During the course of infection, as X molecules of 32-protein become available, they are incorporated together into a replication unit ("replication apparatus") which travels along its DNA template at a rate independent of the concentration of free 32-protein. In principle, if there are only X-1 molecules of 32-protein in a cell, DNA replication should either not begin, or it should be aberrant.

A replicating apparatus which demands exactly X molecules of 32-protein should of course have a unique three-dimensional structure and also require a fixed number of every other protein which functions in it (<u>i.e.</u>, the products of T4 genes 41, 43, 45, and perhaps 44 and 62).

SPECIFIC MODELS

With an <u>in vivo</u> polymerization rate of about 1500 nucleotides sec^{-1} at $37°$ (2), 32-protein must be added to each template strand at a rate of about 150 molecules sec^{-1} in order to keep ahead of the polymerase. As the association constant for 32-protein invasion of a double-helical region is not yet known, the possibility that this is accomplished by random collision with a free intracellular pool of 32-protein molecules cannot yet be ruled out. Nevertheless, it is intriguing to speculate that the X molecules of 32-protein in each replication unit actually remain within that unit as replication proceeds. Regardless of whether DNA replication proceeds continuously on both strands (Fig. 7A) or discontinuously on the second strand (Fig. 7B, Ref. 18), a requirement for circulation of 32-protein within each replicating structure not only puts interesting constraints on the types of replicating structures which are possible (see heading to Fig. 7), but provides a simple rationale for the proposal that replication <u>only</u> proceeds when one particular conformation of the replication complex is attained.

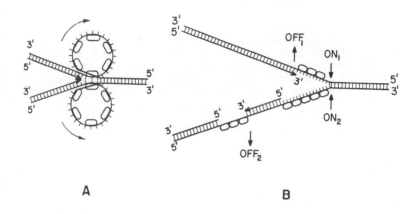

Fig. 7. Schematic models for circulation of 32-protein within the replication apparatus.
A. Continuous synthesis of DNA on both template strands. This model would require a second type of DNA polymerase, permitting polymerization of triphosphates in the 3' to 5' direction on one strand.
B. Discontinuous synthesis of DNA on one template strand (18). Only one type of polymerase is needed. However, the circulation of 32-protein would be more complicated, most likely joining OFF_1 with ON_1, and OFF_2 with ON_2 in a convoluted structure. Note that in both models the total length of single-stranded DNA must be kept constant at all times.

With this viewpoint, it also becomes easier to understand why polymerase and 32-protein are not sufficient by themselves to replicate DNA in the T4 system. The additional gene products needed, thus far identified only by genetic experiments, might be expected to interact with 32-protein, T4 DNA polymerase and/or each other in building the correct spatial relationships between the different regions of the replicating unit. They would also be required for stabilization of the complementary template strands in a complex with 32-protein, since in the absence of other factors, the double helix should be so highly favored under physiological conditions as to allow a structure such as that illustrated in Fig. 5 only a very transient existence.

In concluding, it is instructive to compare this view of the replication apparatus with the ribosome, where an unique three-dimensional structure is also built with proteins around a framework of nucleic acid. As with the ribosome (19), some of the proteins of the replication apparatus are not expected to be bound directly to the nucleic acid component. A crucial difference is, of course, that the ribosome is a stable structure which contains its own RNA while the replication apparatus must be free to slide rapidly along the DNA threaded through it, pushed by the cleavage of triphosphates behind. In such a structure, rather weak protein-protein interactions, such as that observed between T4 DNA polymerase and 32-protein (Fig. 6), might be expected to prevail in place of the much stronger forces which stabilize the ribosome. This could explain why all attempts to isolate a replication apparatus thus far have failed.

OTHER BIOLOGICAL SYSTEMS

If the essential features of T4 DNA replication are shared by other organisms, a protein with properties similar to 32-protein should be found in all replicating systems. In theory, the detection of this type of protein is made difficult in uninfected bacteria by the fact that there are only a few replicating forks in each cell. Recently, we (and others) have found that the gene 5-protein of fd bacteriophage is of the 32-protein type (B.A. and L. Frey, ms in preparation; J. Oey and R. Knippers, personal communication): it binds tightly and cooperatively to single-stranded (but not to double-stranded) DNA; it forms a complex in which the DNA strand is unfolded and fully hyperchromic; and it rapidly denatures poly dAT at 25°. Like 32-protein, the gene 5-protein is synthesized in large amounts (20), and it is required for synthesis of the single-stranded, progeny DNA strands (21).

REFERENCES

1. A. Kornberg, I.R. Lehman and E.S. Simms. Fed. Proc. 15 (1956) 291.

2. D.W. Smith, H.E. Schaller and F.J. Bonhoeffer. Nature 226 (1970) 711.

3. B.M. Alberts, F.J. Amodio, M. Jenkins, E.D. Gutmann and F.L. Ferris. Cold Spring Harbor Symp. Quant. Biol. 33 (1968) 289.

4. B.M. Alberts. Fed. Proc. 29 (1970) 1154.

5. B.M. Alberts and L. Frey. Nature 227 (1970) 1313.

6. J. Marmur and P. Doty. J. Mol. Biol. 5 (1962) 109.

7. F.W. Studier. J. Mol. Biol. 41 (1969) 199.

8. J. Tomizawa, N. Anraku and Y. Iwama. J. Mol. Biol. 21 (1966) 247.

9. R.H. Epstein, A. Bolle, C.M. Steinberg, E. Kellenberger, E. Boy de la Tour, R. Chevalley, R.S. Edgar, M. Susman, G.H. Denhardt and A. Lielausis. Cold Spring Harbor Symp. Quant. Biol. 28 (1963) 375.

10. A. Kozinski and Z.Z. Felgenhauer, J. Virol. 1 (1967) 1193.

11. S. Riva, A. Cascino and E.P. Geiduschek. J. Mol. Biol. 54 (1970) 85.

12. H.R. Warner and M.D. Hobbs. Virology 33 (1967) 376.

13. H.V. Aposhian and A. Kornberg. J. Biol. Chem. 237 (1962) 519.

14. M. Goulian, Z.J. Lucas and A. Kornberg. J. Biol. Chem. 243 (1968) 627.

15. D.P. Snustad. Virology 35 (1968) 550.

16. R. Werner. Cold Spring Harbor Symp. Quant. Biol. 33 (1968) 501.

17. P. Amati. Science 168 (1970) 1226.

18. T. Okazaki and R. Okazaki. Proc. Nat. Acad. Sci., U.S. 64 (1969) 1242.

19. M. Nomura. Bacteriological Reviews 34 (1970) 228.

20. T.J. Henry and D. Pratt. Proc. Nat. Acad. Sci., U.S. 62 (1969) 800.

21. D.A. Marvin and B. Hohn. Bacteriological Reviews 33 (1969) 172.

The experimental work described in this article was supported by grants from the U.S. National Institutes of Health (GM-14927) and The American Cancer Society (E-510).

DISCUSSION

V.G. ALLFREY: Many of the proteins which associate with the proteins of mammalian chromosomes are phosphoproteins. Is either the 32-protein or the f_d protein phosphorylated or reversibly de-phosphorylated?

B.M. ALBERTS: A graduate student, Larry Moran, attempted at one time to label T_4 DNA-binding proteins with ^{32}P. There was no indication that 32-protein contains phosphorus in his experiments.

CHANGES IN CHROMOSOMAL PROTEINS ASSOCIATED WITH GENE ACTIVATION

Nuclear Protein Phosphorylation as a Possible Mechanism for Promoting RNA Initiation

V. G. Allfrey, C. S. Teng and C. T. Teng
The Rockefeller University
New York

Abstract: Interactions between DNA and associated chromosomal proteins are subject to change at times of gene activation. The mechanisms of change are complex, but they often involve reactions - such as phosphorylation, acetylation, and methylation - which alter the charge and conformation of polypeptide chains. Reactions involving histones are taken as a model. These basic proteins, which act to supercoil DNA and restrict its template activity in RNA synthesis, are subject to post-synthetic structural modifications in which acetyl - and phosphoryl - groups are transferred to lysyl - and seryl - residues, respectively. Both acetylation and phosphorylation are increased early in gene activation for RNA synthesis, and they are suppressed in cells in which RNA synthesis is inhibited. These studies of structural alterations in nuclear basic proteins suggest that protein phosphorylation offers a general mechanism for varying the state and function of chromatin.

The model has been extended to consider the "acidic" proteins which appear to be involved in the positive control of RNA synthesis. Methods have been devised for the selective extraction and electrophoretic analysis of a class of nuclear phosphoproteins from various animal tissues. Such proteins occur in the chromatin, and they are distributed differently in different cell types. Their presence in chromatin and their rates of phosphorylation correlate strongly with the RNA synthetic capacity of the nucleus. Phosphoproteins block the suppression of RNA synthesis by histones, and they increase rates of transcription from free DNA in cell-free systems.

The synthesis of certain chromosomal acidic proteins is influenced selectively by hormones (such as cortisol and estradiol) which augment RNA synthesis in the target tissue. The rates of phosphorylation of individual nuclear phosphoproteins vary in a complex way in hormone-stimulated tissues.

DNA-phosphoprotein binding studies indicate a specificity of interaction which may explain the capacity of these proteins to stimulate RNA transcription from some, but not all, DNA's. The relationships between phosphorylation and RNA synthesis in vivo and in vitro suggest that certain chromosomal proteins have functions analogous to those of bacterial and viral Sigma factors. The rapid phosphorylation of chromosomal proteins and the ready reversibility of the process fit a cyclical mechanism for the binding and release of controlling factors which direct the attachment of RNA polymerase(s) to specific sites of initiation on the chromosome.

INTRODUCTION

It is now taken as axiomatic that a prime mechanism for the differentiation of cells - leading to the diversity of specialized tissues in higher organisms - involves the transcription of different regions of the total chromosomal DNA in different cell types, to produce their characteristic RNA populations. This selective, but fractional utilization of the genome occurs despite the presence in most diploid somatic cells of a complete set of chromosomes and a DNA complement which is sufficient to specify the formation of an entire organism. Individual cells suppress the transcription of most of the DNA which they contain (1), while they selectively activate a relatively small number of genes for the synthesis and assembly of the enzymes and structural proteins characteristic of the cell type and the species (1-3). In the course of embryogenesis and development, different sets of genes are activated and repressed in response to programmed signals from the nucleus, the cytoplasm, and the environment.

The two aspects of transcriptional control — suppression of template activity of most of the DNA, and activation of RNA synthesis at particular genetic loci — require the participation of proteins associated with the DNA in the chromatin of interphase cell nuclei. The proteins concerned include the histones - the basic, suppressive components of chromatin - as well as more acidic protein fractions which, as will be seen, show strong indications of positive involvement in the control of RNA synthesis at specific sites.

Interactions between DNA and associated proteins constitute the major topic of this report, which will emphasize recent evidence that the structures of chromosomal proteins are not invariant, but are often subject to enzymatic modification.

Such modifications take place at times when patterns of RNA synthesis are changed. Reactions — such as phosphorylation, acetylation, and methylation of amino acid residues in histones and other nuclear proteins — modify the charge and conformation of the respective polypeptide chains. These changes, which ultimately affect the nature and strength of the interactions between the proteins and DNA, offer informative clues to the molecular events underlying the "scanning" of the genome and the selective activation and repression of particular cistrons in different cell types at different times.

Histone-DNA Interactions : A Model of Negative Control and its Reversal -

The notion that histones might have a suppressive effect on genetic activity was originally advanced in 1950 by Edgar and Ellen Stedman (4) who proposed that " the basic proteins of cell nuclei are gene inhibitors, each histone or protamine being capable of suppressing the activities of certain groups of genes". This surmise, years ahead of its time, received its first experimental verifications in studies of histone effects on RNA synthesis in isolated cell nuclei (5) and in chromatin fractions (6). Histone suppression of RNA synthesis has since been observed in many laboratories, using a variety of DNA templates and RNA polymerases from many sources; there is little doubt that the inhibition is real and reproducible. The interpretation of the effect, however, is complicated by the fact that some histones interact with the RNA polymerase used (7), as well as with the DNA template. However, the finding that RNA synthesis is enormously stimulated when histones are selectively removed from isolated nuclei (5) or from chromatin fractions (8-10) argues in favor of the view that native histone-DNA complexes are much less active as templates for RNA synthesis than the equivalent amounts of free DNA would be. Consequently, the molecular basic of histone suppression has been considered largely in terms of the effects of basic proteins on the physical state of the DNA template. The histones act directly on the long, extended DNA molecule in ways that lead to charge neutralization, supercoiling, and perhaps, cross-linking of DNA helices (e. g. (11)). This type of interaction alters the state of the DNA template, and it may also be expected to influence the binding and the movement of the associated RNA polymerases (7, 12, 13). (Although the interaction between the basic proteins and DNA is usually taken to be largely electrostatic in nature, there are at least five major classes of histones, each of which may be presumed to play some special role in the organization of chromosome structure. Histones differ in their amino acid sequences, in the nature and strength of their binding to DNA, and in their capacity to suppress or modify RNA synthesis. Yet, despite this degree of complexity, it seems unlikely that the limited number of histone species present in the nuclei of higher organisms can account for the multitudinous, selective, and highly specific aspects of transcriptional control which must take place at all genetic loci in the chromosomes of differentiated tissues.)

Apart from these problems concerning the specificity of DNA-histone interactions, the repressed nature of nucleohistone complexes for transcription raises an important related question _ how to alter the association between DNA and histones at times of gene activation. It is evident (as in the case of insect chromosome "puffing" (14)) that the structure and function of chromatin can be varied reversibly at precise regions of the chromosome. It is also well established that the capacity of the nucleus to synthesize RNA is under the control of cytoplasmic factors (15-17), and is influenced by steroid and peptide hormones and by cyclic AMP. When genes are "turned-on", previously repressed regions of the DNA must become accessible. It follows that physiological mechanisms exist for influencing the strength and stability of DNA-histone complexes. How is this brought about?

In recent years it has been discovered that histones are subject to reactions which modify their structure after completion of the polypeptide chain. The reactions include group substitutions - such as acetylation, methylation, and phosphorylation -- which alter the charge and structure of specific amino acid residues in the various polypeptide chains.

The biological significance of all these reactions which alter the structure of previously synthesized chromosomal proteins is not entirely understood, but in the case of histone acetylation, and histone phosphorylation, there is a growing body of evidence that these reactions signal changes in chromosomal structure that frequently <u>precede</u> the activation of repressed gene loci.

<u>Histone Acetylation</u> -   The chemistry of the acetylation reaction has recently been clarified (18, 19). It involves an enzymatic transfer of acetyl groups from acetyl-Coenzyme A to the epsilon-amino groups of specific lysine residues in the histones. The reaction is particularly prominent in the metabolism of the arginine-rich histone fractions, F2a1 and F3.

In molecular terms, the acetylation of lysine residues results in a decrease in the net positive charge of the basic protein. This would be expected to weaken the interactions between the histones and negatively-charged DNA, and would, presumably lead to changes in the fine structure of the chromatin. Because acetylation of histone fraction F2a1 occurs at one particular residue (position-16 in calf thymus F2a1 , (20, 21)) in the most basic portion of the molecule (residues 16 - 20 (20, 21)), a single charge neutralization at this site may have more profound effects than might be anticipated from the small decrease in net positive charge on the whole molecule. In any case, many correlations have now been noted between histone acetylation, chromatin structure, and RNA synthesis. These have been reviewed recently (19, 22), and only a few examples need be cited here.

Histone acetylation has been found to decrease the capacity of the histone to inhibit DNA-directed RNA synthesis in vitro (23). The rate of acetylation and the acetyl content of the histones is greater in those regions of the chromatin that are most active in RNA synthesis (24). The rate of histone acetylation is stimulated in a variety of cell types undergoing a stimulation of RNA synthesis - enhanced acetate uptake has been observed in lymphocytes "transformed" by mitogenic agents (25, 26), in spleen cells stimulated by erythropoietin (27), in liver cells responding to cortisol (19, 24, 28), in early stages of liver regeneration (29), and in mammary epithelial cells exposed to insulin (30).

The response not only involves an increase in the rate of histone acetylation but it may also involve a suppression of histone de-acetylation (19, 29). As a result, the net acetyl content of the histones, i.e. the proportion of the total histone molecules bearing acetyl-lysine residues, is increased during periods of heightened genetic activity.

A suppression of genetic functions is often accompanied by a loss of acetyl groups from the histones. This is seen when RNA synthesis in granulocytes is inhibited by phytohemagglutinin (21), and when nuclear activity is repressed during the maturation of avian red cells (19). In the sperm cells of Arbacia lixula , which are incapable of RNA synthesis, only the non-acetylated form of a histone corresponding to F2a1 can be detected. Embryonic cells of this organism, however, contain both acetylated and non-acetylated forms of this protein (32).

Considering the nature and diversity of the evidence, the conclusion that acetylation represents a mechanism for altering histone-DNA interactions, and that it facilitates transcription is increasingly likely.

Similar conclusions can be drawn from studies of histone phosphorylation.

Histone Phosphorylation - The phosphorylation of histones offers an additional mechanism for charge neutralization of the basic proteins. The reaction was first detected in calf thymus nuclei in vitro (33) and in rat liver in vivo (34). Phosphate-^{32}P-uptake is particularly evident in the very lysine-rich F1 histone fraction, but other histones can also be shown to incorporate ^{32}P-orthophosphate in vivo and in vitro. The major product of the reaction is O-phosphoserine (33, 34), which occurs at a specific site in the polypeptide chain of one of the F1 histones. (The F1 fraction comprises a mixture of closely related proteins, not all of which are phosphorylated.)

The phosphoryl group donor is ATP, and the reaction is catalyzed by histone kinases present in a wide variety of cells and tissues (36-39).

Phosphorylation of the histones takes place after histone synthesis is completed (33, 40). The reaction is reversible, in the sense that phosphate groups incorporated into the histones do not remain there indefinitely, but turn over at rates which vary from one histone fraction to another and which can be altered physiologically (33, 41, 49). A phosphatase which is specific for phosphorylated histones and protamines has been described (42).

There are a number of observations that relate the phosphorylation of histones to genetic activity. For example, lymphocytes "transformed" by phytohemagglutinin increase the rate of phosphorylation of fraction F1 within 15 minutes after addition of the mitogenic agent to the culture medium (41, 43). Thus, the phosphorylation of the lysine-rich histones, like the acetylation of the arginine-rich histones (25), precedes the increase in RNA synthesis in this system. A phosphorylation of histones also occurs in the course of gene activation during regeneration of the liver. Both the rate of phosphorylation and the phosphorus content of histone fraction F1 increase by a factor of two 16-24 hours after partial hepatectomy (44-46, 50). The higher phosphate content of the F1 histone in regenerating liver is due to an increase in the number of F1 molecules phosphorylated (47). The increased proportion of phosphorylated and acetylated forms of the histones is observed at times when the state of the chromatin is altered from a condensed to a more diffuse configuration.

An increased phosphorylation of histones has been noted during regeneration of the pancreas (51), and in a variety of cell types stimulated by hormones. For example, hydrocortisone - which stimulates RNA synthesis in the liver - increases the phosphorylation of the lysine-rich and arginine-rich histones within 30-40 minutes (48). The lysine-rich fraction shows the greatest response. The administration of glucagon to rats, at dosages which are effective in inducing RNA and enzyme synthesis, causes a 15- 25 fold increase in the phosphorylation of the F1 histone within the first hour (35). Insulin, which like glucagon, induces an Actinomycin D-sensitive synthesis of enzymes in the liver, also stimulates histone phosphorylation (35).

The specific kinases which are involved in the phosphoryl-group transfer from ATP to the histones are stimulated by cyclic adenosine-3',5'-monophosphate. A four- to six-fold increase in the rate of histone phosphorylation by a liver enzyme was observed at 10^{-7}M cyclic AMP (36). Similar stimulations have been observed using histone kinases from 15 different bovine tissues (39).

The overall picture that emerges relates phosphorylation of the histones to increased genetic activity. The patterns of phosphorylation of individual lysine-rich histones are species- and tissue-specific (52, 53). This reflects both the complexity of the F1 fraction (i.e. diversity of molecular species (54)), and the differences in the rates of phosphorylation of each species (47).

A particularly interesting and suggestive aspect of histone phosphorylation is that which occurs during spermatogenesis in fish. At the terminal stages of sperm cell differentiation - when histones are replaced by protamines - both classes of chromosomal proteins are phosphorylated (40). The synthesis and phosphorylation of the protamines occurs in the cytoplasm. The newly synthesized protamine is transported into the nucleus and binds to the DNA. At this time in the cell differentiation process, the histones are phosphorylated within the nucleus (40) and they are subsequently degraded. The process of histone release from DNA in the sperm cell nucleus is apparently coupled to the phosphorylation of the basic protein (49). All major histone fractions are phosphorylated at this time, in accord with the view that such structural modifications represent a physiological mechanism for weakening the interactions between histones and DNA.

Measurements have been made of the circular dichroism of histone-DNA complexes, comparing the properties of phosphorylated F1 histones with those of histones which had been enzymatically de-phosphorylated. Clear changes in the structure of the complex were detected (55). Corresponding changes in transcription were observed by Stocken and Stevely (46) who added F1 fractions differing in phosphoserine content to an RNA polymerase system. The capacity of the histone to suppress RNA synthesis was shown to decrease as the phosphate content increased.

Some general conclusions - The variety, extent, and rapidity of the reactions which modify the structure of histones are a sure indication that the biological role of the nuclear basic proteins is far more complex than had previously been assumed. The microheterogeneity which results from one type of substitution, such as phosphorylation, is compounded by alterations of other parts of the same molecule - due to acetylation, various forms of methylation, or (in the case of the F3 histone fraction) to changes in thiol-disulfide ratios.

It may be assumed that these reversible changes in histone structure, brought about by enzymatic reactions, have an effect on DNA-histone interactions in different regions of the chromatin. The weight of evidence now available supports the view that reactions which decrease the net positive charge on the histones, such as phosphorylation and acetylation, correlate with the diffuse state of the chromatin and with increased activity in transcription. In this simple view, such reactions are designed to modify the charge and/or configuration of the chromosomal basic proteins. The resulting changes in their interaction with DNA are readily detectable in vitro, and can be presumed to occur in vivo.

Extrapolation of this basic model to the more acidic proteins associated with DNA in chromatin offers an experimental approach to the problem of how specific genes are turned-on and turned-off by interactions with control proteins, the conformation of the latter depending upon their state of phosphorylation, as suggested below.

"Acidic" Proteins of the Cell Nucleus : Nuclear Phosphoproteins-

A full discussion of the non-histone nuclear proteins is beyond the scope of the present report, and the reader is referred to recent reviews for more detailed treatments (e. g. (56) (57)), but recent developments in the field of the chemistry of the nuclear acidic proteins may be highly relevant to the problem of specificity in genetic transcription in the cells of higher organisms. A brief account of these developments - placing particular emphasis on the role of the nuclear phosphoproteins - will now be presented.

Interest in the non-histone proteins associated with DNA in the chromatin stems from observations that :

1) such proteins have high rates of synthesis and turnover (58-69),

2) that they are present in high concentrations in the chromatin of metabolically active tissues (70-75),

3) that they are preferentially localized in those regions of the chromatin that are most active in RNA synthesis (76, 77),

4) that they are subject to phosphorylation reactions which modify their structure (33, 78-86), particularly at times of gene activation (e.g. (41)), and

5) that their synthesis is selectively influenced by hormones (87, 88, 93).

It has also been reported that nuclear acidic proteins stimulate RNA synthesis in isolated chromatin fractions and augment the template activity of DNA-histone complexes (76, 78, 83, 88, 89). They promote transcription by free DNA (90). There is evidence, based on RNA/DNA hybridization experiments, that the specificity of transcription in re-constituted chromatin fractions depends on the nature of the non-histone proteins present in the mixture (91, 92).

These observations, implying positive control of RNA synthesis and specificity of interaction between the acidic proteins and DNA, should be considered together with the phosphorylation reactions which show that the structures of many chromosomal proteins are subject to reversible change.

Our recent experiments on nuclear phosphoproteins prepared from a variety of animal tissues support the proposition that both aspects of the problem - molecular diversity and phosphate transfer - have a bearing on the control of chromosomal RNA synthesis.

Isolation and properties of nuclear phosphoproteins - A new method for the preparation of nuclear acidic proteins, based on their localization in chromatin, and their characteristic solubility properties, has been applied to the liver, kidney, spleen, brain and other tissues of the rat. Nuclei were first purified by centrifugation through sucrose density barriers (90). A differential extraction of the nuclear proteins was then carried out, using 0.14 M NaCl to remove saline-soluble components, 0.25 N HCl to remove histones, and phenol to solubilize many of the acidic proteins of the residue. (The use of phenol to solubilize proteins associated with nucleic acids has a precedent in the study of plant(94) and bacterial (95) viruses. Phenol has been used to extract proteins with high rates of synthesis from the nuclei of mammalian cells (63, 87).) The phosphoproteins in the phenol phase were dialyzed against 0.1 M acetic acid - 0.14 M 2-mercaptoethanol, and against a series of urea-containing buffers (95) which restore the phenol-soluble proteins to the aqueous phase for further characterization. In the final stage of the preparation, the acidic proteins were dissolved in 0.1 M tris, pH 8.4, containing 0.01 M EDTA, urea, 0.14 M 2-mercaptoethanol (87, 90, 95). Aliquots of the solution were analyzed for total protein content, for alkali-labile phosphate, for phosphoserine and phosphothreonine, and for average amino acid composition (90). Phosphoproteins labeled with ^{32}P were isolated from liver and kidney nuclei of animals which had been injected with ^{32}P-orthophosphate (2mc / 100g body weight) at appropriate times after the administration of cortisol, or saline solutions.

The separation and characterization of individual proteins in the extract was achieved by polyacrylamide gel electrophoresis in the presence of sodium dodecylsulfate (96, 97). A complex banding pattern is obtained in which the distance of migration of individual proteins can be correlated with their molecular weights (97, 98). The relative amounts of individual components has been determined in two ways : 1] by staining of the protein bands with Amidoblack 10B, followed by quantitative densitometry and computer analysis of the densitometer tracing, or 2] by cutting out the bands and extracting the dye from each gel slice in dimethylsulfoxide. (We have shown that the amount of dye recovered from each band is directly proportional to the protein content of that band (90).) The isotope content of the protein in each gel slice was determined by scintillation spectrometry (87, 90).

<u>Tissue specificity</u> -   The complexity of the phenol-soluble protein fraction of rat liver nuclei is indicated by the multiple banding pattern shown in Figure 1 A. Computer analysis of the curve indicates the presence of at least 38 components. A densitometer tracing of the gel photograph is aligned with the banding pattern and shown in the upper part of the figure. The distribution of the bands and their relative proportions are

highly reproducible in different preparations of the phenol-soluble proteins from any given tissue, such as liver, but they vary from one tissue to another. This is clearly indicated by the striking differences between the electrophoretic patterns of kidney acidic proteins (Figure 1 B) and liver acidic nuclear proteins (Figure 1 A).

A.
LIVER

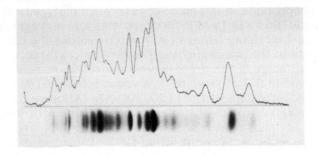

B.
KIDNEY

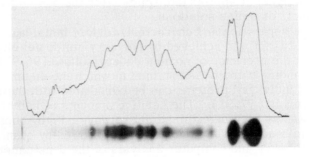

Figure 1. Electrophoretic separation of acidic proteins from rat liver and kidney nuclei. The proteins were stained with Amido Black 10B. A densitometer tracing is aligned over the corresponding banding pattern.

Spleen and brain nucleoproteins show still different electrophoretic patterns.

Thus, tissue specificity, which should be one characteristic of proteins concerned with the control of transcription in differentiated cells, is observed in the phenol-soluble fraction.

Phosphorylation - Tissue-specificity of the nuclear acidic proteins is also indicated in studies of their phosphorylation. The presence of phosphate in the acidic nuclear protein fraction has been demonstrated by analysis of the phenol-soluble proteins for alkali-labile-P content. The average phosphorus content of rat liver phosphoproteins prepared in this way is 0.94% (90). (This figure is to be compared with earlier estimates of 1.14% P in rat liver nuclear phosphoproteins (83) and 1.28 % P in calf thymus nuclear phosphoproteins (79) prepared by a combination of salt extraction and chromatographic purification.) A limited acid hydrolysis of the proteins, followed by chromatography of the hydrolysate on Dowex-50 reveals the presence of phosphoserine and smaller amounts of phosphothreonine. Other tests have shown that the phosphate content of the fractions is not due to contamination by nucleic acids.

The uptake and release of phosphate by nuclear phosphoproteins is a major aspect of their metabolism (33, 79-86). Studies of ^{32}P-orthophosphate incorporation into individual, electrophoretically separated proteins of rat liver and kidney nuclei have revealed clear differences in the extent and kinetics of phosphorylation of different proteins from the same tissue (90, 99), and obvious differences in radio-phosphate distribution in analytical gels from different tissues. Thus, the metabolism, as well as the composition of the phenol-soluble protein fraction indicates tissue-specificity.

Relationship of phosphorylation to gene activation - There is good evidence relating phosphorylation of the nuclear acidic proteins to the RNA synthetic activity of the nucleus. For example, it was shown that "active" and "inactive" chromatin fractions, prepared by the method of Frenster et al. (100), differ in their phosphoprotein contents as well as in their capacity to synthesize RNA. Analyses of calf thymus chromatin sub-fractions were carried out by Langan and Frenster, and the results were expressed in terms of the number of μmoles of alkali-labile P in the protein associated with 100 mg of DNA. The active chromatin contained 13.5 μmoles of P as phosphoprotein, while the clumped, inactive chromatin contained only 3.8 μmoles of P in an equivalent "stretch" of DNA-protein.

In lymphocytes, as in many other cell types, RNA synthesis occurs mainly in the diffuse regions of the chromatin (101), and it is accelerated when lymphocytes are stimulated to enlarge and divide by the addition of phytohemagglutinin (PHA) to the culture medium (25, 41). Changes in chromatin structure and function are reflected in phosphoprotein metabolism. The rate of protein phosphorylation was measured by "pulse-labeling" techniques during the early stages of gene activation in PHA-treated cells. An increase in rate of ^{32}P-incorporation into the nuclear phosphoproteins was detected within 15 minutes after addition of PHA (41). (The change observed was not simply due to changes in the specific activity of the ATP "pool", which decreases very slightly in the PHA-treated lymphocyte.)

An important point is that the increase in protein phosphorylation precedes the increase in RNA transcription in the PHA-treated cells, in accord with the expectation that phosphorylation of the chromosomal proteins may facilitate the utilization of previously unused or repressed DNA - templates. The fact that the "turnover" of previously-incorporated ^{32}P-phosphate groups is also accelerated at times of gene activation suggests a cyclical function of the phosphoprotein (41).

Correlations of protein phosphorylation with gene activation are particularly evident in studies of the effects of hormones on nuclear metabolism. For example, the administration of cortisol to adrenalectomized rats increases RNA synthesis in the liver (102-104) and also influences the rate of synthesis of one of the nuclear acidic proteins (87). The phosphorylation of the nuclear phosphoproteins in liver is stimulated within 5 minutes after administration of the hormone (99). No corresponding changes were seen in ^{32}P-phosphate uptake into nuclear proteins of the kidney - which does not respond to cortisol as does the liver. Kinetic studies of the phosphorylation of individual protein bands in the liver phenol-soluble fraction show that the kinetics of phosphate uptake and release are highly complex. They differ from one protein to another, and the patterns of phosphorylation vary with time after hormone administration. Some indication of the complexity is provided by the data summarized in Figures 2 A and 2 B, which compare the specific ^{32}P-activities of the phosphoproteins of control and cortisol-treated animals at 15 minutes after hormone administration and again at 120 minutes. Stimulations of ^{32}P-uptake into proteins of low mobility (high molecular weight) are evident at early times, but not later. (Compare Figures 2 A and 2 B). Several proteins of relatively high mobility (low molecular weight) show a greater stimulation of phosphate incorporation at later times after hormone administration. Whether the delayed response in the phosphorylation of some nuclear proteins is related to the "programming" of the RNA synthetic response (104) remains to be determined. What is clear is that the differences observed are not likely to be due to fluctuations in nuclear ATP "pool" size, which would be expected to shift the labeling of all the phosphoproteins in a parallel fashion.

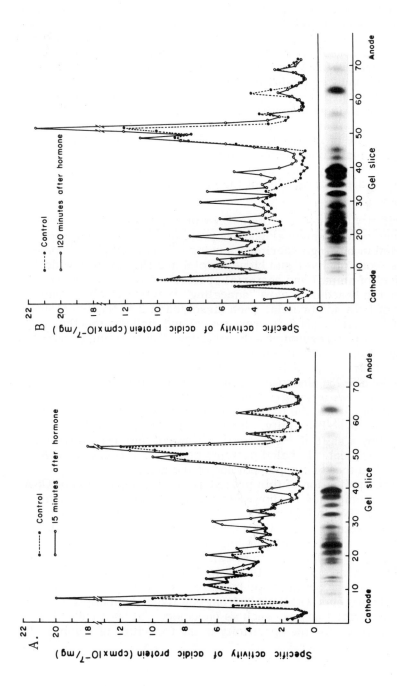

Figure 2. Time-dependence of cortisol stimulation of ^{32}P-phosphate into liver P-proteins.

In any case, the stimulation of phosphorylation of the acidic chromosomal proteins of the liver by cortisol (99), and the stimulation of phosphorylation of the basic chromosomal proteins by cortisol (48), by cyclic AMP (35), by glucagon (35) and by insulin (35) appear to be related phenomena. The hormone-induced modifications of the structure of both major classes of chromosomal proteins presumably reflect the changing state and function of the chromatin in the target cell.

The phosphate content of the acidic nuclear protein fraction and the rate of incorporation of ^{32}P-orthophosphate are much lower in mature nucleated erythrocytes of birds than they are in immature erythroid cells (86). This, too, matches the RNA synthetic capacities of the respective developmental stages.

Specificity of phosphoprotein-DNA interactions - The interactions between nuclear phosphoproteins and DNA have been studied in binding experiments using ^{32}P-labeled phosphoproteins from rat liver and kidney nuclei and DNA from a wide variety of animal and bacterial species. The procedure employed is based on a slow "annealing" reaction in which samples of DNA and protein, each dissolved in 2 M NaCl-5 M urea-0.01 M "tris", pH 8.0, were mixed and dialyzed together against a progression of salt solutions of decreasing concentration (106). After removal of the urea by dialysis against 0.01 M "tris", pH 8.0 - 0.01 M NaCl, the samples were layered over a 5-25% sucrose gradient and centrifuged. The distribution of DNA in the gradient was indicated by its absorption at 260 mμ, while the position of the phosphoproteins was indicated by ^{32}P-activity. Binding of proteins to DNA is accompanied by a shift of ^{32}P-activity downward into denser regions of the gradient. Labeled proteins that are not bound to DNA remain in the light upper portion of the gradient.

The application of this procedure for the separation of DNA-protein complexes is shown for rat liver phosphoproteins and rat liver DNA in Figure 3. A similar binding of rat kidney phosphoproteins to rat liver DNA can be demonstrated (90, 105). However, a comparable binding reaction has not been noted for all DNA's. The failure of calf thymus DNA to bind rat liver nuclear phosphoproteins is shown in Figure 4. Negative results have also been obtained for mixtures of rat liver or rat kidney phosphoproteins with DNA's from dog liver, human placenta, and bacterial sources, such as pneumococcal DNA (90, 105).

The basis for the specificity in the formation of soluble

phosphoprotein-DNA complexes is not understood and further work on the system is required, but the conclusion that liver nuclear phosphoproteins have a capacity to selectively combine with the DNA of the tissue of origin is supported by the recent work of Kleinsmith, Heidema and Carroll (107) who used DNA/cellulose column chromatography to demonstrate that a fraction of the non-histone protein from rat liver chromatin binds to rat DNA but not to DNA's from salmon sperm or from E. coli.

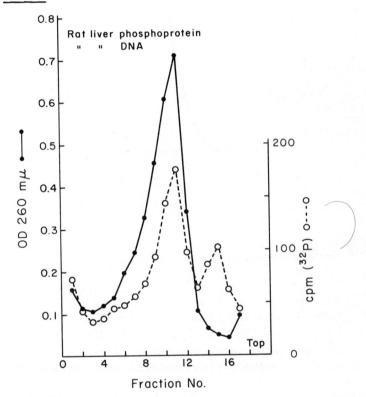

Figure 3. Separation of rat liver nucleoprotein-DNA complexes by sucrose-density gradient centrifugation after "annealing" reaction. The position of DNA in the gradient is indicated by the UV-absorption (solid line). The distribution of the ^{32}P-labeled phosphoprotein is shown by the dashed line. Note the migration of ^{32}P-labeled protein with the DNA peak.

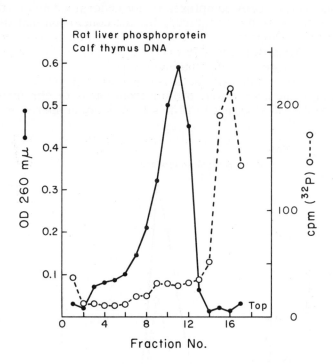

Figure 4. Evidence for selectivity in DNA-binding by rat liver nuclear phosphoproteins. The phosphoprotein fraction, labeled with ^{32}P, was mixed with calf thymus DNA under "annealing" conditions, and then centrifuged on a 5-25% sucrose gradient. Note that the isotopically-labeled proteins (dashed line) fail to form a soluble DNA-protein complex and remain at the top of the gradient.

The question arises as to whether all or only a fraction of the nuclear phosphoproteins combine selectively with DNA in this manner. The problem was investigated by recovering the proteins from the soluble DNA-protein complex isolated from the sucrose-gradient. Electrophoretic analyses showed that many, but not all, of the liver phenol-soluble proteins had combined with DNA under these conditions (90). A similar differential binding reaction, involving different proteins, was found in the study of the complexes formed between rat kidney phosphoproteins and rat DNA (90).

The results show that many of the acidic proteins of mammalian cell nuclei, though they differ from one cell type to another, have a capacity to selectively combine with the appropriate nucleotide sequences of DNA in vitro. This strongly suggests that the nuclear phosphoproteins are closely associated with DNA in vivo. Support for this conclusion is provided by the demonstration that the proteins described can be recovered from isolated chromatin fractions (90), and by autoradiographic studies which show a localization of ^{32}P-labeled proteins along the salivary gland chromosomes of Sciara larvae (108).

Specific enhancement of transcription by nuclear phosphoproteins -

Recent studies in this laboratory (90) indicate that the nuclear phosphoprotein fraction includes components which stimulate transcription in a cell-free, DNA-dependent RNA synthesizing system. The effect is readily demonstrated using the soluble RNA polymerase from E. coli (prepared and assayed according to the method of Burgess (109, 110).), rat liver DNA and phosphoproteins from rat liver and kidney nuclei. The incorporation of ^{3}H-UTP is taken as a measure of RNA synthesis. The results, summarized in Table 1, show that RNA synthesis is increased 78-90% by rat liver nuclear proteins and 72% by rat kidney nuclear acidic proteins, provided that rat DNA is used as the template. No stimulatory effect is seen when the rat nuclear proteins are added to calf thymus DNA under the same "annealing" and assay conditions.

TABLE 1

Selective Stimulation of DNA-Template Activity by Nuclear Acidic Proteins

Conditions	^{3}H-UTP Utilization for RNA Synthesis			
	Expt. 1		Expt. 2	
	cpm/μg DNA	%	cpm/μg DNA	%
rat liver DNA *	440	100	865	100
rat liver DNA + rat liver P-protein	835	190	1530	178
rat liver DNA + rat kidney P-protein	757	172	-	-
calf thymus DNA**	2180	100	5000	100
calf thymus DNA + rat liver P-protein	2200	103	4700	94

* 0.45 units RNA polymerase/ 0.1 ml reaction mixture (90).
** 1.5 units RNA polymerase/ 0.1 ml reaction mixture.

The RNA polymerase results agree with the specificity of phosphoprotein-DNA binding as indicated in Figures 3 and 4.

Conclusions - The nuclear phosphoproteins have many characteristics expected of proteins concerned with the specific control of gene expression in the cells of higher organisms

1] They are localized in active regions of the chromatin.
2] Their distribution in different cell types is tissue-specific.
3] Their activity - particularly with regard to phosphate turnover on the hydroxyl groups of serine residues - reflects the RNA synthetic capacity of the tissue.
4] They combine selectively with the DNA of the species of origin.
5] Transcription from the DNA/phosphoprotein complex, once formed, exceeds that from DNA alone.

We conclude that the nuclear phosphoprotein fraction includes components which are involved in the positive control of RNA synthesis. It is likely that such components recognize and combine with specific polynucleotide sequences to promote transcription at particular gene loci.

The mechanism by which transcription is selectively facilitated is not known, but we propose that phosphorylation is a critical variable in the interaction between the phosphoproteins DNA, and the RNA polymerase. The reversibility of the phosphorylation reaction <u>in vivo</u> suggests a cyclical binding and release mechanism.

In many respects, the nuclear phosphoprotein control system is analogous to the <u>sigma</u> factor control of RNA polymerase activity in bacterial systems (109-113). The <u>sigma</u> component appears to confer specificity on the polymerase by facilitating its attachment to specific initiation (or "promoter") sites on the DNA. After initiation, the <u>sigma</u> factor is released from the enzyme and can participate in a second round of initiation. Phosphorylation of <u>E. coli</u> <u>sigma</u> factors by protein kinases from animal tissues has been found to stimulate RNA synthesis (114). The <u>sigma</u> phosphorylation reaction is itself stimulated by cyclic AMP (114). (This effect may constitute a basis for the induction of <u>lac</u> mRNA and beta-galactosidase synthesis in bacterial systems by cyclic AMP (see (115, 116) and the appropriate inducers.)

The resemblance of the <u>sigma</u> phosphorylation system - particularly with respect to its hormonal (cyclic AMP) response- to the phosphorylation mechanisms known to exist in the chromosomes of higher organisms is strikingly close. Further developments in this area promise to reveal protein phosphorylation as a general mechaism for the control of chromosomal activity in prokaryotic and eukaryotic cells.

REFERENCES

1. V. G. Allfrey and A. E. Mirsky, Proc. Nat. Acad. Sci. U. S., 48 (1962) 1590.
2. J. Paul and R. S. Gilmour, J. Mol. Biol., 16 (1966) 242.
3. B. J. McCarthy and B. H. Hoyer, Proc. Nat. Acad. Sci. U. S., 52 (1964) 915.
4. E. Stedman and E. Stedman, Phil. Trans. Roy. Soc. (London), B 234 (1951) 565.
5. V. G. Allfrey, V. C. Littau and A. E. Mirsky, Proc. Nat. Acad. Sci., U. S., 49 (1963) 414.
6. R. C. Huang and J. Bonner, Proc. Nat. Acad. Sci. U. S., 48 (1962) 1216.
7. T. C. Spelsberg and L. S. Hnilica, Biochim. Biophys. Acta, 195 (1969) 55.
8. K. Marushige and J. Bonner, J. Mol. Biol., 15 (1966) 160.
9. J. Hindley, Abstr. Intern. Congr. Biochem. 7th (Tokyo), 1 (1964) 82.
10. G. P. Georgiev, L. N. Ananieva and Y. V. Kozlov, J. Mol. Biol., 22 (1966) 365.
11. B. M. Richards and J. F. Pardon, Exp. Cell Res., 62 (1970)184.
12. T. Y. Shih and J. Bonner, J. Mol. Biol., 48 (1970) 469.
13. Y. V. Kozlov and G. P. Georgiev, Nature, 228 (1970) 245.
14. W. Beerman, J. Exptl. Zool., 157 (1964) 49.
15. J. B. Gurdon and D. D. Brown, J. Mol. Biol., 12 (1965) 27.
16. L. R. Thompson and B. J. McCarthy, Biochem. Biophys. Res. Commun., 30 (1968) 166.
17. H. Harris, J. F. Watkins, C. E. Ford and G. I. Schoefl, J. Cell Sci., 1 (1966) 1.
18. E. L. Gershey, G. Vidali and V. G. Allfrey, J. Biol. Chem., 243 (1968) 5018.
19. V. G. Allfrey, Federation Proc., 29 (1970) 1447.
20. R. J. Delange, D. M. Fambrough, E. L. Smith and J. Bonner, J. Biol. Chem., 244 (1969) 319.
21. Y. Ogawa, G. Quagliarotti, J. Jordan, C. W. Taylor, W. C. Starbuck and H. Busch, J. Biol. Chem., 244 (1969) 4387.
22. V. G. Allfrey, in "The Histones and Nucleohistones" (D. M. P. Phillips, Editor),(Plenum Press, London, 1971) p. -
23. V. G. Allfrey, R. M. Faulkner and A. E. Mirsky, Proc. Nat. Acad. Sci. U. S., 51 (1964) 786.

24. V. G. Allfrey, Canadian Cancer Conf., 6 (1964) 313.
25. B. G. T. Pogo, V. G. Allfrey and A. E. Mirsky, Proc. Nat. Acad. Sci. U. S., 55 (1966) 805.
26. T. Ono, H. Terayama, F. Takaku and K. Nakao, Biochim. Biophys. Acta, 179 (1969) 214.
27. F. Takaku, K. Nakao, T. Ono and H. Terayama, Biochim. Biophys. Acta, 195 (1969) 396.
28. V. G. Allfrey, Cancer Research, 26 (1966) 2026.
29. B. G. T. Pogo, A. O. Pogo, V. G. Allfrey and A. E. Mirsky, Proc. Nat. Acad. Sci. U. S., 59 (1968) 1337.
30. W. F. Marzluff and K. S. McCarty, J. Biol. Chem., 245, (1970) 5635.
31. B. G. T. Pogo, V. G. Allfrey and A. E. Mirsky, J. Cell Biol., 35 (1967) 477.
32. L. Wangh, A. Ruiz-Carrillo and V. G. Allfrey, manuscript in preparation.
33. L. J. Kleinsmith, V. G. Allfrey and A. E. Mirsky, Proc. Nat. Acad. Sci. U. S., 55 (1966) 1182.
34. M. G. Ord and L. A. Stocken, Biochem. J., 98 (1966) 5P.
35. T. A. Langan, Proc. Nat. Acad. Sci. U. S., 64 (1969) 1276.
36. T. A. Langan, Science, 162 (1968) 579.
37. E. Schiltz and C. E. Sekeris, Z. physiol. Chem., 350 (1969) 317.
38. T. A. Langan, J. Biol. Chem., 244 (1969) 5763.
39. J. F. Kuo, B. K. Krueger, J. R. Sanes and P. Greengard, Biochim. Biophys. Acta, 212 (1970) 79.
40. K. Marushige, V. Ling and G. H. Dixon, J. Biol. Chem., 244 (1969) 5953.
41. L. J. Kleinsmith, V. G. Allfrey and A. E. Mirsky, Science, 154 (1966) 780.
42. M. H. Meisler and T. A. Langan, J. Biol. Chem., 244 (1969) 4961.
43. M. E. Cross and M. G. Ord, Biochem. J., 118 (1970) 191.
44. M. G. Ord and L. A. Stocken, Biochem. J., 107 (1968) 403.
45. W. S. Stevely and L. A. Stocken, Biochem. J., 100 (1966) 20c.
46. L. A. Stocken and W. S. Stevely, Biochem. J., 110 (1968) 187.
47. R. H. Buckingham and L. A. Stocken, Biochem. J., 117 (1970) 157.
48. L. D. Murthy, D. S. Pradhan and A. Sreenivasan, Biochim. Biophys. Acta, 199 (1970) 500.
49. M. T. Sung and G. H. Dixon, Proc. Nat. Acad. Sci. U. S., 67 (1970) 1616.
50. R. M. Gutierrez and L. S. Hnilica, Science, 157 (1967) 1324.

51. P. J. Fitzgerald, W. H. Marsh, M. G. Ord and L. A. Stocken, Biochem. J., 117 (1970) 711.
52. B. Jergil, M. Sung and G. H. Dixon, J. Biol. Chem., 245 (1970) 5867.
53. T. A. Langan, Science, in press (1971)
54. J. M. Kinkade and R. D. Cole, J. Biol. Chem., 241 (1966) 5798.
55. H. A. Adler, B. Schlafhausen, T. A. Langan and G. D. Fasman, Proc. Nat. Acad. Sci. U. S., in press.
56. L. S. Hnilica, Progr. Nucl. Acid. Res. Mol. Biol., 7 (1966)25.
57. R. H. Stellwagen and R. D. Cole, Ann. Rev. Biochem., 38 (1969) 951.
58. M. M. Daly, V. G. Allfrey and A. E. Mirsky, J. Gen. Physiol., 36 (1952) 173.
59. R. M. S. Smellie, W. M. McIndoe and J. N. Davidson, Biochim. Biophys. Acta, 11 (1953) 559.
60. V. G. Allfrey, M. M. Daly and A. E. Mirsky, J. Gen. Physiol., 38 (1955) 415.
61. V. G. Allfrey, A. E. Mirsky and S. Osawa, J. Gen. Physiol., 40 (1957) 451.
62. O. P. Samarina, Biokhimiya, 26 (1961) 61.
63. W. J. Steele and H. Busch, Cancer Research, 23 (1963)1153.
64. A. L. Dounce and C. A. Hilgartner, Exp. Cell Research, 36 (1964) 228.
65. L. S. Hnilica, H. A. Kappler, and V. S. Hnilica, Science, 150 (1965) 1470.
66. V. Holoubek, L. Fanshier, T. T. Crocker and L. S. Hnilica, Life Sci., 5 (1966) 1691.
67. H. Hyden and B. S. McEwen, Proc. Nat. Acad. Sci. U. S., 55 (1966) 354.
68. V. Holoubek and T. T. Crocker, Biochim. Biophys. Acta, 157 (1968) 352.
69. C. S. Teng and T. H. Hamilton, Proc. Nat. Acad. Sci. U. S., 63 (1969) 465.
70. A. E. Mirsky and H. Ris, J. Gen. Physiol., 34 (1951) 475.
71. C. W. Dingman and M. B. Sporn, J. Biol. Chem., 239 (1964) 3483.
72. K. Marushige and G. H. Dixon, Dev. Biol., 19 (1969) 397.
73. K. Marushige and H. Osaki, Dev. Biol., 16 (1967) 474.
74. R. A. Paoletti and R. C. Huang, Biochemistry, 8 (1969)1615.
75. V. Seligy and M. Miyagi, Exp. Cell Research, 58 (1969) 28.
76. J. H. Frenster, Nature, 206 (1965) 680.
77. F. Dolbeare and H. Koenig, Proc. Soc. Exp. Biol. Med., 135 (1970) 636.

78. T. A. Langan, in " Regulatory Mechanisms for Protein Synthesis in Mammalian Cells. eds. A. SanPietro, M. R. Lamborg and F. T. Kenney. (Academic Press, New York, 1968) p. 101.
79. L. J. Kleinsmith and V. G. Allfrey, Biochim. Biophys. Acta, 175 (1969) 123.
80. L. J. Kleinsmith and V. G. Allfrey, Biochim. Biophys. Acta, 175 (1969) 136.
81. J. N. Davidson, S. C. Frazer and W. C. Hutchinson, Biochem. J., 49 (1951) 311.
82. R. N. Johnson and S. Albert, J. Biol. Chem., 200, (1953) 335
83. T. A. Langan, in " Regulation of Nucleic Acid and Protein Biosynthesis". eds. V. V. Koningsberger and L. Bosch. Elsevier, Amsterdam, 1967) p. 233.
84. W. Benjamin and A. Gellhorn, Proc. Nat. Acad. Sci., U. S., 59 (1968) 262.
85. G. Patel, V. Patel, T. Y. Wang and C. R. Zobel , Arch. Biochem. Biophys., 128 (1968) 654.
86. E. L. Gershey and L. J. Kleinsmith, Biochim. Biophys. Acta, 194 (1969) 519.
87. K. S. Shelton and V. G. Allfrey, Nature, 228 (1970) 132.
88. C. S. Teng and T. H. Hamilton, Biochem. Biophys. Res. Commun. 40 (1970) 1231.
89. T. Y. Wang, Exp. Cell Research, 61 (1970) 455.
90. C. S. Teng, C. T. Teng and V. G. Allfrey, J. Biol. Chem., in press (1971).
91. T. C. Spelsberg and L. S. Hnilica, Biochem. J., 120 (1970) 435.
92. R. J. Gilmour and J. Paul, J. Mol. Biol., 40 (1969) 137.
93. G. Stein and R. Baserga, J. Biol. Chem., 245 , (1970) 6097.
94. A. Gierer and G. Schramm, Nature, 177 (1956) 702.
95. E. Vinuela, I. D. Algranati and S. Ochoa, Eur. J. Biochem., 1 (1967) 3.
96. D. F. Summers, J. V. Maizel and J. E. Darnell, Proc. Nat. Acad. Sci. U. S., 54 (1965) 505.
97. A. L. Shapiro, E. Vinuela and J. V. Maizel, Biochem. Biophys. Res. Commun., 28 (1967) 815.
98. K. Weber and M. Osborn, J. Biol. Chem., 244 (1969) 4406.
99. C. S. Teng, C. T. Teng and V. G. Allfrey, J. Biol. Chem. (submitted) (1971).
100. J. H. Frenster, V. G. Allfrey and A. E. Mirsky, Proc. Nat. Acad. Sci. U. S., 50 (1963) 1026.
101. V. C. Littau, V. G. Allfrey, J. H. Frenster and A. E. Mirsky, Proc. Nat. Acad. Sci. U. S., 52 (1964) 93.

102. M. Feigelson, P. R. Gross and P. Feigelson, Biochim. Biophys. Acta, 55 (1962) 495.
103. F. T. Kenney and F. J. Kull, Proc. Nat. Acad. Sci. U. S., 50 (1963) 493.
104. F. L. Yu and P. Feigelson, Biochem. Biophys. Res. Commun. 35 (1969) 499.
105. C. T. Teng, C. S. Teng and V. G. Allfrey, Biochem. Biophys. Res. Commun., 41 (1970) 690.
106. I. Bekhor, G. M. Kung and J. Bonner, J. Mol. Biol., 39 (1969) 351.
107. L. J. Kleinsmith, J. Heidema and A. Carroll, Nature, 226 (1970) 1025.
108. W. B. Benjamin and R. M. Goodman, Science, 166 (1969) 629.
109. R. R. Burgess, J. Biol. Chem., 244 (1970) 6160.
110. R. R. Burgess, J. Biol. Chem., 244 (1970) 6168.
111. A. A. Travers and R. R. Burgess, Nature, 222 (1969) 537.
112. E. K. F. Bautz and F. A. Bautz, Nature, 226 (1970) 1219.
113. D. Berg and M. Chamberlin, Biochemistry, 9 (1970) 5055.
114. O. J. Martelo, S. L. C. Woo, E. M. Reimann and E. E. Davie, Biochemistry, 9 (1970) 4807.
115. G. Zubay, D. Schwartz and J. Beckwith, Proc. Nat. Acad. Sci. U. S., 66 (1970) 104.
116. H. C. Varmus, R. L. Perlman and I. Pastan, J. Biol. Chem., 245 (1970) 6366.

This research was supported in part by grants from the American Cancer Society (E-519), the National Foundation/March of Dimes (CRBS-223) and the United States Public Health Service (GM-17383-01).

DISCUSSION

B.L. HORECKER: Do you have any idea what enzymes are involved in phosphorylation and dephosphorylation of these acidic phosphoproteins? You implied that phosphorylation was not mediated by the protein kinase.

V.G. ALLFREY: It is not the histone kinase. For one thing, the phosphorylation of these proteins, as studied in isolated nuclei or in isolated preparations of chromatin or in isolated phosphoproteins, is not stimulated by cyclic AMP, as is phosphorylation of the F_1 histone. The other is that the release of phosphate as we have studied it in isolated nuclei results in a release of inorganic phosphate. I would guess that it is due to a simple phosphatase type reaction. However, if one studies this in intact nuclei *in vivo*, the release of phosphate as well as the uptake of ^{32}P shows signs of energy dependence: so the whole system may have to be turning over to release the phosphate under natural conditions.

R. WARREN: Bonner and co-workers working on pea seedlings have described a 3.2S RNA of about 60 nucleotides associated with the basic proteins from this chromatin. Would your fractionation procedure allow you to see this, or did you see any indication of this type of RNA.

V.G. ALLFREY: When we began to get specific binding, we tested to see whether such RNA's were involved in the recognition of specific nucleotide sequences in DNA. We analyzed for RNA in the phosphoprotein preparation that was used for the binding experiment. We labelled with a great deal of RNA precursor such as ^{14}C-orotic acid and found no label in the protein fraction as isolated. There was no RNA detectable by chemical analysis by the orcinol reaction. There was no RNAase effect detectable in tests where we attempted to destroy the reaction by pretreating the phosphoprotein with ribonuclease; so by three tests we were unable to find any RNA in these preparations.

D. NOVELLI: You said that these acidic nuclear proteins are synthesized much more rapidly than histones; do you have any idea of the half-life of these proteins and whether the half-life is affected by the injection of cortisone?

V.G. ALLFREY: I can make a rough estimate of the half-life of the particular protein (MW 41,000) stimulated by cortisol. That had a half-life of 7 hours.

D. NOVELLI: Is that the average half-life of the mixture?

V.G. ALLFREY: No. The average half-life of the mixture depends on the tissue. Half-lives up to 90 hours have been noted for some acidic proteins of the liver and kidney.

B. ALBERTS: In looking over the older literature on mammalian DNA-binding proteins, it was always my impression that what were called insoluble acidic proteins were only insoluble because the histones had first been extracted with acid. It is not clear to me whether the insolubility that you are dealing with is also due to that factor or whether the acidic proteins you are talking about are actually insoluble in aqueous solution in their native state.

V.G. ALLFREY: They can be solubilized in high salt, and that was the original extraction procedure devised by Langan and used by Kleinsmith. The proteins go into high salt, and after removing the histone and nucleic acid, the acidic proteins are further purified by chromatographic procedures. When they are prepared that way they have very low solubility in neutral buffers. Once you get them off by themselves they precipitate readily; it is for that reason that isoelectric focusing shows all those bands as turbid areas. They are really quite insoluble. When they are prepared by phenol extraction they also have very low solubilities. One of our problems is that they tend to aggregate.

B. ALBERTS: You are saying that the native proteins are soluble at high salt concentrations when you extract them from the DNA, but when you lower the salt concentration they come out of solution?

V.G. ALLFREY: Right.

S.B. WEISS: Can you tell me what is the specificity of the phosphoproteins for the various DNA's if 1) the DNA's are denatured and 2) if the phosphoproteins are dephosphorylated?

V.G. ALLFREY: I wish I could answer that. The dephosphorylation experiment is under way. We have been able to take 80% of the dephosphorylated protein in DNA binding and in the RNA polymerase assay. If the DNA is heat-denatured, we do not get the formation of a readily-discernible soluble complex. Instead, a kind of a random aggregate forms in which all of the protein is recovered, but binding is not selective. What we need is a better method for measuring the DNA-binding affinities of individual phosphoproteins.

S. FRIEDMAN: I'd like to follow up this question; did you do other experiments on the characterization of the soluble DNA protein complex, beside the centrifugation experiment you showed? I'd like to know if you find any conformation changes associated with the formation of the complex or a change in the hyperchromicity of the DNA upon melting the complex.

V.G. ALLFREY: No. We haven't done that; some of this study of DNA-protein complexes and their conformation changes is going to be done in collaboration with Dr. G. Fasman at Brandeis University. We haven't done it yet.

P. BYVOET: Although I am afraid that it is somewhat superfluous, I would like to point out that if you are measuring half-lives in a tissue, for example, in the liver, you have to be careful with the interpretation of the data. For example, if one injects an animal with a precursor for DNA and measures the half-life of liver DNA starting shortly after administration of the label, one finds a short half-life. If, on the other hand, one waits a while after the injection and then starts measuring, one finds a long half-life, because during the intervening period all the label is lost that was incorporated in rapidly turning-over cell populations of the liver. A precursor will be incorporated most actively into those macromolecules which turn over fastest, and it is often hard to tell from which cell type they originate or even if they originate from a single cell type. The liver contains many different cell types and the half-life of 90 hours for the mixture of acidic proteins might well pertain to proteins of the rapidly turning-over cell populations in the liver. A half-life of 90 hours would be similar to that of the DNA in those cell populations.

V.G. ALLFREY: The experiments that I was citing were based on long term turn-over experiments where there was a great deal of pre-labeling, with ^{15}N-glycine. (This was before C^{14} was available, and what we did was measure the retention of ^{15}N-glycine in proteins in the liver residual nuclear proteins as a function of time, losing half of the counts in about 96 hours.) I don't have any other data on them, except for recent work on that particular band in the acrylamide gels. What you say about the mixed population of the cells in the liver is true. Some cells are turning over and this will confuse any estimate of average half-life of nuclear proteins.

D. BOROVSKY: Do you have any idea what kind of chemical bond you have between the phosphate protein and the DNA?

V.G. ALLFREY: There's no evidence for covalent linkage of any sort. The complex is readily dissociated by salt and by relatively low salt concentration. The problem is that the protein, having been isolated in phenol and treated with urea, was originally denatured. What we are counting on is renaturation of the protein to conform to the DNA; how effective that process is, I don't know. I don't know whether this is as strong a bond as occurs naturally or not; I doubt it. In any case, it is a covalent bond because salt readily dissociates the complex.

D. BOROVOSKY: But if the protein is denatured, it just might migrate with the DNA without having any specificity; and in the cell, when the protein is not denatured, there might be some specificity between binding of protein to DNA. So in your experiments when you denatured a protein you just defeated your purpose.

V.G. ALLFREY: Well, I'm not going to deny that the denaturation of the protein is a disadvantage. The point is that even after it has occurred it's possible to demonstrate specificity of a sort with respect to DNA binding and with respect to template activity. What you are saying is true that it seems a mistake to start out by denaturing the protein if one is trying to study its natural properties, but when we began that was not our aim. Our aim was to separate the proteins from the nucleic acids and from other proteins in the nucleus so that we could do the chemistry on them, study their phosphate uptake and turn-over, determine sites of phosphorylation, and so on. This is still a method of choice because we now do not have to worry about contamination by nucleic acids.

QUATERNARY STRUCTURE AND SUBSTRATE BINDING OF THE AMINOACYL-tRNA SYNTHETASES, PARTICULARLY TRYPTOPHANYL-tRNA SYNTHETASE OF ESCHERICHIA COLI

KARL H. MUENCH and DAVID R. JOSEPH
Division of Genetic Medicine
Departments of Medicine and Biochemistry
University of Miami School of Medicine

Abstract: The aminoacyl-tRNA synthetases may be placed in three categories: enzymes consisting of a single polypeptide chain, enzymes consisting of subunits of equal mass, and enzymes consisting of subunits of unequal mass. Tryptophanyl-tRNA synthetase of *Escherichia coli* can be placed in the second category. Its subunits are equal in mass and charge. Studies suggest the enzyme has a binding site for ATP and tryptophan on each subunit. An unusual property of the enzyme is its synthesis of tryptophanyl-ATP ester (Trp-ATP). Sucrose density gradient centrifugation demonstrates a 1:1 tRNA:enzyme complex but does not exclude the likelihood of a 2:1 complex.

INTRODUCTION

The aminoacyl-tRNA synthetases now present a complex array of structures which offer various possibilities for substrate interactions. For the present purpose we shall define three groups of aminoacyl-tRNA synthetases on the basis of quaternary structure: enzymes consisting of a single polypeptide chain, enzymes consisting of two or four subunits of equal mass, and enzymes consisting of subunits of unequal mass. Examples of enzymes in each group are listed in Table 1.

Single polypeptide chains. With the exception of Gln-tRNA synthetase of *Escherichia coli* all of the enzymes in the first category have molecular weights near 100,000 daltons. The Gln-tRNA synthetase is the smallest of these enzymes so far reported. The evidence for a single polypeptide chain is most convincing for Ile-tRNA synthetase (3,4) and Val-tRNA synthetase (4) of *E. coli*. For the other enzymes listed for *E. coli* at least one method of reduction and denaturation has failed to reveal subunits. The Lys- and Val-tRNA synthetases of yeast are placed only tentatively in this category, and the possibility of their having subunits remains (12). All of these enzymes studied so far have the expected single substrate binding site as demonstrated by a variety of procedures.

Subunits of identical mass. There are now numerous examples (Table 1) of aminoacyl-tRNA synthetases with subunits of identical mass, usually near 50,000 daltons. We shall refer to these subunits as monomers to avoid confusion, for some of the enzymes

TABLE 1
Quaternary structure and substrate binding sites of some aminoacyl-tRNA synthetases

Enzyme[a]	Molecular weight	Subunits	Binding sites[b]
Gln	69,000 (1)	None (1)	One, B(1)
Ile	112,000 (2)	None (2,3,4)	One, A(5), C(6), E(4)
Leu	105,000 (7)	None (7,8)	
Lys	100,000 (9)	None (10)	One, A(11)
Lys A	100,000 (12)		One, A(12), B(12,13), C(13)
Val	110,000 (14)	None (4)	One, A(14), B,C(15), D(16), E(4)
Val A	112,000 (17)		One, B(18)
Ala	140,000 (19)	Two (19)	
Glu B	190,000 (20)	Four (20)	
His C	100,000 (21)	Two (22)	
Met	190,000 (23)	Four (24)	
Met	96,000 (25)	Two (25)	Two, D,E(26)
Phe	181,000 (27)	Four (27)	
Pro	100,000 (8,28)	Two (28)	
Ser	100,000 (29)	Two (29)	One, B(30)
Tyr	95,000 (31)	Two (8)	
Trp	74,000	Two	Two, A
Trp D	108,000 (32)	Two, (32)	Two, A(32)
Gly	230,000 (33)	$\alpha_2\beta_2$ (33)	
Phe A	237,000 (34)	$\alpha_2\beta_2$ (34)	One, B(35)

[a] Unless indicated the enzymes, designated by amino acid, are from *E. coli*. Other sources: A, yeast; B, *Micrococcus cryophilus*; C, *Salmonella typhimurium*; D, bovine pancreas.
[b] Methods used for determination of binding sites are designated: A, gel filtration of a substrate-enzyme complex; B, sucrose density gradient centrifugation of a substrate-enzyme complex; C, binding of a substrate-enzyme complex to a nitrocellulose filter; D, quenching of enzyme fluorescence with tRNA; E, equilibrium dialysis.

may be either dimers or tetramers, for example the Met-tRNA synthetase (23-26). No monomer has been shown to possess catalytic activity, either in amino acid-dependent ATP-PP$_i$ exchange or in aminoacyl-tRNA formation. Moreover, for two enzymes, Pro-tRNA synthetase (28) and Phe-tRNA synthetase (27) of *E. coli*, the monomers are clearly inactive catalytically, and for Phe-tRNA synthetase the dimer is inactive, too. That demonstration is possible because these enzymes, unlike any of the others so far reported, can be converted to monomers reversibly. The monomers, stable under conditions of enzyme assay, can be shown to regain activity upon dimerization (Pro-tRNA synthetase) or tetramerization (Phe-tRNA synthetase).

Although activity of monomers remains to be demonstrated, binding sites for methionine, ATP, and tRNA^{Met} or tRNA^{fMet} have been demonstrated on each monomer of Met-tRNA synthetase by equilibrium dialysis and fluorescence quenching (26). A subunit with mass intermediate to the monomer and dimer has enzymatic activity. This subunit is formed by exposure of the tetramer to a macromolecular factor present in extracts of *E. coli* (36).

Evidence indicates one other subunit enzyme has substrate binding sites on each monomer. The Trp-tRNA synthetase of bovine pancreas is a dimer able to complex two molecules of tryptophan and two molecules of ATP (32). We shall here present evidence for two substrate binding sites on Trp-tRNA synthetase of *E. coli*.

The Met-tRNA synthetase of *E. coli* demonstrates that enzymatic activity possibly resides both in dimers and tetramers (23-25). In fact the enzyme may be purified in either form. However, the Met-tRNA synthetase tetramer may not be active but may change into an active dimer during assay at 37° (24). The Phe-tRNA synthetase has no activity as a dimer and must be in the tetrameric form to be active (27).

The monomers of Met-tRNA synthetase meet rigid criteria of identity, not only in mass but in charge (24), in N-terminal sequence, and in an octapeptide sequence at the active site (26). The Phe-tRNA synthetase monomers are identical in charge by polyacrylamide gel electrophoresis in the presence of 8 M urea and give identical tryptic peptides (27). The fact that sodium dodecyl sulfate-polyacrylamide gel electrophoresis in presence of 1% 2-mercaptoethanol yields a predominant subunit of 90,000 daltons which becomes 43,000 daltons after performic acid oxidation may indicate presence of a disulfide bond in the dimer (27).

Subunits of unequal mass. The third category of aminoacyl-tRNA synthetases consists of enzymes composed of subunits of unequal mass. The only two examples, each with an $\alpha_2\beta_2$ quaternary structure, are the Gly-tRNA synthetase of *E. coli* (33) and the Phe-tRNA synthetase of yeast (34). In the former the mass of the α subunits is 33,000 daltons and that of the β subunits is 80,000 daltons, whereas in the Phe-tRNA synthetase of yeast the α subunit mass is 56,000 daltons and the β subunit mass is 63,000 daltons. The existence of nonidentical subunits raises the question of differential function, which so far has not been demonstrated.

Whatever the significance of different quaternary structures may be, enzymes from different species but with the same amino acid specificity do not always possess the same quaternary structure, a fact demonstrated by the Phe-tRNA synthetases of yeast and $E.\ coli$.

The Trp-tRNA synthetase from a mammalian source, bovine pancreas, was the first aminoacyl-tRNA synthetase to be highly purified and characterized (37). We have purified and partially characterized the Trp-tRNA synthetase of $E.\ coli$, and now we shall present evidence to place the enzyme with that of bovine pancreas in the second category of Table 1.

RESULTS AND DISCUSSION

Quaternary structure. We purified Trp-tRNA synthetase of $E.\ coli$ 1000-fold in 21% yield by column chromatography on DEAE-cellulose, hydroxylapatite, and Amberlite CG-50 (38). The purified enzyme is homogeneous by the following criteria: absence of a measurable quantity of any of the other aminoacyl-tRNA synthetases (38), constant specific activity on Sephadex G-200 gel filtration (8), migration as a single band on polyacrylamide gel electrophoresis (8), and migration as a single band on polyacrylamide gel electrophoresis in the presence of 8 M urea (38).

The molecular weight of the active form of the enzyme is 74,000 daltons by gel filtration on standardized Sephadex G-150 and G-100 columns and 81,000 daltons by sucrose density gradient centrifugation in the presence of four standards, the $s_{20,w}$ being 5.0 S (38).

The presence of subunits of identical mass, 37,000 daltons, was demonstrated by gel electrophoresis in the presence of sodium dodecyl sulfate and 2-mercaptoethanol (8) and by gel filtration in the presence of 8 M urea and 2-mercaptoethanol on a standardized agarose column (38). Evidence for identical charge of these monomers was their migration as a single band on polyacrylamide gel electrophoresis in the presence of 8 M urea (38).

In all kinetic studies so far reported for subunit enzymes there exists no evidence for interaction of the active sites. Similarly, kinetic analysis of Trp-tRNA synthetase revealed the classical linear patterns, suggesting complete independence of active sites. The K_m values for all three substrates are typical, but the turnover number is 1200, the highest reported for any aminoacyl-tRNA synthetase (38).

Fluorescence quenching. Although amino acid analysis revealed only four tryptophan residues per dimer and 18 tyrosine residues per dimer, the enzyme gave a strong fluorescence spectrum for tryptophan and no detectable tyrosine fluorescence. For fluorescence quenching studies (16,26) conditions were found to stabilize the enzyme, tRNA$_a^{Trp}$, and tRNA$_i^{Trp}$ at 23°*. However, under these conditions (50 mM potassium phosphate buffer, pH 6.9, 5 mM MgCl$_2$) the substrate, tRNA$_a^{Trp}$, did not quench the enzyme fluorescence.

Trp-ATP-enzyme complex. The enzyme could be isolated in a complex with tryptophan and ATP by gel filtration on Sephadex G-75 (8). Formation of the complex with tryptophan depended on the presence of ATP, and formation of the complex with ATP depended on the presence of tryptophan. In contrast the Trp-tRNA synthetase of bovine pancreas forms a complex with tryptophan alone (32) or with ATP alone (40). The complex with ATP and tryptophan formed by the Trp-tRNA synthetase of bovine pancreas transfers the tryptophanyl residue to tRNA and hydroxylamine (40). Such transfer, with varying efficiency, is the rule for all other complexes reported (5,11-14), because they contain aminoacyl-AMP. However, the complex with ATP and tryptophan formed by the Trp-tRNA synthetase of *E. coli* does not contain Trp-AMP but rather Trp-ATP (tryptophanyladenosine triphosphate), an ester first described as a product of the Trp-tRNA synthetase of bovine pancreas (41) and later shown to be formed by the *E. coli* enzyme (42). Proof of the Trp-ATP-enzyme structure for the *E. coli* enzyme complex came from its inability to form ^{32}P-ATP in the presence of added ^{32}P-PP$_i$ (38), from its failure to transfer the tryptophanyl residue to tRNA (38), and from isolation of Trp-ATP on DEAE-cellulose chromatography after treatment of the complex with 8 M urea (38). The Trp-ATP isolated by DEAE-cellulose chromatography was labeled in the γ-phosphate with ^{32}P and in the tryptophanyl residue with ^{14}C and exhibited stoichiometry of 1:1 (38).

Stoichiometry of binding of Trp-ATP to enzyme was 1.8:1.0 on the basis of enzymatic activity recovered from the gel column, or 1.1:1.0 on the basis of enzyme (as protein) placed on the column (38).

Although formation of detectable amounts of free Trp-ATP *in vitro* has required concentrations of ATP greater than 1 mM (42), the Trp-ATP-enzyme complex forms in 1 mM ATP (38). The intracellular concentration of ATP in *E. coli* is 1 mM or higher (43). Therefore, formation of a Trp-ATP-enzyme complex *in vivo* seems possible, and the question of a physiologic role for Trp-ATP remains open.

Sulfhydryl groups. Titration of the enzyme with DTNB, 5,5'-dithiobis(2-nitrobenzoic acid) (44), revealed two sulfhydryl groups per molecule of enzyme dimer, and both were protected by the simultaneous presence of tryptophan, ATP, and Mg^{2+}. Except in the presence of those substrates incubation of the enzyme for 60 min at 25° with 2 X 10^{-4} M DTNB gave complete inactivation.

---

*tRNA$_a^{Trp}$ is the active and tRNA$_i^{Trp}$ the inactive conformation of tRNA specific for tryptophan (39).

The presence of only two sulfhydryl groups, both protected by substrates and both necessary for activity of an enzyme consisting of two subunits of equal mass and charge is highly indicative that those sulfhydryl groups reside at or near the active sites. In a parallel example the Ile-tRNA synthetase of $E.\ coli$ has a single sulfhydryl group that is necessary for activity and reacts rapidly with N-ethylmaleimide unless isoleucine and ATP are present (45). Eight other sulfhydryl groups in this enzyme have different characteristics (2,45). The unusual sulfhydryl group presumably resides at or near the single active site (45). Nevertheless, these studies do not definitively locate the sulfhydryl groups at the active sites. Conceivably both sulfhydryl groups of the Trp-tRNA synthetase even though residing on a subunit lacking an active site, could be necessary for activity and protected by substrate binding.

The Trp-tRNA synthetase of bovine pancreas has eight titratable sulfhydryl groups, four of which are essential for activity and are protected from DTNB by the presence of ATP and tryptophan (46).

The low thiol content of $E.\ coli$ Trp-tRNA synthetase is unusual, Lys-tRNA synthetase of $E.\ coli$ being the only other aminoacyl-tRNA synthetase reported in that range (47). That enzyme is unique among aminoacyl-tRNA synthetases in not being inactivated by DTNB or para-chloromercuribenzoate. Lysine, ATP, and Mg^{2+} do not protect the sulfhydryl groups of Lys-tRNA synthetase from reaction with DTNB (47).

Sedimentation of tRNA-enzyme complex. Sucrose density gradient centrifugation has been used extensively to study tRNA-enzyme complex formation (1,12,13,15,18,30,35) in spite of certain theoretical and practical objections. The high pressure gradients developed in high-speed centrifuge tubes have enormous effects on equilibrium constants of macromolecular interactions unless there is no change in molar volumes (48). This objection has been discussed at length by Knowles *et al.* (30). Because the binding of tRNAs to their specific aminoacyl-tRNA synthetases is generally enhanced under acidic conditions, most of the studies of tRNA-enzyme complexes in sucrose density gradients have been done at pH values below 6.5. Although such complexes are specific, most of the aminoacyl-tRNA synthetases have relatively low catalytic activity below pH 6.5, and the relationship of such complexes to the functional complex in aminoacyl-tRNA formation remains obscure.

Table 1 shows that sucrose density gradient centrifugation has confirmed the results of other methods for determining the single binding site on the aminoacyl-tRNA synthetases composed of a single polypeptide chain. However, in no case has sucrose density gradient centrifugation revealed more than one binding site on the enzymes composed of subunits, for example Ser-tRNA synthetase of $E.\ coli$ (30) and Phe-tRNA synthetase of yeast (35). For the

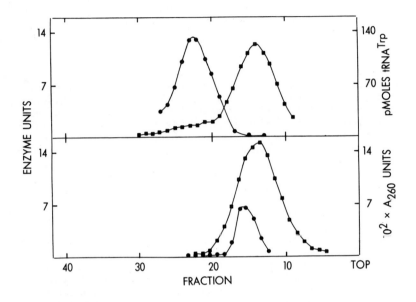

Fig. 1. Sucrose density gradient centrifugation of Trp-tRNA synthetase in the presence of excess tRNA$_a^{Trp}$ or of tRNA lacking tRNA^{Trp}. Pure Trp-tRNA synthetase (13 μg, 170 pmoles) and either 1.0 A$_{260}$ unit of tRNA containing 970 pmoles of tRNA$_a^{Trp}$ (upper panel) or 1.2 A$_{260}$ unit of tRNA containing less than 0.6 pmole tRNA$_a^{Trp}$ (lower panel) in 0.10 ml of 50 mM potassium phosphate buffer, pH 5.8, 5 mM MgCl$_2$, 0.2 mM EDTA, 1 mM dithiothreitol and bovine serum albumin 50 μg/ml was layered on 5 to 20% linear sucrose gradients in 4.6 ml of the same solution. The gradients were then centrifuged 16 hours at 39,000 rpm in a Beckman SW 39 rotor (49). The tubes were punctured, and 5-drop (0.11-ml) fractions were collected and assayed for Trp-tRNA synthetase (●), tRNA$_a^{Trp}$ (■, upper panel) and A$_{260}$ (■, lower panel) as previously described (42). Recovery of tRNA$_a^{Trp}$ was 100% and of A$_{260}$ was 95%. Recovery of enzyme at the time centrifugation was stopped was not determined in this experiment. In parallel experiments it ranged from 83 to 100%.

Purified tRNA$_a^{Trp}$ activated immediately before use by heating to 60° with Mg^{2+} (39) was prepared by chromatography on BD-cellulose and hydroxylapatite, and tRNA free of tRNA^{Trp} was prepared by chromatography on BD-cellulose (38).

aminoacyl-tRNA synthetases with subunits sucrose density gradient centrifugation has not been compared to other methods of determining the number of binding sites per enzyme molecule. Therefore, we have begun a study of complex formation between tRNA^{Trp} and Trp-tRNA synthetase by sucrose density gradient centrifugation.

As shown in the lower panel of Fig. 1, if the $s_{20,w}^{0.725}$ of tRNA^{Trp} is 4.6 S (49) then Trp-tRNA synthetase has a relative value, 5.3 S. If we assume that the enzyme and tRNA undergo no conformational changes in forming the complex, then a 1:1 complex of tRNA:enzyme should sediment at 7.9 S (fraction 24), and a 2:1 complex of tRNA:enzyme should sediment at 10 S (fraction 30). As shown in the upper panel of Fig. 1, in the presence of a 5.7-fold excess tRNA$_a^{Trp}$ the enzymatic activity sedimented at 7.4 S. The quantity of tRNA in the 7.4 S peak was 0.7 mole per mole of enzyme. These data were confirmed in repeated experiments with a 2-fold excess of tRNA$_a^{Trp}$. Whereas the data in Fig. 1 indicate the presence of one binding site per enzyme molecule they do not exclude the presence of another.

As Lagerkvist and Rymo (18) have discussed, formation of a stable 1:1 complex and the total absence of free enzyme in the presence of excess tRNA probably can be explained by the nearly equal sedimentation velocities of free tRNA and free enzyme, such that dissociation of the 1:1 complex during sedimentation does not preclude reassociation. By an extension of that reasoning reassociation of the components, tRNA and tRNA-enzyme, of a 2:1 complex would be disfavored by their unequal sedimentation velocities. Therefore, we suggest that sucrose density gradient sedimentation may not be adequate as a method to demonstrate more than one binding site for tRNA on an aminoacyl-tRNA synthetase. Repeated attempts to demonstrate a 2:1 complex by this method and with larger amounts of enzyme (460 pmoles) showed no peak of enzyme activity in the 10 S region.

In summary, Trp-tRNA synthetase of *E. coli* is a dimer consisting of monomers which are probably identical. Evidence so far indicates that each dimer has two substrate binding sites, but further confirmation is needed. Sucrose density gradient centrifugation demonstrates only one binding site for tRNA^{Trp} on each dimer. The method has provided the same result for two other aminoacyl-tRNA synthetases with subunits, and the findings can best be viewed as a limitation of the method, which does not exclude presence of multiple binding sites. We predict that all aminoacyl-tRNA synthetases with identical subunits will be shown to have a binding site for each substrate on each subunit. The aminoacyl-tRNA synthetases with $\alpha_2\beta_2$ structures pose a different problem, and substrate binding data on these enzymes are eagerly awaited.

REFERENCES

1. W.R. Folk, in press.
2. A.N. Baldwin and P. Berg, *J. Biol. Chem.*, 241 (1966) 831.
3. D.J. Arndt and P. Berg, *J. Biol. Chem.*, 245 (1970) 665.
4. F. Berthelot and M. Yaniv, *Europ. J. Biochem.*, 16 (1970) 123.
5. A.T. Norris and P. Berg, *Proc. Nat. Acad. Sci. U.S.*, 52 (1964) 330.
6. M. Yarus and P. Berg, *J. Mol. Biol.*, 28 (1967) 479.
7. H. Hayashi, J.R. Knowles, J.R. Katze, J. Lapoint, and D. Söll, *J. Biol. Chem.*, 245 (1970) 1401.
8. K.H. Muench, A. Safille, M. Lee, D.R. Joseph, D. Kesden, and J.C. Pita, Jr., *Proc. Second Miami Winter Symp.*, W.J. Whelan and J. Schultz, eds. North-Holland, Amsterdam, 1970. Vol. 1 p. 52.
9. J. Waldenström, *Europ. J. Biochem.*, 3 (1968) 483.
10. R.D. Marshall and P.C. Zamecnik, *Biochim. Biophys. Acta*, 181 (1969) 454.
11. J. Waldenström, *Europ. J. Biochem.*, 5 (1968) 239.
12. L. Rymo, U. Lagerkvist, and A. Wonacott, *J. Biol. Chem.*, 245 (1970) 4308.
13. S.A. Berry and M. Grunberg-Manago, *Biochim. Biophys. Acta* 217 (1970) 83.
14. M. Yaniv and F. Gros, *J. Mol. Biol.*, 44 (1969) 1.
15. M. Yaniv and F. Gros, *J. Mol. Biol.*, 44 (1969) 17.
16. C. Helene, F. Brun, and M. Yaniv, *Biochem. Biophys. Res. Commun.*, 37 (1969) 393.
17. U. Lagerkvist and J. Waldenström, *J. Biol. Chem.*, 242 (1967) 3021.
18. U. Lagerkvist and L. Rymo, *J. Biol. Chem.*, 244 (1969) 2476.
19. M. Lazar, M. Yaniv, and F. Gros, *C.R. Acad. Sci. Paris*, 266 (1968) 531.
20. N.L. Malcolm, *Biochim. Biophys. Acta*, 190 (1969) 347.
21. F. DeLorenzo and B.N. Ames, *J. Biol. Chem.*, 245 (1970) 1710.
22. F. DeLorenzo, Personal communication.
23. R.L. Heinrikson and B.S. Hartley, *Biochem. J.*, 105 (1967) 17.
24. F. Lemoine, J.-P. Waller, and R. van Rapenbusch, *Europ. J. Biochem.*, 4 (1968) 213.

25. C.J. Bruton and B.S. Hartley, *Biochem. J.*, 108 (1968) 281.
26. C.J. Bruton and B.S. Hartley, *J. Mol. Biol.*, 52 (1970) 165.
27. M.H.J.E. Kosakowski and A. Böck, *Europ. J. Biochem.*, 12 (1970) 67.
28. M. Lee and K.H. Muench, *J. Biol. Chem.*, 223 (1969) 244.
29. J.R. Katze and W. Konigsberg, *J. Biol. Chem.*, 245 (1970) 923.
30. J.R. Knowles, J.R. Katze, W. Konigsberg, and D. Söll, *J. Biol. Chem.*, 245 (1970) 1407.
31. R. Calendar and P. Berg, *J. Biol. Chem.*, 5 (1966) 1681.
32. G. Lemaire, R. van Rapenbusch, C. Gros, and B. Labouesse, *Europ. J. Biochem.*, 10 (1969) 336.
33. D.L. Ostrem and P. Berg, *Proc. Nat. Acad. Sci. U.S.*, 67 (1970) 1967.
34. F. Fasiolo, N. Befort, Y. Boulanger, and J.-P. Ebel, *Biochim. Biophys. Acta*, 217 (1970) 305.
35. N. Befort, F. Fasiolo, C. Bollack, and J.-P. Ebel, *Biochim. Biophys. Acta*, 217 (1970) 319.
36. D. Cassio and J.P. Waller, *Europ. J. Biochem.*, 5 (1968) 33.
37. E.W. Davie, V.V. Koningsberger, and F. Lipmann, *Arch. Biochem. Biophys.*, 65 (1956) 21.
38. D.R. Joseph and K.H. Muench, unpublished data.
39. K.H. Muench, *Biochemistry*, 8 (1969) 4880.
40. A.V. Parin, E.P. Savelyev, and L.L. Kisselev, *FEBS Letters*, 9 (1970) 163.
41. S.B. Weiss, H.G. Zachau, and F. Lipmann, *Arch. Biochem. Biophys.*, 83 (1959) 101.
42. K.H. Muench, *Biochemistry*, 8 (1969) 4872.
43. H.A. Cole, J.W.T. Wimpenny, and D.E. Hughes, *Biochim. Biophys. Acta*, 143 (1967) 445.
44. G.L. Ellman, *Arch. Biochem. Biophys.*, 82 (1959) 70.
45. M. Iaccarino and P. Berg, *J. Mol. Biol.*, 42 (1969) 151.
46. M. DeLuca and W.D. McElroy, *Arch. Biochem. Biophys.*, 116 (1966) 103.
47. R. Stern, M. DeLuca, A.H. Mehler, and W.D. McElroy, *Biochemistry*, 5 (1966) 126.
48. G. Kegeles, L. Rhodes, and J.L. Bethune, *Proc. Nat. Acad. Sci. U.S.*, 58 (1967) 45.
49. R.G. Martin and B.N. Ames, *J. Biol. Chem.*, 236 (1961) 1372.

We acknowledge the technical assistance of Mrs. Alicia Safille. This work was supported by United States Public Health Service (National Institutes of Health) Grants 5-P01-AM-09001-06 and GM-02011-01, by American Cancer Society Grants PRA-21 and IN 51J, and by a special equipment grant from the Florida Division of the American Cancer Society.

DISCUSSION

K.B. JACOBSON: In looking at the formation of these complexes in other systems, rather low pH is used to get enzyme-tRNA complexes. I'd be interested in knowing what pH, relative to that for optimum conditions, you are using to prepare the enzyme-tryptophan, the enzyme-ATP, and the enzyme-tRNA complex.

K.H. MUENCH: The Trp-ATP-enzyme complex was prepared at pH 8.8, which is the pH optimum for both tryptophan activation and Trp-tRNA formation. The tRNA-enzyme complex was prepared and sedimented at pH 5.8, where the enzyme has very little activity. Therefore, the relationship of the complex to the functional interaction of tRNA and enzyme remains an open question. However, the tRNA-enzyme complex is specific, as shown by control experiments in which no complex forms with $tRNA_i^{Trp}$

U.Z. LITTAUER: Did you try to detect the formation of your complex by some other techniques, such as the Millipore filter binding?

K.H. MUENCH: Yes, but we were unable to demonstrate the complex in this way. It may be that we haven't found the right conditions.

S. WEISS: You raised the specter of a compound that I thought we had buried a long time ago, and that was tryptophanyl-ATP. I recall, though, at the time when we were working with animal adenylates, we had come to the conclusion that AMP-tryptophan was highly reactive and that it would react not only with ATP but with any other nucleoside or nucleotide. We actually found tryptophanyl-ATP, tryptophanyl-ADP and even tryptophanyl-CTP if CTP was added. I just wondered whether you had seen this also and whether anyone else had reported an aminoacyl type linkage other than with tryptophan.

K.H. MUENCH: This type of compound has not been reported for any amino acid other than tryptophan, and we ourselves have not yet done the survey. As far as other acceptors are concerned, these studies are difficult to do. For example, deoxy ATP is an inhibitor of the enzyme and therefore cannot be added in sufficient concentration to test it as an acceptor.

A. KORNBERG: Under conditions where catalytic amounts of enzyme are used, do you see any accumulation of this ester of tryptophan and ATP?

K.H. MUENCH: Yes. We have been able to show, for instance, that its formation is more than stoichiometric with enzyme and is dependent on the ATP concentration.

AMINOACYL-tRNA SYNTHETASES: THEIR BIOLOGICAL STABILITY AND REGULATION IN ESCHERICHIA COLI

F. C. NEIDHARDT[*] and J. J. ANDERSON[**]
Department of Biological Sciences
Purdue University, Lafayette, Indiana 47907

Abstract: Recent studies on the regulation of aminoacyl-tRNA synthetases in Escherichia coli are discussed in relation to the several physiological functions of these indispensable enzymes, and to their apparent instability under certain conditions. A new mutant is described that has led to the findings that E. coli is programmed to maintain a functional excess of valyl-tRNA synthetase, and that the mutant enzyme is inactivated as a result of its participation in the aminoacylation of tRNA.

INTRODUCTION

We have recently been concerned with the regulation in bacteria of one set of enzymes that undergo a striking interaction with nucleic acids - the aminoacyl-tRNA synthetases (1). Our interest has been primarily that of the physiologist intrigued by the way cells coordinate macromolecule synthesis during growth under different conditions. The participation of aminoacyl-tRNA synthetases in several of these processes led us to study these enzymes, largely through the isolation of mutants with defective synthetases. Analysis of these mutants has uncovered interesting properties of the synthetases and some of these properties seem to be consequences of the unique interaction of these enzymes with transfer RNA.

In this presentation I would like to discuss what is known about the way bacterial cells, specifically Escherichia coli,

---

[*] Present address: Department of Microbiology, The University of Michigan Medical School, Ann Arbor, Michigan 48104

[**] Present address: Department of Biological Sciences, Dartmouth College, Hanover, New Hampshire 03755

regulate the synthesis and activity of their complement of twenty aminoacyl-tRNA synthetases. It will be obvious before long that our information is still at a primitive state, and that few important aspects of synthetase regulation have really been solved. Nevertheless discussion is in order if only to identify the extent of our ignorance.

PHYSIOLOGY OF AMINOACYL-tRNA SYNTHETASES

Let us first consider a few features of aminoacyl-tRNA synthetases that may have had an influence on the evolution of mechanisms to regulate these enzymes.

1. The best known cellular function of aminoacyl-tRNA synthetases is the generation of aminoacyl-tRNA for protein synthesis, and this function is an indispensable one. No matter what the chemical nature of the medium, the rate of protein synthesis creates a demand for charged tRNA that can be met only by the activity of the cell's complement of aminoacyl-tRNA synthetases. In this respect these enzymes differ from the amino acid biosynthetic enzymes, which are required for growth only in media deficient in the respective amino acid. The rapid growth seen in rich media demands a relatively high activating activity for each amino acid in the same circumstance that makes biosynthetic enzymes redundant impedimenta. One would not, therefore, expect the structural genes for activating enzymes to be part of the operon or operons specifying the enzymes of the biosynthetic pathways; nor would one expect the formation of synthetases to be regulated by the same signals that regulate the biosynthetic enzymes.

2. Because aminoacyl-tRNA synthetases are proteins required for protein synthesis, the loss of any one synthetase activity would be lethal for the cell. Therefore one would not expect cells to adjust the level of these enzymes to match coordinately the cell's demands for aminoacyl-tRNA, or else this level might become dangerously low during prolonged periods of slow growth. Instead, one might expect to see a rather high minimum level of these enzymes.

3. Thirteen of the E. coli synthetases can be purified to a reasonably high degree of homogeneity by fractionations that accomplish a 300-700 fold increase in protein specific activity [alanine (2), arginine (3), glycine (4), isoleucine (5), leucine (6), methionine (7), phenylalanine (8,9), proline (10), serine (11), threonine (12), tryptophan (13), tyrosine (14), and valine (15)]. Each synthetase must then represent on the order of 0.2% of the cell's total protein, and the twenty synthetases must

account for approximately 4% of the cell's total protein. It is difficult to evaluate whether this is sufficiently large to exert a force toward the evolution of a flexible control on the synthesis of these enzymes.

4. In the cases of isoleucine (16), leucine (17), histidine (18), threonine (19), and valine (20), the synthetases appear to be involved in the process by which amino acids repress their own biosynthetic enzymes. Restriction of amino acid activation either by the addition of an inhibitor of the synthetase or by thermal inactivation of mutant synthetases derepresses these biosynthetic pathways. This somewhat strange situation is not well understood. One of its interesting consequences is that if the product of the synthetase reaction, aminoacylated tRNA, is the signal for repression of a biosynthetic pathway, then it is unlikely that it is a very effective signal for repression of the cognate synthetase. If it were an effective signal, then amino acid supplementation could hardly bring about maximum repression of the biosynthetic pathway while simultaneously either derepressing or having no effect on the synthetase level.

5. By a process which appears also to involve guanosine tetraphosphate (MS I) (21), amino acid activation is involved somehow in the control of stable RNA synthesis and accumulation. Restricting either the supply of an amino acid or its activation leads to a preferential reduction in stable RNA accumulation in normal, stringent cells (22,23,24). It is not easy to guess whether this additional involvement of synthetases imposes special restrictions on the regulation of synthetase level.

6. One of the substrates for each synthetase is tRNA. It is an unusual substrate in that it is present in the cell in concentrations not greatly different from that of the enzyme. As a rough approximation we may assume that 4% of the cellular protein consists of synthetase molecules of 120,000 M.W.; a mass of cells containing 1 gram of protein will therefore contain approximately 2×10^{17} molecules of synthetase. Total RNA is equal to approximately 30% of the total protein, and 20% of this is tRNA of 25,000 M.W.; therefore approximately 1.4×10^{18} molecules of tRNA would be present in this mass of cells, or an average of approximately 7 tRNA molecules per synthetase molecule. Most synthetases are responsible for charging more than one species of tRNA, and in several cases the tRNA's for an amino acid are not present in equimolar amounts; therefore for many tRNA - synthetase couples, there must be close to a 1:1 molar ratio in the cell, and probably for all couples the ratio is less than 10:1. This fact suggests that a change in the amount of a synthetase in the cell will have an effect on the fraction of both uncharged and charged tRNA that is complexed with synthetase molecules. Such a change could conceivably

affect not only the rate of protein synthesis but also repression of the biosynthetic pathway and the regulation of RNA synthesis. Regulation of synthetases might therefore be important to guard against overproduction as well as underproduction.

REGULATION OF AMINOACYL-tRNA SYNTHETASES

From these features of synthetase physiology one might guess that synthetases should be regulated separately from the amino acid biosynthetic enzymes, that their level should be maintained slightly above the minimum necessary for rapid growth in rich medium, and that this level should not become reduced coordinately when growth slows down. Such behavior perhaps could be achieved if synthetases were not subject to repression, but instead were "constitutive" enzymes made at a nearly invariant differential rate as determined by the properties of their respective promoters, preferably coordinately with transfer RNA. Alternatively, they could be governed by some device that is potentially capable of varying their differential rate of formation, but which ordinarily maintains a nearly constant level of each synthetase.

To see how accurate these guesses are, let us examine what has been learned about synthetase regulation during the past several years:

1. Synthetases show little variation in levels in cells grown at different rates in media of different chemical composition. Addition of the cognate amino acid to cells growing in minimal medium usually changes neither the growth rate nor the level of the respective synthetase. Addition of a rich supplementation of amino acids and growth factors causes an increase in growth rate without a concomitant increase in synthetase level. A sampling of some of this data for two enzymes is shown in Table 1. The variation of the valine enzyme at 30°C in this strain actually shows the greatest coordination with growth rate we have observed; synthetases for other amino acids, notably arginine, isoleucine, histidine, and phenylalanine have been examined in our laboratory and elsewhere and for each of them there is less than a 25% variation under different "normal" growth conditions (25,26, unpublished experiments, and references cited in 26).

These results, to a first approximation, are consistent with the arguments presented above that one would not expect the formation of synthetases to be regulated by the same signals that regulate the biosynthetic enzymes, and that strict coordination of level with growth rate would appear to be risky because of the vital nature of these enzymes and because of possible ill

TABLE 1

Effect of medium composition and temperature on the level of two synthetases in E. coli strain NP2.*

Medium	Temp. (°C)	Growth rate constant (hr^{-1})	Specific Activity of tRNA-synthetase (units/mg)	
			Arginine	Valine
Minimal	30	0.25	0.13	0.28
	40	0.54	0.13	0.37
Minimal + val + ile	30	0.25	0.12	0.29
	40	0.54	-	0.37
Minimal + 20 a.a.	30	0.48	0.13	0.36
	40	0.92	-	-
Tryptone + yeast extract + glucose	30	1.1	0.13	0.40
	40	2.1	-	0.38

* These data were compiled in the author's laboratory by Dr. James J. Anderson, Dr. Luther S. Williams, and Mr. Galen McKeever. Methods of assay are cited in Williams and Neidhardt (26).

---

effects if these enzymes were made non-coordinately with transfer RNA. Also consistent is the finding that the structural genes for synthetases are not found as part of the biosynthetic operons (27).

2. Synthetase levels do vary from their "normal" values during certain special growth conditions, most of which affect the internal supply of their amino acid substrates (25,26). Examples of this variation are shown in Table 2. They fall into two categories: Those in which the level is below "normal" and those in which they are above "normal". Interestingly, both conditions seem to derive from the same general stress on the cell - a limitation of the cognate amino acid. This proved puzzling to us for some time, but was apparently solved when we learned that aminoacyl-tRNA synthetases undergo inactivation during growth under amino acid-limited conditions (26). The story is somewhat complicated, and involves a still not well understood difference between the growth of cells in a chemostat and the growth of bradytrophic strains. In essence, however,

TABLE 2

Effect of amino acid limitation on level of synthetases.*

Strain	A.A.-tRNA synthetase	Growth condition	Growth rate (hr^{-1})	Relative Sp. Act.**
Chemostat experiments:				
HP1802	arg	unrestricted	0.90	1.00
		arg. limitation	0.42	0.48
NP29028	ile	unrestricted	0.38	1.00
		ile. limitation	0.16	0.30
AB1132	his	unrestricted	0.65	1.00
		his. limitation	0.40	0.35
Non-chemostat experiments:				
HP180201	arg	unrestricted	0.90	1.00
		arg. limitation (bradytroph)	0.40	2.14
NP2	ile	unrestricted	0.60	1.00
		ile. limitation (val inhibition)	linear	1.93
AB1132	his	unrestricted	0.68	1.00
		his. limitation (β-alanylhistidine)	0.18	3.10

* These data are taken from the Ph.D. thesis of Dr. Luther S. Williams, (Purdue University, 1968) except for the experiment with strain NP2 (25).

** Normalized in each case to the unrestricted condition.

---

density-labelling enabled us to discover that aminoacyl-tRNA synthetases are subject to inactivation in vivo, and that this process frequently masks an increase in the differential rate of synthetase formation during amino acid limitation. In Table 3 are summarized some measurements of the differential rate of synthetase formation as measured by density labelling.

3. Not only do synthetase levels vary over a considerable range when amino acid restrictions are applied to the cells, but

TABLE 3

Physiological adjustment of arginyl-tRNA synthetase formation.*

Growth condition**	Differential rate of enzyme formation (units/mg protein)
(a.) Derepressing	0.290
(b.) Normal	0.135
(c.) Repressing	<0.005

* Data compiled from Williams and Neidhardt (26).

** Derepressing growth condition: slow growth of a leaky arginine auxotroph in arginine-free minimal medium.
Normal growth condition: growth in minimal medium without arginine limitation.
Repressing growth condition: the temporary condition for the first generation of growth after adding arginine to cells that had been grown for a long period under the derepressing growth condition.

---

transient rates of synthetase formation show an even greater range. These transient rates are observed whenever normal growth conditions are restored to a cell which has been caused to increase or decrease the level of a synthetase from its "normal" value by means of amino acid restriction. The adjustment back to a normal level usually occurs by a very marked alteration in the rate of synthetase formation. A cell with an excess of synthetase almost totally halts production until dilution by growth lowers the level to normal (see "repressing condition" in Table 3); similarly, a cell with a deficit of some synthetase will accelerate its formation many fold (26).

INACTIVATION OF AMINOACYL-tRNA SYNTHETASES

The discovery of in vivo lability of aminoacyl-tRNA synthetase activity introduces a new factor that is potentially of utmost importance for synthetase regulation. If an indispensable enzyme is subject to inactivation, then it is of obvious survival value to be able to increase its differential rate of synthesis when its activity becomes subnormal. For this reason it becomes less satisfactory for synthetases to be "constitutive" enzymes than for their level to be maintained by an adjustable mechanism that is sensitive to the level of these enzymes.

That amino acid activating enzymes should be subject to inactivation _in vivo_ is a bizarre situation; such enzymes should have been selected for stability. Instability must therefore be an unavoidable property of proteins that catalyze the aminoacylation of tRNA, or else inactivation of synthetases has some role in the physiology of protein synthesis. There is no known evidence to support the latter alternative, but the former is quite believable from what is known about synthetases. These enzymes must choose with great selectivity one amino acid and an appropriate tRNA molecule and catalyze an ATP-mediated reaction between them. Such functional requirements for synthetase activity must impose enormous restriction on the type of structure evolved, and must guarantee a certain minimum size and complexity. It is possible that the functioning of a synthetase is such an elaborate and complex process that the enzyme is forced to assume a configuration that has a significant probability of decaying to an irreversibly inactive form.

Much evidence supports this view. First of all, the conformation of several well-studied synthetases can be profoundly altered by interactions with their several ligands. Mehler and Mitra (28) have shown that $tRNA_{arg}$ must bind to the arginyl-tRNA synthetase before that enzyme can bind arginine; furthermore, the binding of $tRNA_{arg}$ confers stability to the enzyme. The binding of $tRNA_{ile}$ to the isoleucyl-tRNA synthetase changes that enzyme's substrate specificity (Loftfield and Eigner, 29), and the binding of isoleucine promotes the rates of release and binding of acylated and non-acylated $tRNA_{ile}$ (Yarus and Berg, 30). Some of the sulfhydryl groups of valyl-tRNA synthetase are protected from oxidation by the binding of valyl-AMP [Yanif and Gros (15), George and Meister (31)], and the enzyme has increased thermostability when valyl-AMP is bound to it [Yanif and Gros (32)]. Similarly, mutant tryptophanyl-tRNA synthetases are stabilized by tryptophanyl-AMP [Doolittle and Yanofsky (33)]. The intensity of tryptophane fluorescence by the valyl-tRNA synthetase changes upon addition of val-ol-AMP (34) or of $tRNA_{val}$ [Helene and Yanif, cited in (15)], and Ohta _et al_ (35) demonstrated by optical methods a conformational change in tyrosyl-tRNA synthetase upon interaction with tRNA.

Secondly, synthetases are unusually susceptible to mutations that increase their thermolability. With no special selective procedures after mutagenesis other than the elimination of nutritionally bypassable mutants, valyl- and phenylalanyl-tRNA synthetase mutants can be detected in nearly every set of temperature sensitive mutants isolated from many E. coli strains, and mutants altered in other synthetases have been isolated with very little difficulty. "Mutants" in the glycine enzyme have in fact been discovered in several supposedly normal "wild" strains (36,37).

Recent work with a new temperature-sensitive mutant in our laboratory suggests that synthetase inactivation is related in an interesting manner to the catalysis of aminoacylation by these enzymes (J. J. Anderson and F. C. Neidhardt, manuscripts in preparation). Escherichia coli NP29 is a well-studied mutant strain that possesses a thermolabile valyl-tRNA synthetase. The mutant enzyme operates almost normally in vivo at 30°C, but is rapidly inactivated at 40°C. The enzyme decays within minutes of preparing cell-free extracts, no matter what the temperature. Strain NP2907 was isolated as a spontaneous mutant of strain NP29, and its valyl-tRNA synthetase retains enzymic activity in vitro; nonetheless it still differs from the wild type enzyme in stability and apparent K_m for ATP. The new mutant locus co-transduces with pyrB at the same frequency as does the structural gene, valS, for this synthetase. Strain NP2907 is therefore regarded as differing from the wild strain only in the structure of its valyl-tRNA synthetase.

The new mutant grows normally at 30°C, and upon a shift to 40°C growth quickly accelerates exactly as for normal cells. Exponential growth, however, cannot be maintained at 40°C, but proceeds linearly with time beginning at a point which depends on the rate of growth of the cells in the particular medium employed. In very rich medium the transition occurs almost immediately upon the shift to 40°C, in glucose medium enriched with amino acids it occurs almost a generation later, and in glucose minimal medium it occurs after 1.5 generations of growth.

Irrespective of the growth medium, net synthesis of valyl-tRNA synthetase ceases in the new mutant as soon as the temperature is raised to 40°C. In fact, as long as valine (and isoleucine, to overcome valine inhibition) is present in the medium, the amount of synthetase per ml of culture during prolonged incubation at 40°C remains constant with time whether the cells are prevented from growing by any of a number of means, or whether they are permitted to grow slowly (minimal medium) or rapidly (rich medium). This behavior is the classic symptom when the synthesis of a protein is temperature-sensitive and the protein itself is stable at the inhibitory temperature. On this basis mutants are usually designated as having a "temperature-sensitive synthesis" of the protein in question.

Nevertheless there is an alternative explanation for such behavior. If at the restrictive temperature the enzyme were made at a normal rate but inactivated at an exactly balancing rate,

then simple dilution of the enzyme would occur during growth of
the cells and there would be neither a net gain nor loss of enzyme activity per ml of culture. This explanation is usually
discounted because of the necessary postulation of an exactly
balanced synthesis and destruction of enzyme activity, and is
considered really implausible if the balance must occur over a
range of growth rates, i.e., if one must postulate that the rate
of destruction is high when the rate of synthesis of the enzyme
is low, and that it ceases altogether when enzyme synthesis
halts.

Much to our surprise, the appropriate test of whether valyl-
tRNA synthetase was being made at 40°C in strain NP2907 yielded
an affirmative answer; growth of the cells at 30°C in medium
containing D_2O, followed by a shift to normal H_2O medium at 40°C
revealed that valyl-tRNA synthetase was made at 40°C but that no
net accumulation could occur because of an exactly equal rate of
inactivation of the enzyme.

The most interesting feature of this behavior, of course, is
that the rate of destruction of valyl-tRNA synthetase activity
must somehow vary in proportion to the rate of synthesis of the
enzyme in different media at 40°C, and fall to near zero in the
absence of growth. Additional experiments revealed that the
enzyme decayed in vivo if either the intracellular level of
valine or of ATP were lowered.

We have postulated an explanation based on the notion that
since this enzyme functions in protein synthesis, its rate of
functioning will be proportional to the overall rate of protein
synthesis. The essence of the model is that the aminoacylation
of tRNA by valyl-tRNA synthetase in this mutant entails a con-
figurational change in the enzyme that can lead to its inactivation; the more rapidly the enzyme works, the faster it is inactivated.

The work of Yarus and Berg (30) with isoleucyl-tRNA synthetase is particularly important to this problem, and because of
many kinetic similarities between the valyl and isoleucyl enzymes (15), we have chosen to express our explanation in terms
of their picture of synthetase function. In Figure 1 is presented a slightly modified statement of their model. Four steps
are pictured in the cycle of enzyme function in vivo. In the
first step, a synthetase molecule containing bound valyl adenylate interacts with an appropriate tRNA molecule to form the
ternary complex: enzyme-valyl AMP-tRNA. In the second step, the
valyl residue is transferred to the bound tRNA, releasing AMP.
In the third step, valine and ATP bind to the enzyme and react
to form valyl AMP. In the fourth and rate-limiting step, valyl-
tRNA dissociates from the enzyme, leaving it free to bind to

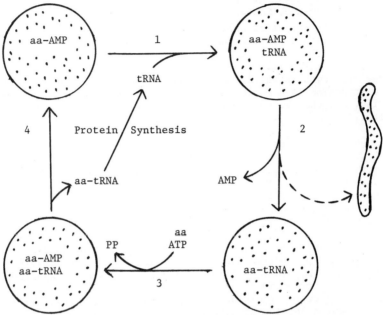

Fig. 1. A model of the aminoacylation of tRNA by the valyl-tRNA synthetase of E. coli mutant NP2907 during growth. The flattened figure to the right depicts a synthetase molecule inactivated during the conformational change in the enzyme as the aa-AMP complex breaks down in step 2. The overall cycle is a modified form of that proposed by Yarus and Berg (30).

another tRNA molecule.

The predominant functional state of the enzyme in vivo and, under usual assay conditions, in vitro, is the ternary complex: enzyme - valyl-AMP - valyl-tRNA. When protein synthesis is blocked, as by chloramphenicol, this is probably the only state possible, and we know that the enzyme of strain NP2907 is perfectly stable under this condition. If protein synthesis is permitted at 40°C, the generation of uncharged tRNA permits the cycle to function and the enzyme undergoes continuous changes in its conformation. It is this flexing of the enzyme as the valyl-AMP complex is broken down and the tRNA acylated that we view as destructive for this enzyme. The faster the rate of protein synthesis, the faster the enzyme cycles through these steps, and the faster it is inactivated. At low temperatures the enzyme can withstand this distortion as it functions, but as the temperature is raised each turn of the cycle takes a greater toll of the enzyme. At 40°C the rate constants are such that one synthetase molecule becomes inactivated, on the average, for each one formed. This means that a valyl-tRNA synthetase molecule in NP2907 at 40°C will survive, on the average, somewhere in the neighborhood of 20,000 cycles of aminoacylation.

Full experimental details that support this model are to be published elsewhere, but we should mention an important confirmation of the model by some observations made on the behavior of the enzyme *in vitro*. Extracts of NP2907 cells can be incubated in the presence of valine and ATP for prolonged periods at 40°C without damaging the enzyme. The mere addition of tRNA to such a mixture induces a rapid inactivation of the enzyme.

In summary, the paradox that valyl-tRNA synthetase is stable in strain NP2907 at 40°C under non-growth conditions, but unstable during growth, is viewed as a consequence of the enzyme becoming inactivated by its own functioning. The valyl-tRNA synthetase of this strain is of course not a normal one, but the rapid inactivation of normal synthetases during intermittant growth in the chemostat (26) suggests that the mutant enzyme may simply have an amplified property of normal synthetases.

Thus it seems plausible that synthetase formation is subject to effective and flexible regulation at least partly in order to minimize the danger of death through depletion of some synthetase during a metabolic crisis.

AMINOACYL-tRNA SYNTHETASE LEVEL

The same mutant, NP2907, provided a unique opportunity to learn whether the normal level of valyl-tRNA synthetase that is usually maintained by cells growing in any medium is an excess of what is needed. One possible explanation for the apparent failure of the cell's regulatory system to match synthetase level with growth rate is that these enzymes are inefficient in cells growing in minimal media, and therefore that the level of synthetase being maintained is actually the minimum appropriate level.

This notion is testable because one can lower the level of valyl-tRNA synthetase in strain NP2907 by growing the cells for an appropriate time at 40°C. Therefore, without changing the chemical composition of the medium one can observe the rate of protein synthesis that the cells can maintain over a range of valyl-tRNA synthetase level. These experiments have been performed, and it appears from the results that the level of enzyme maintained in strain NP2907 is just sufficient to support growth at its maximum rate in very rich medium, and that this level is a three-fold excess with respect to overt growth requirements in glucose minimal medium and also with respect to the requirement for valine activation in biosynthetic enzyme repression (J. J. Anderson and F. C. Neidhardt, manuscripts in preparation). The activity in NP2907, as measured in standard in vitro assays, is approximately half that in non-mutant cells, either because there are fewer molecules of enzyme or because each is less active.

If it is generally true in normal cells that the regulation of aminoacyl-tRNA synthetases is not designed to match their level to immediate functional demands in protein synthesis, then novel biochemical signals may be expected in the regulatory process. The overall picture of synthetases being maintained at levels higher than absolutely necessary for protein synthesis may be related either to other roles they play, or to their potential instability.

CONTROL OF AMINOACYL-tRNA SYNTHETASE ACTIVITY

Our discussion thus far has concerned one aspect of synthetase regulation, control of synthetase formation, and its possible relation to synthetase inactivation brought about by conformational changes of the enzyme.

Another control has been postulated by Brenner, DeLorenzo and Ames (38). Measurements of the effect of energy charge,

$$[(ATP) + 1/2(ADP)]/[(ATP)+(ADP)+(AMP)]$$

on the functioning of histidyl-tRNA synthetase in vitro reveals that this enzyme behaves as a fairly typical ATP-utilizing system in that it is inhibited by low values of energy charge. This result suggests to these workers that synthetases may respond in vivo to energy-poor conditions by decreasing their rate of aminoacylation of tRNA.

No further information is available to aid in assessing the correctness of this suggestion; in particular, no in vivo studies have been reported that shed light on the matter, so further judgement will have to wait. One potential difficulty with the notion that an ATP deficiency would be met by reduced aminoacylation of tRNA is that a senseless derepression of biosynthetic pathways (histidine, isoleucine, leucine and valine) that are regulated through synthetase function might result.

SUMMARY

The formation of aminoacyl-tRNA synthetases appears to be carefully regulated in bacterial cells. The rate of synthesis of these enzymes can be varied independently of each other and independently of the amino acid biosynthetic enzymes over a very wide range, from near zero to several times higher than the normal rate. Nevertheless, the level of each synthetase seems not to be adjusted to match different demands for aminoacylated-tRNA, but rather to be held at an apparently arbitrarily fixed value which is more than sufficient to meet the maximum demands for aminoacylated tRNA during rapid growth, and is a considerable excess during slow growth. Under certain growth conditions aminoacyl-tRNA synthetases are subject to high rates of inactivation. In one mutant this inactivation seems related in an interesting manner to the catalysis of tRNA acylation by these enzymes. It is possible that the susceptibility of these indispensable enzymes to inactivation has played a role in the evolution of the control mechanisms regulating their synthesis. The genetic basis of this regulation and the biochemical signals involved have not yet been identified.

REFERENCES

1. F. C. Neidhardt, Bacteriol. Revs., 30 (1966) 701.

2. A. Bock, Arch. Mikrobiol., 68 (1969) 165.

3. S. K. Mitra and A. H. Mehler, J. Biol. Chem.,242 (1967) 5490.

4. D. O. Ostrem and P. Berg, Proc. Natl. Acad. Sci., 67 (1970) 1967.

5. A. N. Baldwin and P. Berg, J. Biol. Chem.,241 (1966) 831.

6. H. Hayashi, J. R. Knowles, J. R. Katze, J. Lapointe, and D. Soll, J. Biol. Chem.,245 (1970) 1401.

7. C. J. Bruton and B. S. Hartley, Biochem. J.,108 (1968) 281.

8. M. P. Stulberg, J. Biol. Chem., 242 (1967) 1060.

9. M. H. J. E. Kosakowski and A. Bock, Europ. J. Biochem., 12 (1970) 67.

10. M. Lee and K. H. Muench, J. Biol. Chem., 244 (1969) 223.

11. J. R. Katze and W. Konigsberg, J. Biol. Chem., 245 (1970) 923.

12. D. I. Hirsh, J. Biol. Chem., 243 (1968) 5731.

13. K. H. Muench, A. Safille, M. Lee, D. R. Joseph, D. Kesden, and J. C. Pita, Jr., in: Proc. Miami Winter Symp., eds. W. J. Whelan and J. Schultz (North-Holland, Amsterdam, 1970) p. 52.

14. R. Calendar and P. Berg. Biochem., 5 (1966) 1681.

15. M. Yanif and F. Gros. J. Mol. Biol., 44 (1969) 1.

16. A. Szentirmai, M. Szentirmai and H. E. Umbarger, J. Bacteriol., 95 (1968) 1672.

17. J. M. Calvo, M. Freundlich and H. E. Umbarger, J. Bacteriol., 97 (1969) 1272.

18. J. R. Roth and B. N. Ames, J. Mol. Biol., 22 (1966) 325.

19. G. Nass, K. Poralla and H. Zahner, Biochem. Biophys. Res. Comm., 34 (1969) 84.

20. L. Eidlic and F. C. Neidhardt, Proc. Natl. Acad. Sci., 53 (1965) 539.

21. M. Cashel and J. Gallant, Nature, 221 (1969) 838.

22. W. L. Fangman and F. C. Neidhardt, J. Biol. Chem., 239 (1964) 1844.

23. L. Eidlic and F. C. Neidhardt, J. Bacteriol., 89 (1965) 706.

24. G. Edlin and P. Broda, Bacteriol. Revs., 32 (1968) 206.

25. G. Nass and F. C. Neidhardt, Biochim. Biophys. Acta, 134 (1967) 347.

26. L. S. Williams and F. C. Neidhardt, J. Mol. Biol., 43 (1969) 529.

27. A. L. Taylor, Bacteriol. Revs., 34 (1970) 155.

28. A. H. Mehler and S. K. Mitra, J. Biol. Chem., 242 (1967) 5495.

29. R. B. Loftfield and E. A. Eigner, J. Biol. Chem., 240 (1965) 1482.

30. M. Yarus and P. Berg, J. Mol. Biol., 42 (1969) 171.

31. H. George and A. Meister, Biochim. Biophys. Acta, 132 (1967) 165.

32. M. Yanif and F. Gros, J. Mol. Biol., 44 (1969) 31.

33. N. F. Doolittle and C. Yanofsky, J. Bacteriol., 95 (1968) 1283.

34. D. Cassio, F. Lemoine, J. P. Waller, E. Sandrin, and R. A. Boissonnas, Biochem., 6 (1967) 827.

35. T. Ohta, I. Shimada and K. Imahori, J. Mol. Biol., 26 (1967) 519.

36. A. Bock and F. C. Neidhardt, Zeits. fur Vererbungsl., 98 (1966) 187.

37. J. Carbon, P. Berg and C. Yanofsky, Cold Spring Harbor Symp. Quant. Biol., 31 (1966) 487.

38. M. Brenner, F. DeLorenzo and B. N. Ames, J. Biol. Chem., 245 (1970) 450.

ACKNOWLEDGEMENTS

This work was supported in part by research grants to the senior author from the National Science Foundation (GB 6062) and from the U. S. Public Health Service (GM 08437).

DISCUSSION

N. STEBBING: According to the earlier observation you described, the level of synthetase varies as the cells grow, and in this respect there was a difference between the wild type cells and the auxotrophic cells. Now, there is one difference between these two types of cells. Karlstrom & Gorini (1969, J. Mol. Biol. <u>39</u>, 89) have shown that in auxotrophs different rates of growth will greatly vary the level of the pool, although this may not be true of the wild type to the same extent. So I think there is a possibility of explaining the difference between the two on this basis. I think that one could ask whether there is more direct evidence for the level of the pool of amino acids controlling the level of the synthetase. What one would have to do is to measure the level of the pool of amino acids. Have you done this?

F.C. NEIDHARDT: I am not sure that I followed your reasoning. The wild cells that have been used in our studies can only be used if there is a way in which one can make a particular amino acid restricting for growth. There are few tricks one can use; one of them is to employ an inhibitor of the biosynthesis of the particular amino acid, as we have used for valine, to inhibit the biosynthesis of isoleucine. We have assumed that the growth rate of the cells under these conditions is an indirect measure of some intracellular pool of the amino acids. I think you have a valid point, if you are concerned with what pool it is that is important in serving as a signal. There are no good measurements of either the pool of free amino acids, or of the pool of amino acids which is enzyme-bound, or of the pool of amino acids which is attached to tRNA under the conditions of growth that we have used.

N. STEBBING: You are however in a fortunate position, and that is that you do find a difference in the level of synthetase in the wild type under different conditions of growth; so you don't have to, in any way, jigger around the exogenous environment of the cell to achieve a change; you have observed it already.

F.C. NEIDHARDT: No. The synthetases of a normal cell - the four or five at least that have been looked at extensively, and another half dozen that have been looked at superficially - do not change during growth in media of different composition, nor do they vary much with temperature. This is the primary observation which, for many years, led people to consider that synthetases are constitutive, i.e. that they are always made at a constant rate. The variation with growth rate is very slight: 10-25%, at most, over a 10-fold range in growth rate.

N. STEBBING: Yes, but that is precisely the sort of experiment that one does not want to do. One does not want to alter the composition of a medium, because the consequences could be many. You have a situation where you don't have to change the composition of a medium but merely the growth rate in a chemostat, and as a result, you do find the change in the synthetase levels. So one can simply ask, how do the pools of all the amino acids vary in that situation?

F.C. NEIDHARDT: Perhaps it will clarify the discussion if I just remind you that in all of those cases where one sees a variation in the level of a synthetase, one sees it as a result of tampering with the supply of that amino acid to the cell, whether by means of auxotrophic growth in a chemostat, by bradytrophic growth in batch culture, or by the use of an inhibitor. And in all of these experiments if you look at irrelevant synthetases such as the methionine and the tryptophan enzyme, when you have varied the arginine supply, you find their differential rate of formation has not been affected.

K. MUENCH: Just a minor clarification. Concerning the ratio of the seven tRNA's per enzyme (aminoacyl tRNA synthetase), I had thought that the ratio was more like one to one. That is the case, at least, for $tRNA^{Trp}$ and Trp-tRNA synthetase. The first work I know of on this was that of Callender and Berg on the $tRNA^{Tyr}$ and Tyr-tRNA synthetase of both *E. coli* and *B. subtilis* and, it seems to me, the ratio there was 1:1. Then, very recently DeLorenzo and Ames published the concentrations of $tRNA^{His}$ and His-tRNA synthetase in *S. typhimurium*.

F.C. NEIDHARDT: You are quite right. This makes the story even more dramatic. I was taking the most conservative information, information that you get just by lumping all of the synthetases and all of the tRNA, and in fact, in the cases where information is known precisely about individual amino acids the ratio is even lower, approaching almost one to one. You are quite right.

M.T. SKARSTEDT: I have two questions. First: Do you know of a case in which more than one synthetase has been demonstrated for a given amino acid?

F.C. NEIDHARDT: Yes, there are several cases - most of them obtained through work with eukaryotic cells. I would prefer, however, to hear from one of the experts from Oak Ridge; Dr. Jacobson may, in fact, have information about that.

K.B. JACOBSON: We have been working with Neurospora phenylalanine-tRNA synthetase. In this case, there is but one synthetase, but it exists in three chromatographically different forms whose concentration varies as a function of the growth stage of the organism. In early growth, there are three peaks. I will call them (1), (2) and (3). The third peak exists exclusively at early growth stages but older cultures have primarily the first peak. Furthermore, peak 3 can be converted to peak 1 by incubating an early homogenate. We feel that these three chromatographic forms are all one synthetase under one gene's control, but undergoing some sort of modification after their primary structure is determined. So I think that we are really left with very little evidence, as far as I am aware, that there are multiple synthetases for any given tRNA, perhaps others can comment on this.

F.C. NEIDHARDT: Well, for *E. coli*, the combination of biochemical and genetic information that one has for a dozen of the enzymes now would certainly make it difficult to maintain that there are two synthetases for any of those amino acids.

D.G. NOVELLI: With regard to eukaryotic cells there have been definitely demonstrated multiple synthetases, but these are organelle specific. That is, the mitochondrial enzyme, of which there is only one for each amino acid, may be quite different from the cytoplasmic enzyme in many cases. As far as I know, there is no really convincing case of multiple synthetases in *E. coli* or any other cell.

F.C. NEIDHARDT: The enzyme that I have been talking about, the valine enzyme, upon T-even phage infection acquires a new subunit, or at least it acquires a polypeptide factor made under the control of a phage gene. The factor binds tightly to the valyl-tRNA synthetase of the host, modifying many properties of the enzyme. The two forms of the enzyme are easily separated from each other. This situation is, of course, quite special. To answer your question, the mounting evidence is that there is one synthetase per amino acid in *Escherichia coli*.

M.T. SKARSTEDT: Thank you. My other question is: Could you summarize briefly how the tRNA levels within cells vary according to growth conditions?

F.C. NEIDHARDT: In a few hours I probably could. There is an enormous amount of information about it, but unfortunately, none of it is easily put together in one simple answer, partially because people working in this area have used many organisms and many different growth conditions, not all of which are easily reproducible. A few gross generalizations can be made: during amino acid restriction tRNA is not made; tRNA and ribosomal RNA vary as a function of growth rate irrespective of the chemical nature of the medium, but the variation is not a simple mathematical function; it definitely varies from one strain of *coli* to another. I think one point you may be getting at is what happens if one tampers with the rate of synthesis of an activating enzyme; does this influence the rate of formation or the level of the cognate tRNA molecules? We have not seen such variation during the growth of the mutants that I described this morning. To cite just one example, when the valine enzyme of strain NP2907 decreases during growth at 40°C the amount of valine acceptor tRNA remains coordinate with the total tRNA of the cell. We have not achieved a specific change in the rate of formation of the tRNA.

THE ROLE OF A SPECIFIC ISOACCEPTOR tRNA IN GENETIC SUPPRESSION AND IN ENZYME REGULATION IN DROSOPHILA*

K. BRUCE JACOBSON and E. H. GRELL
Biology Division, Oak Ridge National Laboratory
Oak Ridge, Tennessee 37830

Abstract: The vermilion mutant, v, of Drosophila melanogaster is unable to convert tryptophan to the brown eye pigment, ommochrome. The biochemical lesion in the vermilion mutant is a deficiency in tryptophan pyrrolase. In the v, su(s)^2 double mutant the activity of the tryptophan pyrrolase is restored partially and the brown eye pigment is produced. The suppressor mutant, su(s)^2, was described in 1932 by Schultz and Bridges. We have examined the su(s)^2 mutant for changes in tRNA because suppression in some bacteria occurs through an alteration of the anticodon of tRNA.

We have found that 1) the suppressor mutant fails to produce one of the three isoacceptor forms of tyrosine tRNA that occur normally, 2) the tryptophan pyrrolase of the vermilion mutant is inactive but is restored to 75% of the wild-type activity by addition of RNase T_1, 3) the "activated" tryptophan pyrrolase of vermilion is inhibited by wild-type tRNA but not by suppressor tRNA, and 4) the tyrosine tRNA that the suppressor fails to produce is the species that causes the inhibition of the tryptophan pyrrolase of the vermilion mutant.

We propose that one of the tyrosine isoacceptors is involved in the normal function of tryptophan pyrrolase and that the vermilion mutation alters the protein in such a way that the tRNA becomes inhibitory. Suppression of this inhibition is accomplished by preventing the occurrence of this form of tyrosine-tRNA.

INTRODUCTION

Suppression of a defective mutation allows an organism to express a normal phenotype even though the defective mutation remains present in the genome. In bacteria a genetically altered tRNA can "correct" both nonsense and missense codons that occur at the site of the defective mutation. An alteration of a base in the anticodon allows the tRNA to

---

*Research sponsored by the United States Atomic Energy Commission under contract with the Union Carbide Corporation.

introduce an amino acid in a growing peptide chain in response to a nonsense codon (1, 2, 3, 4) or to a missense codon (5, 6). Alteration of a species of tRNA by a base change at a site other than the anticodon also can cause suppression (7).

Since all tRNA-mediated suppression has been demonstrated exclusively in microorganisms, we inquired whether such a mechanism occurs in higher organisms. We examined the genetically characterized suppressor of vermilion eye color in Drosophila melanogaster.

The vermilion locus, $\underline{v}$, occurs on the X chromosome and is recessive (8). The failure of the homozygous recessive fly to synthesize brown eye pigment is due to the inability to convert tryptophan to kynurenine, an intermediate in pigment synthesis (9, 10, 11). Tryptophan pyrrolase, the enzyme that catalyzes this conversion, is present at greatly reduced concentration in homogenates of the vermilion mutant, 0–25% in various reports (12, 13, 14).

The suppressor of vermilion, $\underline{su(s)^2}$, also occurs on the X chromosome and is recessive. Its locus is widely separated from that of vermilion (15). The $\underline{v}$, $\underline{su(s)^2}$ double mutant exhibits a partial restoration of tryptophan pyrrolase concentration (12, 13), and a reduction in the accumulation of nonprotein tryptophan (16). Three other pigment mutants — sable, speck, and purple — also are suppressed by the $\underline{su(s)^2}$ locus (17).

TRANSFER RNA IN SUPPRESSOR AND WILD-TYPE DROSOPHILA

To compare flies of the suppressor mutant to wild-type (Samarkand) strain, large scale cultures of each were established. Transfer RNA was extracted from 20 grams of adult flies by homogenization in phenol, precipitation with ethanol, and adsorption to DEAE-cellulose at 0.3 M NaCl and elution with 1 M NaCl. The yield was 1 mg tRNA per gram of flies, the A_{260}/A_{280} was 1.9–2.0, the contamination by DNA or protein was less than 1%, and by ribosomal RNA was less than 10%.

Crude enzyme preparations were made by homogenization of 25 grams of adult flies in 10 mM Tris-Cl (pH 7.5), 10 mM ß-mercaptoethanol, 10 mM $Mg(OAc)_2$, and 10% glycerol, centrifugation at 79,000 X g for 90 min, adsorption on DEAE-cellulose, and elution with 0.3 M NaCl. The enzyme was stable at $-196°C$ for six months in 50% glycerol and the above buffer solution. A detailed account of the preparation of the tRNA and the enzyme fraction will appear elsewhere (18).

To compare suppressor and wild-type tRNA's by cochromatography the aminoacyl-tRNA was produced using [^{14}C]amino acid and [^3H]amino acid, respectively. The conditions for charging were optimized for pH and for ATP and Mg^{++} concentration and then were shown to result in a linear relationship between tRNA added and aminoacyl-tRNA formed. The aminoacyl-tRNA was separated from all other components of the reaction mixture by adsorption to DEAE-cellulose and subsequent elution Samples to be compared were then mixed and applied to the reversed phase II (Freon) chromatography system of Weiss and Kelmers (19) and eluted with a linear gradient of NaCl that contained also 10 mM NaOAc (pH 4.5), 10 mM $MgCl_2$, 1 mM EDTA, and 5 mM ß-mercaptoethanol. Fractions were assayed for acid-precipitable radioisotopes of carbon and hydrogen.

Four amino-acid-specific tRNA's are compared in Fig. 1. There are no differences in the chromatographic profiles of phenylalanyl-, seryl-, or leucyl-tRNA from the wild-type Samarkand strain and from the v, su(s)^2 double mutant. On the other hand, there is a slight forward displacement of the tyrosyl-tRNA from the suppressor mutant. When the column tubing is treated with silane, the wild-type tyrosyl-tRNA can be resolved into

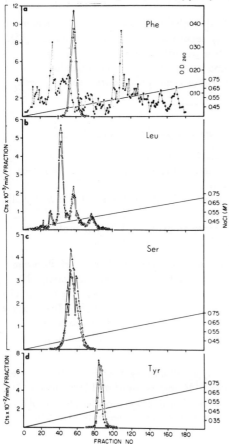

Fig. 1. Cochromatography on reversed phase II column of aminoacyl-tRNA's from v, su(s)^2 double mutant and wild-type flies. ^{3}H-labeled amino acid was employed to charge suppressor tRNA (Δ) and ^{14}C-labeled amino acid for wild type (O). The A_{260} is shown as (----) in 1a. a) Phenylalanyl-tRNA, b) leucyl-tRNA, c) seryl-tRNA, d) tyrosyl-tRNA.

three peaks and, as seen in Fig. 2a, the suppressor lacks tRNA$_2^{Tyr}$. Since the same amount of tRNA (5 A$_{260}$ units) was added from the wild type as from the suppressor mutant, the increase in peak 1 of tRNA^{Tyr} indicates that the total amount of tyrosine tRNA is the same in the two types of flies. Indeed, when su(s)^2/su(s)^2, and su(s)^2/+ genotypes were compared to wild type we found that each contained 13–14 pmoles tyrosyl-tRNA per A$_{260}$ of tRNA.

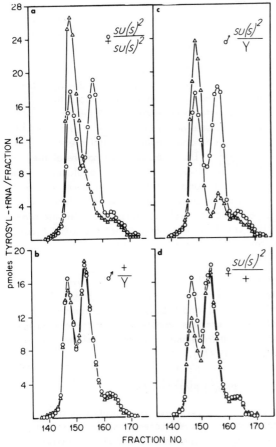

Fig. 2. Comparison of tyrosyl-tRNA from various genotypes of Drosophila. [^{14}C]tyrosine and [^{3}H]tyrosine were used to charge tRNA from the test genotype and wild type, respectively. tRNA's from the following genotypes (Δ) were cochromatographed with wild-type tRNA (O): a) su(s)^2/su(s)^2 females, b) +/Y males, c) su(s)^2/Y males, d) su(s)^2/+ females of second cross (female progeny from the first cross gave a similar elution pattern).

To demonstrate that the alteration in tyrosyl-tRNA isoacceptors is controlled by a locus on the X chromosome, a double reciprocal cross was made. Hemizygous males, su(s)^2/Y were mated to wild-type females, and conversely wild-type males were mated to homozygous recessive females, su(s)^2/su(s)^2. The progeny had genotypes of +/Y, +/su(s)^2 and su(s)^2/+,

su(s)2/Y, respectively.   Males and females from each cross were separated and the tRNA's were extracted and compared by cochromatography.
Fig. 2b shows that the tRNA^{Tyr} from the male +/Y and from mixed sexes of the wild-type strain chromatograph identically.   Fig. 2c shows that the hemizygous male, su(s)2/Y, is deficient in tRNA$_2^{Tyr}$ but not so much as the homozygous recessive female, su(s)2/su(s)2 (Fig. 2a).   Fig. 2d indicates that the control of tRNA$_2^{Tyr}$ is recessive, since the heterozygote, su(s)2/+, is not different from the wild-type strain.   Therefore the regulation of the production of tRNA$_2^{Tyr}$ occurs on the X chromosome.   The vermilion locus is ruled out because the tyrosyl-tRNA from the vermilion mutant cochromatographs identically to wild-type tyrosyl-tRNA.

The final establishment of the role of the su(s)2 locus in regulating production of tRNA$_2^{Tyr}$ was accomplished by inducing several new suppressor mutants.   Flies from a non-suppressor strain shown to contain tRNA$_2^{Tyr}$ were treated with ethyl methanesulfonate to produce several newly induced suppressor mutants that mapped at the same position as su(s)2.   The tRNA^{Tyr} from one of these, su(s)e1, was examined and shown to be deficient in tRNA$_2^{Tyr}$ in the same manner as su(s)2.   Thus the recessive nature of the su(s)2 phenotype is paralleled by the recessive nature of the regulation of the production of tRNA$_2^{Tyr}$, both traits are found on the X chromosome, and a newly isolated suppressor mutant also exhibits the characteristic deficiency for tRNA$_2^{Tyr}$.   So the su(s)2 locus is quite probably responsible for the production of tRNA$_2^{Tyr}$.   Since it is recessive we presume that the su(s)2 locus is not the structural gene for the tRNA but for an enzyme that converts a precursor molecule to tRNA$_2^{Tyr}$.

Various controls were performed.   The enzyme and tRNA from wild type and suppressor were interchanged in the preparation of tyrosyl-tRNA, the tRNA was heated to 85°C for 6 min, and the tRNA was chromatographed on Sephadex G-100; in each case there was no alteration in the chromatographic profile of tRNA^{Tyr} due to the treatment.   Detailed information on these tRNA studies will appear shortly (18).

EFFECT OF RNase T_1 ON TRYPTOPHAN PYRROLASE

The mechanism of suppression of the vermilion mutant must be quite different from that of the defective mutants of bacteria mentioned in the introduction, since the tRNA change in Drosophila involves a recessive gene and the disappearance of a tRNA species rather than the appearance of a new form.   Therefore, we postulated that tRNA$_2^{Tyr}$ is directly involved with the function of tryptophan pyrrolase.

Support for this hypothesis was found first in the observation that the addition of RNase T_1 stimulates tryptophan pyrrolase activity in the vermilion extract to a level approaching that of the wild type (Fig. 3).   No significant change in the activity of the enzyme from wild-type Drosophila results from addition of RNase T_1. There are several mutant alleles known at the vermilion locus. The v mutant is suppressed by $\underline{su(s)_2^2}$ but the v^{36f} mutant is not. When RNase T_1 is added to extracts from v^{36f} there is no increase in tryptophan pyrrolase activity.

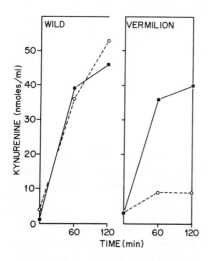

Fig. 3. Activation of tryptophan pyrrolase of the vermilion mutant by RNase T_1. The 15,000 X g supernatant of a 25% homogenate in 0.05 M potassium phosphate (pH 7.5) was assayed for tryptophan pyrrolase. The appearance of kynurenine was measured by a modified Bratton-Marshall assay (20) when 0.1 ml of Drosophila extract was added to 0.005 M tryptophan, 0.001 M 2-mercaptoethanol, and 0.1 M potassium phosphate (pH 7.5). RNase T_1 is either present at 2.5 units (21) per ml (●), or absent (O).

To determine whether the RNase T_1 removes some RNA that is bound to the tryptophan pyrrolase or simply hydrolyzes unassociated RNA, the vermilion extract was subjected to centrifugation in a sucrose density gradient. We observed an enzyme peak at a position on the gradient that corresponds to 9S, but activity was obtained only by including RNase T_1 in the assay mixture or in the homogenate prior to centrifugation. If tRNA were not bound to the enzyme it would remain at the 4S portion of the gradient. We could detect no difference in the sedimentation rates of tryptophan pyrrolase with and without treatment with RNase T_1 prior to centrifugation. These experiments using RNase T_1 indicate that some form of RNA inhibits the tryptophan pyrrolase of the vermilion mutant and that if it is tRNA it is bound to the enzyme.

EFFECT OF tRNA ON TRYPTOPHAN PYRROLASE

To test whether tRNA itself inhibits the enzyme, a crude extract of the vermilion mutant was treated with RNase T_1 to give maximum activity, and the RNase was then separated from the "activated" tryptophan pyrrolase by chromatography on G-75 Sephadex. Unfractionated tRNA from wild-type Drosophila was added to a tryptophan pyrrolase reaction mixture in which the enzyme concentration limited the velocity of kynurenine formation. As shown in Fig. 4, the tRNA inhibits the enzyme. Maximal inhibition occurs when the absorbance at 260 nm of the tRNA in the reaction mixture is 0.012. If we assume that there are 20 A_{260} units/mg of tRNA, that the molecular weight of tRNA is 30,000, and that $tRNA_2^{Tyr}$ is 2% of the total, then the concentration of $tRNA_2^{Tyr}$ is $4 \times 10^{-10} M$ when maximal inhibition of tryptophan pyrrolase is achieved. In many enzyme assays the concentration of enzyme is approximately 10^{-9} to 10^{-10} M. Therefore we might guess that there is a simple stoichiometry for the interaction of the enzyme and tRNA, with perhaps 1–2 tRNA's complexing with a molecule of tryptophan pyrrolase.

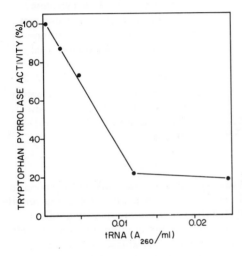

Fig. 4. Inhibition of "activated" tryptophan pyrrolase by unfractionated tRNA. Using a rate-limiting amount of the extract from vermilion that was pretreated with RNase T_1, the standard enzyme assay was performed with and without tRNA from wild-type flies. The amount of kynurenine produced in 30 min was determined.

To determine whether $tRNA_2^{Tyr}$ specifically is responsible for the enzyme inhibition we then fractionated the wild-type tRNA. Using a reversed phase II chromatography column (19), we fractionated 200 A_{260} units of tRNA, so that the profile of absorbance at 260 nm appeared as shown in Fig. 5a. An aliquot of each tRNA fraction was added to a tryptophan pyrrolase reaction mixture in which the "activated" enzyme from the vermilion mutant was rate-limiting and the final absorbance of the tRNA at 260 nm was 0.004. The major inhibitory fraction emerged from the column at 0.42–0.50 M NaCl. This fraction was pooled and dialyzed against 0.005 M sodium phosphate (pH 4.6) and subsequently chromatographed on a column (1 X 6 cm) of hydroxylapatite. As shown in Fig. 5b, three absorbance peaks emerged, two of which accepted tyrosine; but only one of the tyrosine tRNA peaks inhibited the "activated tryptophan pyrrolase of the vermilion mutant.

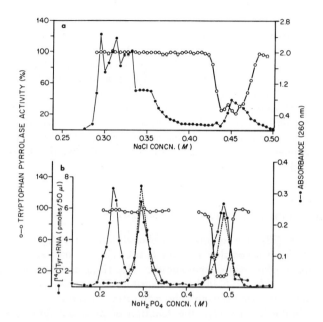

Fig. 5. Inhibition of "activated" tryptophan pyrrolase by tRNA fractions. a) Wildtype tRNA was chromatographed on a reversed phase II column as described in the text. The eluting gradient contained 4 l and varied linearly from 0.25 to 0.70 M NaCl. b) The 0.42 to 0.50 M fraction of (a) was chromatographed on a hydroxylapatite column as described in the text. The eluting gradient contained 300 ml of 0.05 M potassium acetate and varied from 0.10 to 0.60 M potassium phosphate. Aliquots (50 µl) were added to the standard tryptophan pyrrolase assay to determine their inhibitory effect.

The purity of the tyrosine tRNA peaks is probably greater than 80% since the second and third peak of Fig. 5b accepted 1835 and 1340 pmoles/A_{260} respectively. Therefore the third peak is predominantly $tRNA^{Tyr}$, probably corresponds to the $tRNA_2^{Tyr}$ observed on reversed phase II chromatography, and is the major component of the wild-type tRNA that inhibits the tryptophan pyrrolase of the vermilion mutant.

Fractionation of tRNA from the vermilion mutant $\underline{v}$ also resulted in isolation of an inhibitory $tRNA^{Tyr}$ analogous to that shown in Fig. 5b. On the other hand, when tRNA extracted from the double mutant $\underline{v}$, $\underline{su(s)}^2$ was tested it did not inhibit the "activated" tryptophan pyrrolase at any concentration up to 0.2 A_{260}/ml.

These findings are consistent with the postulate that $tRNA_2^{Tyr}$ is associated in some way with tryptophan pyrrolase in the wild-type fly. The vermilion mutant $\underline{v}$ then causes an alteration in protein structure so that the bound tRNA inhibits the enzyme. This inhibition of tryptophan pyrrolase is prevented in the double mutant $\underline{v}$, $\underline{su(s)^2}$ by preventing the formation of $tRNA_2^{Tyr}$, thus accomplishing a suppression of the $\underline{v}$ mutation. A mechanism of suppression of vermilion that does not involve transcription or translation has been proposed by Marzluf (22) and by Tartof (14). The tryptophan pyrrolase of the nonsuppressible vermilion mutant $v\underline{^{36f}}$ presumably is altered in some other manner so that it is inactive no matter whether tRNA is bound to it or not. It should be noted that uncharged tRNA was used to inhibit tryptophan pyrrolase.

ASSOCIATION OF tRNA WITH OTHER ENZYMES

Duda <u>et al</u>. (23) have reported that phenylalanine tRNA (either charged or uncharged) forms a complex with the first enzyme in the bacterial pathway for synthesis of aromatic amino acids. Hatfield and Burns (24) have found that leucyl-tRNA is bound to an immature form of the bacterial enzyme, threonine deaminase, and prevents the maturation of the enzyme to its active form. In the latter study, the function of an isoacceptor form of $tRNA^{Leu}$ is suggested. In the present study we find that a single isoacceptor form of $tRNA^{Tyr}$ (uncharged) is responsible for inactivating tryptophan pyrrolase.

A more detailed report of the inhibition studies described here will appear elsewhere (25).

REFERENCES

1. D. L. Englehardt, R. E. Webster, R. C. Wilhelm and N. D. Zinder, <u>Proc. Nat. Acad. Sci. U.S.A.</u>, 54 (1965) 1791.

2. J. D. Smith, J. N. Abelson, B. F. C. Clark, H. M. Goodman and S. Brenner, <u>Cold Spring Harbor Symp. Quant. Biol.</u>, 31 (1966) 479.

3. R. C. Wilhelm, <u>Cold Spring Harbor Symp. Quant. Biol.</u>, 31 (1966) 496.

4. M. R. Cappechi and G. N. Gussin, <u>Science</u>, 149 (1965) 417.

5. J. Carbon, P. Berg and C. Yanofsky, <u>Proc. Nat. Acad. Sci. U.S.A.</u>, 56 (1966) 764.

6. N. K. Gupta and H. G. Khorana, <u>Proc. Nat. Acad. Sci. U.S.A.</u>, 56 (1966) 772.

7. D. Hirsh, Nature, 228 (1970) 57.
8. T. H. Morgan and C. B. Bridges, Carnegie Inst. Wash. Pub. 237 (1916) 27.
9. A. H. Sturtevant, Proc. Sixth Int. Cong. Genet., 1 (1932) 304.
10. G. W. Beadle and B. Ephrussi, Proc. Nat. Acad. Sci. U.S.A., 22 (1936) 536.
11. A. Butenandt, W. Weidel and E. Becker, Naturwissenschaften, 28 (1940) 63.
12. C. Baglioni, Heredity, 15 (1960) 87.
13. S. Kaufman, Genetics, 47 (1962) 807.
14. K. D. Tartof, Genetics, 62 (1969) 781.
15. J. Schultz and C. B. Bridges, Amer. Natur., 66 (1932) 323.
16. M. M. Green, Genetics, 34 (1949) 564.
17. D. L. Lindsley and E. H. Grell, Carnegie Inst. Wash. Pub. 627 (1968) 471 pp.
18. D. R. Twardzik, E. H. Grell and K. B. Jacobson, J. Mol. Biol., in press.
19. J. F. Weiss and A. D. Kelmers, Biochemistry, 6 (1967) 2507.
20. G. A. Marzluf, Z. Vererbungslehre, 97 (1965) 10.
21. K. Takahashi, J. Biochem. (Tokyo), 49 (1961) 1.
22. G. A. Marzluf, Genetics, 52 (1965) 503.
23. E. Duda, M. Staub, P. Venetianer and G. Denes, Biochem. Biophys. Res. Commun., 32 (1968) 992.
24. G. W. Hatfield and R. O. Burns, Proc. Nat. Acad. Sci. U.S.A., 66 (1970) 1027.
25. K. B. Jacobson, Nature, in press.

DISCUSSION

M.T. SKARSTEDT: Have you examined the kinetics of the inhibition which is imposed by $tRNA^{Tyr}$ on the enzyme? That might tell you something about the nature of the site that is inhibited.

K.B. JACOBSON: Yes, we'd like to do that. Actually, the assay procedure for tryptophan pyrrolase can be described as just about miserable. You can tell whether the enzyme is active, 50% active or inactive, but for getting down to fine points it is a pretty inadequate kind of assay. We shall have to "tune up" the enzyme assay before we try asking any semi-sophisticated questions like this. I might emphasize, by the way, that it is the tRNA^{Tyr} that is involved in the tryptophan metabolizing pathway; I have no explanation for this. It doesn't offer any guidance in predicting which tRNA to look for as a regulator of any other enzyme you might look at.

B.L. HORECKER: You have not proposed a function for the tRNA in the wild type enzyme?

K.B. JACOBSON: No, we simply postulate it to make a consistent hypothesis for explaining the effect on the vermilion enzyme. There are other mutants of vermilion at this locus that are non-suppressible as opposed to this one which is suppressible. The non-suppressible allele is not activated by ribonuclease. So I presume there is a defect in the protein structure at a site that results in an inactive enzyme whether tRNA is bound to it or not.

R. ARMSTRONG: Have you tried any other sources of tRNA, say *E. coli* tRNA for the inhibition? Is this specific for the *Drosophila* tRNA?

K.B. JACOBSON: I don't know, we haven't looked at that.

REGULATION OF GENE EXPRESSION IN ESCHERICHIA COLI BY CYCLIC AMP

I. PASTAN[*], B. DE CROMBRUGGHE[*], B. CHEN[*], W. ANDERSON[*],
J. PARKS[*], P. NISSLEY[*], M. STRAUB[**], M. GOTTESMAN[*], and
R. L. PERLMAN[**]

[*] National Cancer Institute
[**] National Institute of Arthritis and Metabolic Disease
National Institutes of Health, Bethesda, Md. 20014

To cope with the unexpected, all cells have genetic information which is not usually expressed but can be readily activated. In E. coli and some other bacteria, some of these genes are under the control of 3',5'-adenosine monophosphate (cAMP) (1).

When E. coli is confronted with a variety of substances it can use as a source of energy (these substances include a number of sugars, amino acids and nucleic acids), the organism will induce the synthesis of the proteins necessary to transport these substances into the cell and convert them into a form that can be readily utilized by pre-existing constitutive enzymes. The substance about which most information is available is lactose which induces the synthesis of the lac permease that transports lactose into the cell and β-galactosidase that converts it to glucose, which the cell is equipped to metabolize, and galactose, which in turn induces the synthesis of a second set of enzymes, those of the gal operon.

Even though the inducer is present, cells make inducible enzymes at a very low rate if they are grown in glucose. This ability of glucose to repress the synthesis of inducible enzymes (termed the glucose effect or catabolite repression) has long puzzled investigators. The finding by Makman and Sutherland (2) that glucose lowered cyclic AMP levels and our finding that cyclic AMP overcame glucose repression of the synthesis of various inducible enzymes indicated that the presence of cAMP was required for the synthesis of such proteins. These include the proteins of the lac, gal, ara, and mal operons and the glycerol regulon, tryptophanase, D-serine deaminase, thymidine phosphorylase and flagellar protein (2). Further, mutants unable to make cyclic AMP due to loss of the enzyme adenyl cyclase are also unable to make inducible enzymes unless grown in the presence of cyclic AMP (3).

Even stronger evidence for the role of cAMP in regulating the synthesis of inducible enzymes comes from studies in broken cells. Cell-free extracts of E. coli are able to make some β-galactosidase when supplemented with large amounts of λh80dlac DNA (4,5) and galactokinase when supplemented with λpgal DNA (6), but appreciable amounts of these enzymes are made only when cAMP is added (Table 1).

TABLE 1

Synthesis of β-galactosidase and galactokinase in cell-free extracts

Cell extract genotype	DNA	CRP 5 ug	cAMP (10^{-4} M)	β-galacto-sidase units/ml	Galacto-kinase (mU/ml)
lac Δ crp$^+$	λh80dlac	-	-	2.0	-
lac Δ crp^{-1}	λh80dlac	-	+	10.0	-
lac Δ crp^{-1}	λh80dlac	+	-	1.4	-
lac Δ crp^{-1}	λh80dlac	+	+	7.0	-
gal Δ crp$^+$	λpgal	+	-	-	60
gal Δ crp$^+$	λpgal	+	+	-	250

Legend to Table 1: S-30 were prepared and incubated as described previously. Data in Part B taken from Parks et al. (6).

In addition to the adenyl cyclase mutants, another class of mutants which are unable to synthesize β-galactosidase, galactokinase and other inducible enzymes has been found. These mutants have high levels of cAMP. They are missing a protein which has a high affinity for cAMP and has therefore been called the cyclic AMP receptor protein (CRP). This protein has been purified to homogeneity, has a molecular weight of 45,000 and is composed of two subunits each with a molecular weight of 22,500. The protein binds cyclic AMP reversibly (K_d 1.1 x 10^{-5} M) and binding of cAMP is competitively inhibited by cyclic GMP (K_I 0.7 x 10^{-5} M). Cell-free extracts from such mutants are unable to make β-galactosidase or galactokinase unless supplemented with cyclic AMP and purified CRP (Table 1 and Ref. 6).

In intact cells, cyclic AMP stimulates the rate of synthesis of lac and gal mRNA when the appropriate inducers are present, lactose or its analogue IPTG (isopropyl thiogalactoside) for the lac operon (7) and galactose or its analogue fucose for gal (8).

Recently, we have begun to investigate how cyclic AMP and the cyclic AMP receptor protein act to control transcription in purified cell-free systems. In these studies, we have employed highly purified RNA polymerase and CRP. As shown in Table 2, lac mRNA is made when DNA containing lac or gal genes, RNA polymerase, CRP and cyclic AMP are present. By hybridization methods, this RNA has been found to correspond to transcription; off the correct strand of the bacterial gene. Further, in the lac system, purified lac repressor prevents lac transcription IPTG overcomes this repression. Thus, the model proposed by Jacob and Monod (9) for the control of lac transcription has been experimentally confirmed.

TABLE 2

Synthesis of lac mRNA in cell-free system

DNA	CRP 7 ug/ml	cAMP 10^{-4} M	lac repressor 3 ug/ml	IPTG 10^{-2} M	cpm in lac mRNA
λ h80dlac p^S	−	−	−	−	29
λ h80dlac p^S	+	−	−	−	50
λ h80dlac p^S	−	+	−	−	71
λ h80dlac p^S	+	+	−	−	688
λ h80dlac p^S	+	+	+	−	97
λ h80dlac p^S	+	+	+	−	647

Legend to Table 2: Each reaction mixture contains in addition to the components shown, RNA polymerase 25 ug/ml, rho 5 ug/ml, ATP, GTP, UTP at 1.15×10^{-4} M, CTP at 0.75×10^{-4} M, $MgCl_2$ 0.15 M, Tris-HCl (pH 7.9) 0.02 M, and dithiothreitol 1×10^{-4} M.

How do CRP and cAMP act? Do they modify RNA polymerase, allowing it to bind to the lac promoter, or do they bind to the promoter, allowing RNA polymerase to bind? We favor the latter model since CRP will bind to lac DNA in the presence of cyclic AMP, and the binding of CRP is displaced more effectively by λh80dlac DNA than by λh80 DNA (Table 3). In these studies, we have employed an unusual lac mutant (λh80dlac p^s) which has a mutation in the promoter which appears to allow it to be transcribed more efficiently (10).

TABLE 3

Binding of cAMP-CRP complex to ^{3}H-λh80dlac p^s DNA

cAMP (10^{-4} M)	CRP (1 ug)	Competing DNA (ug/ml)		cpm bound
		λh80dlac p^s (ug/ml)	λh80 (ug/ml)	
-	-	-	-	100
+	-	-	-	106
-	+	-	-	205
+	+	-	-	411
+	+	.3	-	260
+	+	.6	-	230
+	+	-	.3	398
+	+	-	.6	388

Legend to Table 3: Each reaction mixture contained all the components of Table 1 except RNA polymerase. Incubations were for 10 min at 23°C and the complex was trapped on Millipore filters and the content of the ^{3}H determined. The initial concentration of λh80dlac p^s DNA was 0.5 ug/ml and 800 cpm were added to each reaction mixture.

Bautz and Bautz (11) have shown with bacteriophage DNA that RNA polymerase will bind to the promoter site in the absence of nucleotides to form a rifampicin resistant complex. We find in studying lac transcription that only when λh80dlac p^s DNA, RNA polymerase, CRP and cAMP are incubated together is such a complex found. Since the cAMP-CRP complex binds to lac DNA in the absence of RNA polymerase, cAMP and CRP must modify the DNA so that RNA polymerase can bind and be ready to initiate transcription when nucleotides are added. This modification of DNA is reversible since cyclic GMP which inhibits the binding of cAMP to CRP rapidly dissociates the rifampicin resistant complex. The nature of the modification in the lac promoter made by the cAMP-CRP complex is still unclear, but controlled transcription of the lac and gal operons using only purified substances has been achieved.

REFERENCES

1. I. Pastan and R. Perlman. Science 169 (1970) 339.

2. R. S. Makman and E. W. Sutherland. J. Biol. Chem. 240 (1965) 1309.

3. R. Perlman and I. Pastan. Biochem. Biophys. Res. Comm. 37 (1969) 151.

4. D. A. Chambers and G. Zubay. Proc. Nat. Acad. Sci. 63 (1969) 118.

5. B. de Crombrugghe, H. E. Varmus, R. L. Perlman and I. Pastan. Biochem. Biophys. Res. Comm. 38 (1970) 894.

6. J. S. Parks, M. Gottesman, R. L. Perlman, and I. Pastan. J. Biol. Chem., in press (1970).

7. H. Varmus, R. Perlman and I. Pastan. J. Biol. Chem. 245 (1970) 2259.

8. Z. Miller, H. E. Varmus, J. Parks, R. L. Perlman and I. Pastan. Submitted to J. Biol. Chem. (1970).

9. F. Jacob and J. Monod. J. Molec. Biol. 3 (1961) 318.

10. B. de Crombrugghe, B. Chen, M. Gottesman, I. Pastan, H. E. Varmus, M. Emmer, R. L. Perlman. Nature, in press (1971).

11. E. F. Bautz and F. A. Bautz. Nature 226 (1970) 1219.

DISCUSSION

W.J. WHELAN: Does your receptor protein have any effect on the cyclic AMP activation of protein kinases?

I. PASTAN: We sent a sample of CRP to Ed Krebs two months ago. I haven't heard from him about the result so I assume that it doesn't work.

R. NOVAK: Does the phosphorylation of the sigma subunit in RNA polymerase have any effect on your system?

I. PASTAN: You're referring to the experiments of Davies and coworkers in which they showed that a protein kinase from muscle which was prepared by Krebs would stimulate transcription from a T_4 template. Using that protein kinase we do not find any correction of extracts from crp^- mutants when measuring translation. We haven't gone back to the transcription systems to re-do that experiment, but there is no reason that we would expect that it should work.

M. LIPSCOMB: You indicated that chloramphenicol acetyltransferase was controlled by cyclic AMP, but isn't this enzyme constitutive in *E. coli*; perhaps I'm wrong on that?

I. PASTAN: Yes, it appears to be constitutive in *E. coli*. Somewhere in the beginning, I tried to hedge and said that, in general, inducible enzymes are under cyclic AMP control. One doesn't know but there might be an internal inducer for the chloramphenicol acetylase; but I think, like all generalizations it is bound to break down. I think Bill Shaw has a comment.

W.V. SHAW: The question of the constitutive synthesis of chloramphenicol acetylase is interesting. It is always described as a constitutive enzyme in strains of *E. coli* bearing episomes containing the structural gene. However, we have done some measurements of the specific activity throughout growth from the early exponential phase through to stationary phase. There was a definite increase in specific activity throughout the exponential phase of growth, suggesting the definition of constitutive as "the continuous synthesis of enzyme at a fixed level in the absence of inducer" is not necessarily so for some enzymes. We have also kept open the thought that the chloramphenicol acetylase may well be induced by some endogenous metabolite.

DNA—RNA POLYMERASE INTERACTION: THE EFFECTS OF THE FORMATION OF A SINGLE PHOSPHODIESTER BOND[1]

A. G. So,[2] K. M. Downey and B. S. Jurmark
Raymy Haber Karp Hematology Research Laboratories
Departments of Medicine and Biochemistry
University of Miami School of Medicine

Abstract: The effects of the formation of a single phosphodiester bond on both the stability of the DNA-RNA polymerase complex to dissociation by high ionic strength and on the sensitivity of the enzyme to the action of the antibiotic Rifampicin have been studied using poly [d(A-T)] as a template for the RNA polymerase of Escherichia coli. The formation of a single phosphodiester bond between ApU and ATP or between UpA and UTP has been shown to result in the stabilization of the d(A-T)n-enzyme complex to dissociation by high salt concentrations. This bond formation, and the resultant stabilization of the complex has been shown to require sigma factor. The formation of the trinucleotide ApUpA, but not the formation of UpApU, rendered the enzyme resistant to Rifampicin inhibition. With poly [d(A-T)] as a template, the enzyme catalyzed a pyrophosphate exchange reaction with UTP but not with ATP, and pyrophosphorolysis of phosphodiester bonds was observed only when the 3'-terminal nucleotide was UMP.

INTRODUCTION

The DNA-enzyme-RNA complex which exists during the synthesis of RNA by the DNA-dependent RNA polymerase may be distinguished from the DNA-enzyme complex which forms in the absence of RNA synthesis, both by its relative stability to dissociation by high ionic strength, and by its lack of sensitivity to the antibiotic Rifampicin.

The formation of an initiation complex consisting of DNA, RNA polymerase, and purine ribonucleoside triphosphates has been reported to stabilize the DNA-enzyme complex to dissociation by high salt concentrations (1,2) and to protect the enzyme against inhibition by Rifampicin (3). In these studies it is not clear whether the formation of the initiation complex results from the binding of a purine ribonucleoside triphosphate to the DNA-enzyme complex, or whether phosphodiester bond formation is a pre-requisite for the formation of a stable complex. The present studies were undertaken to resolve this question.

In these studies of the interaction of DNA and RNA polymerase, we have chosen poly [d(A-T)] as a template. This double-helical, alternating copolymer of deoxyadenylate and thymidylate cannot act as a template for the synthesis of poly A or poly U, and thus, the effects of single ribonucleoside triphosphates, either purine or pyrimidine, can be studied in the absence of phosphodiester bond formation. Moreover, the two dinucleotides complementary to the poly [d(A-T)] template, i.e. ApU and UpA, can be utilized both to study the effects of pre-formed phosphodiester bonds on the DNA-enzyme complex, and as a tool to limit RNA synthesis to the formation of a single phosphodiester bond.

EXPERIMENTAL

RNA polymerase was prepared from E. coli cells by the method previously described (4). The binding assay, the assay for pyrophosphate exchange and the polymerization assay were as previously reported (4). Sigma factor was prepared by the method of Burgess et al (5).

RESULTS AND DISCUSSION

The binding of RNA polymerase to a DNA template results in the formation of a DNA-enzyme complex which is retained on a Millipore filter, whereas the enzyme and DNA, separately, are not retained. This complex is readily dissociated, and its formation prevented, by high ionic strength (6). In contrast, once the synthesis of RNA has taken place, the DNA-enzyme-RNA complex is relatively stable to dissociation by high salt (4,7).

The effects of increasing ionic strength on the stability of the DNA-enzyme complex are shown in Figure 1.

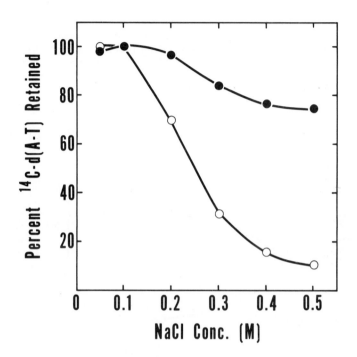

Fig. 1. Stabilization of the ^{14}C-poly[d(A-T)] - RNA polymerase complex by poly(A-U) synthesis.[3] The reaction mixtures were incubated for 4 minutes at 25°. Washing buffers contained the indicated concentrations of NaCl.

In this experiment, ^{14}C-labelled poly[d(A-T)] and RNA polymerase were incubated in the presence (solid circles) and absence (open circles) of the substrates ATP and UTP. The reaction mixtures were filtered on Millipore filters, and washed with buffer containing increasing concentrations of NaCl.

In the absence of poly(A-U) synthesis, the poly[d(A-T)] - enzyme complex is quite unstable, being almost completely dissociated at 0.4 M NaCl. Once poly(A-U) synthesis has taken place, however, the DNA-enzyme-RNA complex is relatively resistant to dissociation. The addition of pre-formed poly(A-U) had no stabilizing effect on the DNA-enzyme complex, suggesting that the formation of a stable complex requires active RNA synthesis.

With poly[d(A-T)] as a template, the two dinucleotides complementary to the template, ApU and UpA, may be used to initiate RNA synthesis, although these dinucleotides cannot be incorporated into internal positions in the RNA chains. The effects of the complementary dinucleotides on both chain initiation and elongation are shown in Figure 2.

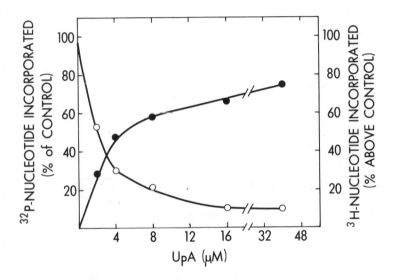

Fig. 2. Effect of UpA on chain initiation and polymerization.[4] The reaction mixtures were incubated for 10 minutes at 37°, and the reaction stopped by the addition of cold 5% TCA.

In this experiment, tritium labelled ribonucleoside triphosphates were used to measure the overall rate of RNA synthesis, (solid circles), and ATP labelled with ^{32}P in the γ-phosphate was used to measure the rate of chain initiation (open circles). Initiation with poly[d(A-T)] as a template has been shown to occur almost exclusively with ATP (8). Increasing concentrations of the complementary dinucleotide UpA stimulated the overall rate of poly(A-U) synthesis, while ATP-initiation as measured by the incorporation of ^{32}P was inhibited. Similar results were obtained with the other complementary dinucleotide ApU. Control experiments have shown that noncomplementary dinucleotides such as ApA or UpU had no effect on either chain initiation or elongation. These experiments indicate that the complementary dinucleotides compete with ATP to initiate RNA chains and are incorporated into the 5'-ends of the newly synthesized poly(A-U).

By using as substrates a complementary dinucleotide and a single ribonucleoside triphosphate, it is possible to synthesize a single phosphodiester bond. The incubation of ApU with ATP in the presence of poly[d(A-T)] and RNA polymerase results in the formation of the trinucleoside diphosphate ApUpA. Chain elongation cannot take place in the absence of the alternating substrate, UTP. Similarly, UpA and UTP will form the corresponding trinucleotide UpApU, and, in this case, chain elongation cannot take place in the absence of ATP. The enzyme catalyzed formation of a single phosphodiester bond is sufficient to stabilize the DNA-enzyme complex to dissociation by high salt, as shown in Table 1.

TABLE 1

Stabilization of the ^{14}C-poly[d(A-T)]-Enzyme Complex by the Formation of a Single Phosphodiester Bond[3]

Components Added	Amount of ^{14}C-Poly[d(A-T)] Retained on Filter (CPM)
Control	330
ATP	400
ApU	340
ApU, ATP	1300
UTP	450
ATP, UTP	2130
UpA	420
UpA, UTP	1400
ApU, UTP	1300
UpA, ATP	460

The reaction mixtures were incubated for 2 minutes at 25°. The washing buffer contained 0.4 M NaCl.

In this experiment, ^{14}C-labelled poly[d(A-T)] was incubated with RNA polymerase under each of the experimental conditions indicated. The reaction mixtures were then filtered on Millipore filters, and washed with buffer containing 0.4 M NaCl. Under these conditions, the poly[d(A-T)]-enzyme complex dissociates and is not retained on the filter, as shown by the control experiment. Incubation of the d(A-T)$_n$-enzyme complex with single ribonucleoside triphosphates, either ATP or UTP, did not stabilize the complex, although when both substrates were present, i.e., when there was synthesis of poly(A-U), the complex was resistant to dissociation by high salt. Preformed dinucleotides, either ApU or UpA, were not able to stabilize the complex, however, the enzyme-catalyzed formation of a phosphodiester bond, either between ApU and ATP, or between UpA and UTP, did result in stabilization of the poly[d(A-T)]-enzyme complex.

Incubation of the dinucleotide UpA with ATP did not result in phosphodiester formation, demonstrating the substrate specificity imposed on the enzyme by the base sequence of the DNA template. Similarly, the trinucleotide ApUpU cannot be formed with this template. However, incubation of the poly[d(A-T)]-enzyme complex with ApU and UTP does result in the stabilization of the complex. It is proposed that stabilization, in this case, is not the result of the formation of the trinucleotide ApUpU, but rather it is the result of pyrophosphorolysis of the dinucleotide ApU, followed by reformation of the phosphodiester bond.

The formation of a single phosphodiester bond either between ApU and ATP (ApUpA), or between UpA and UTP (UpApU) is dependent on the presence of sigma factor, the subunit of RNA polymerase required for chain initiation (5). The requirement for sigma factor is shown in Table 2, where the formation of a phosphodiester bond is measured by the formation of a ^{14}C-poly[d(A-T)]-enzyme complex which is resistant to dissociation by 0.4 M NaCl.

TABLE 2

Requirement for Sigma Factor for the Formation of a Single Phosphodiester Bond.

Components Added	Amount of (^{14}C)-d$(A-T)_n$ Retained on Filter (cpm)
Core enzyme	300
Core enzyme + sigma factor	310
ApU + ATP + core enzyme	320
ApU + ATP + sigma factor	0
ApU + ATP + core enzyme + sigma factor	1180
UpA + UTP + core enzyme	210
UpA + UTP + core enzyme + sigma factor	1140

The reaction mixtures were incubated for 2 minutes at 25° and washed with 0.4 M NaCl.

These results suggest that the formation of the phosphodiester bond between a complementary dinucleotide and a single ribonucleoside triphosphate represents RNA chain initiation rather than chain elongation. This hypothesis is further supported by the sensitivity of the reaction to inhibition by Rifampicin, an antibiotic that has been shown to inhibit RNA chain initiation, but has little effect when added after RNA synthesis has begun (9). As shown in Table 3, Rifampicin inhibits phosphodiester bond formation between ApU and ATP, UpA and UTP, and the synthesis of poly(A-U) as measured by the formation of a stable DNA-enzyme complex.

TABLE 3

Inhibition of Phosphodiester Bond Formation by Rifampicin[3]

Components Added	Amount of ^{14}C-Poly[d(A-T)] Retained on Filter (CPM)
Control	330
ApU, ATP	1490
ApU, ATP, Rifampicin	690
UpA, UTP	1300
UpA, UTP, Rifampicin	560
ATP, UTP	2110
ATP, UTP, Rifampicin	640

Incubation conditions were described in Table 1. The control contained only enzyme and ^{14}C-poly[d(A-T)].

Preincubation of the poly[d(A-T)]-enzyme complex with either ATP, UTP, ApU or UpA did not protect the enzyme against the action of Rifampicin, although protection was observed following the formation of a single phosphodiester bond, as shown in Table 4.

TABLE 4

Effect of Phosphodiester Bond Formation on Rifampicin Inhibition

Components Added During Preincubation	Components Added During Subsequent Incubation	Percent Activity in the Presence of Rifampicin
Experiment 1		
None	UpA, ^{3}H-ATP, UTP	52
UpA	^{3}H-ATP, UTP	53
UpA, UTP	^{3}H-ATP	59
UpA, ATP	^{3}H-UTP	53
UTP	UpA, ^{3}H-ATP	52
Experiment 2		
None	ApU, ATP, ^{3}H-UTP	41
ApU	ATP, ^{3}H-UTP	49
ApU, ATP	^{3}H-UTP	94
ApU, UTP	^{3}H-ATP	48
ATP	ApU, ^{3}H-ATP	49

The reaction mixtures were preincubated for 4 minutes at 37°, followed by the addition of Rifampicin and the other components indicated. The reaction mixtures were incubated for 6 minutes at 25°. Percent activity was calculated from control experiments in which no Rifampicin was added during the incubation period.

In these experiments, poly[d(A-T)] and RNA polymerase were incubated with the components listed in the first column. Following the pre-incubation, Rifampicin was added to the reaction mixtures, and finally, the missing components necessary for poly(A-U) synthesis were added back. The degree of protection against Rifampicin inhibition was measured by the subsequent synthesis of poly(A-U) in the presence of the antibiotic.

The formation of a phosphodiester bond between ApU and ATP during the pre-incubation period protected the enzyme against Rifampicin inhibition, so that the subsequent rate of poly(A-U) synthesis was 94% of that observed in a control experiment in which no Rifampicin was added. The presence of either single ribonucleoside triphosphates or pre-formed dinucleotides during the pre-incubation period afforded no protection against Rifampicin inhibition, nor did protection result from the formation of the trinucleotide UpApU. Although the poly$[d(A-T)]$-enzyme complex is stabilized by the formation of a phosphodiester bond between either ApU and ATP (ApUpA) or UpA and UTP (UpApU), only the formation of the trinucleotide ApUpA protected the enzyme against inhibition by Rifampicin. The lack of protection by the formation of UpApU may be related to the specificity of the pyrophosphate exchange reaction that is observed with this enzyme when $d(A-T)_n$ is used as a template.

RNA polymerase has been shown to catalyze a PP_i exchange reaction under conditions where the synthetic reaction is limited by low substrate concentrations (10,11). As shown in Table 5, significant PP_i exchange is observed when the $d(A-T)_n$-enzyme complex is incubated with UpA and UTP, or when there is formation of a trinucleotide with a 3'-terminal uridine nucleotide.

TABLE 5

Requirements for Pyrophosphate Exchange

Additions	PP_i Incorporated ($\mu\mu$moles)
None	70
ATP	110
UTP	380
ATP (0.32 mM), UTP	440
ATP (0.08 mM), UTP	1490
ApU	100
ApU, ATP	110
ApU, UTP	690
UpA	130
UpA, UTP	2150
UpA, ATP	90

The reaction mixtures were preincubated in the absence of ATP or UTP and $^{32}P-PP_i$ for 5 minutes at 37°, followed by incubation at 37° for 10 minutes.

A small amount of PP_i exchange is observed with UTP alone, and the rate is stimulated at low concentrations of ATP. Higher ATP concentrations, which favor poly(A-U) synthesis, inhibit the exchange reaction. Considerable exchange is also observed when ApU and UTP are incubated with the $d(A-T)_n$-enzyme complex, suggesting that pyrophosphorolysis can occur with a preformed phosphodiester bond, followed by reformation of ApU and repetition of the cycle. These results are consistent with the observation that incubation of the $d(A-T)_n$-enzyme complex with ApU and UTP results in the formation of a complex resistant to dissociation by high salt.

Little or no PP_i exchange is observed when the poly[d(A-T)]-enzyme complex is incubated with either ATP, UpA and ATP, or ApU and ATP, suggesting that no PP_i exchange occurs with ATP or with a 3'-terminal adenine nucleotide, when $d(A-T)_n$ is used as a template. However, PP_i exchange with ATP is observed with this enzyme when poly dT is used as a template (11), suggesting that the secondary structure of the template plays an important role in determining the substrate specificity of the enzyme.

It is perhaps significant that the formation of a trinucleotide that does not allow pyrophosphorolysis to occur (ApUpA) protects the enzyme against inhibition by Rifampicin; while the formation of a trinucleotide which readily undergoes pyrophosphorolysis (UpApU) affords no such protection. Whether or not there is a relationship between these two phenomena is not, as yet, clear.

REFERENCES

1. D. D. Anthony, E. Zeszotek, and D.A. Goldthwait, Proc. Nat. Acad. Sci. U.S., 56 (1966) 1026.

2. N.W. Stead and O.W. Jones, Biochim. Biophys. Acta, 145 (1967) 679.

3. E. diMauro, L. Snyder, P. Marino, A. Lamberti, A. Coppo, and G. P. Tocchini-Valentini, Nature (London), 222 (1969) 533.

4. A. G. So and K. M. Downey, Biochemistry, 9 (1970) 4788.

5. R. R. Burgess, A.A. Travers, J.J. Dunn, and E.K.F. Bautz, Nature (London), 221 (1969) 43.

6. O.W. Jones and P. Berg, J. Mol. Biol. 22 (1966) 199.

7. J.P. Richardson, Progr. Nucleic Acid Res. Mol. Biol. 9 (1969) 75.

8. U. Maitra and J. Hurwitz, Proc. Natl. Acad. Sci. U.S. 54 (1965) 815.

9. A. Sippel and G. Hartmann, Biochim. Biophys. Acta. 157 (1968) 218.

10. I. H. Goldberg, M. Rabinowitz, and E. Reich, Proc. Nat. Acad. Sci. U.S. 49 (1963) 226.

11. J.S. Krakow and E. Fronk, J. Biol. Chem. 244 (1969) 5988.

FOOTNOTES

1. Supported by Grants from the National Institutes of Health (NIH-AM-09001) and the American Heart Association (69-640)

2. Established Investigator of the American Heart Association.

3. Reprinted from Biochemistry 9, November, 1970, 4788. Copyright by the American Chemical Society and reprinted by permission of Copyright owner.

4. Reprinted from Biochemistry 9, June, 1970, 2520, Copyright by the American Chemical Society and reprinted by permission of Copyright owner.

DISCUSSION

M. MICHAELIS: Is anything known about the mode of action of Rifampicin - how does it interfere with the enzyme reaction? On which site of the enzyme is it bound?

K. DOWNEY: I'm not aware of any studies on this, but Rifampicin binds to the RNA polymerase, presumably to the β subunit. I don't know whether this is an ionic interaction or if other forces are involved.

B.L. HORECKER: Is there any effect of preformed trinucleotides?

K. DOWNEY: We didn't look at the trinucleotides but I would assume not, as preformed poly r(A-U) did not have any effect.

S. FRIEDMAN: Does the binding of the polymerase to your templates poly d(A-T) protect the latter from muclease attack or haven't you looked at that?

K. DOWNEY: We haven't looked at that, although Kresin has reported that the binding of the enzyme to the template does protect the enzyme against temperature effects and proteases. They were using more complicated templates rather than poly d(A-T).

PART 2

NUCLEIC ACID SYNTHESIS IN VIRAL INFECTION

Editor: J. Schultz

TRANSLATIONAL CONTROL MECHANISMS IN BACTERIOPHAGE DEVELOPMENT

SAMUEL B. WEISS
Department of Biochemistry, University of Chicago, and the
Argonne Cancer Research Hospital (operated by the University
of Chicago for the United States Atomic Energy Commission),
Chicago, Illinois

INTRODUCTION

The development of bacteriophage in infected cells occurs in a sequential and orderly fashion. Phage multiplication can be divided roughly into four periods: synthesis of early proteins, synthesis of phage nucleic acid, synthesis of late proteins and self-assembly. In the DNA-containing bacteriophages, the nucleic acids serve as a template both for its own replication and for the transcription of phage mRNA which is then translated into protein. Although the biochemical machinery of the host is used for phage biosynthesis, it is not clear how the host machinery adapts itself toward this purpose; nor do we understand the control mechanisms that regulate the proper sequence of replicative events. The elegant theories of Jacob and Monod (1) covering the control of groups of bacterial genes have received important experimental support from the isolation of repressors for the lac operon (2) and for lambda phage (3). As yet, however, no strong evidence has been produced to indicate that regulatory genes producing specific repressors operate in the development of the T-phages. Although regulator gene mechanisms are not excluded, it is possible that the control of phage development could be established by phage-induced factors that operate at the level of DNA-transcription (i.e., sigma and rho) as well as at the level of mRNA translation. I should like to review some experimental information that favors the existence of translational control mechanisms during the replication of certain bacteriophages.

It has been generally thought that the translation of phage messengers is accomplished by utilizing the biochemical apparatus of the host, which remains unaltered during the infectious process. The finding that bacteriophages may alter the translational machinery of the host leads us to suspect the validity of this assumption. The types of changes observed in E. coli after bacteriophage infection involve changes in tRNAs and the enzymes that modify them, as well as modifications in the translational properties of the ribosomes themselves.

PHAGE INDUCED MODIFICATIONS OF TRANSFER RNAs

Several groups of workers have shown that following phage infection certain host tRNA species exist in a modified form. Shortly after T2 infection of E. coli, both qualitative and quantitative alterations occur in host leucine tRNA (4-6) as well as induced changes in enzymes that modify tRNA molecules, such as tRNA methylases (7) and tRNA sulfur transferases (8). In our own laboratory, we first observed that pulse-labeling of T4-infected cells with radioactive inorganic sulfate resulted in the formation of a labeled RNA, that showed a distinctly different chromatographic profile on methylated albumin kieselguhr (MAK) from the pulse-labeled ^{35}S-RNA isolated from uninfected cells (8). Both uninfected and infected ^{35}S-RNA were low-molecular-weight in character, but labeled RNA from normal cells showed 2 primary peaks on MAK chromatography, while labeled RNA from T4-infected cells demonstrated 3 distinct peaks. Significant quantitative differences were also observed in the same chromatographic profiles. The newly-induced ^{35}S-RNA material appearing after T4 infection was found to hybridize with T4 DNA and the extent of hybridization was not diminished in the presence of either E. coli tRNA or T4 mRNA. It was concluded, therefore, that the low-molecular-weight ^{35}S-RNA appearing after T4 infection was transcribed, de novo, from T4 DNA and was not related to T4 mRNA by sequence homology.

By adjusting annealing conditions so that aminoacylated tRNAs could be used in hybridization reactions, it became possible to demonstrate that the T4 genome codes for the synthesis of several different tRNA species. Under appropriate annealing conditions, tRNAs isolated from T4-infected cells and charged with radioactive amino acids formed specific radioactive hybrids with T4 DNA (9). Similar reports from other laboratories have confirmed this finding (10, 11). The procedure for the identification of phage tRNAs in infected extracts has been modified so that multiple tRNA species may be detected in a single annealing reaction. Hybridization is carried out with tRNA charged with a mixture of (^{3}H) amino acids, and the radioactive amino acids associated with the tRNA-DNA hybrid are dissociated by heating and subsequently identified by ion-exchange column chromatography (12). At the present time we have identified five different molecular species of tRNA coded by the T4 genome, and 14 new species of tRNA induced by T5 bacteriophage infection (Table 1). Although we consider that it is most likely that T4 phage carries information for only a limited number of tRNAs, it is quite possible that all 20 tRNA species are coded by the T5 phage genome. So far, no phage-specific tRNAs have been detected in extracts from E. coli infected with T7 phage or in B. subtilis infected with SPO1 phage.

TABLE 1

Transfer RNAs coded by bacteriophage T4 and T5

Transfer RNA	T4 Phage	T5 Phage
Arginine	+	+
Proline	+	+
Glycine	+	+
Isoleucine	+	+
Leucine	+	+
Serine	−	+
Valine	−	+
Methionine	−	+
Tyrosine	−	+
Phenylalanine	−	+
Lysine	−	+
Histidine	−	+
Glutamic acid	−	−
Alanine	−	+
Cysteine	−	−
Aspartic acid	−	+
Tryptophan	−	not assayed
Threonine	−	not assayed
Glutamine	not assayed	not assayed
Asparagine	not assayed	not assayed

PHAGE INDUCED MODIFICATIONS OF RIBOSOMES

Several different groups observed that the replication of RNA-containing bacteriophages could be blocked by superinfection of cells with T-even phages (13-15). Later, experiments by Hattman and Hofschneider (16) suggested that the defect might lie at the level of translation, although the nature of the defect was not defined. After the description of bacteriophage-coded tRNAs, it seemed possible that T-even tRNAs might be involved in shutting off RNA phage development. Cell-free protein synthesizing systems were set up to examine this possibility but no differences were detected between the _in vitro_ translation of MS2 RNA when either normal or T4 tRNAs were employed. During the course of these studies, however, we observed a striking difference between the ability of normal and T4-infected cell ribosomes to translate various template

RNAs (17). Ribosomes from T4-infected cells were equally as efficient as ribosomes from normal cells for the translation of poly U and T4 mRNA, but were inhibited (60-90%) in their ability to translate E. coli and MS2 RNA. This alteration in the translational property of E. coli ribosomes was shown to occur early after T4 infection (prior to T4 DNA synthesis) and did not manifest itself when infection was carried out in the presence of chloramphenicol or with T4 ghosts, suggesting that the phenomenon required the synthesis of some phage-induced protein.

Our current information leads us to conclude that host ribosomes are modified after infection by a phage-induced factor that restricts the ribosomes with respect to the messengers they can translate. A protein factor (T4 factor) has been isolated from T4-infected cell ribosomes, which imparts to uninfected ribosomes a similar property of "selective" mRNA translation. In the presence of T4 factor, E. coli ribosomes efficiently translate T4 and T5 template RNAs but are markedly restricted in their ability to translate E. coli and MS2 RNAs. In addition, we now know that the T4 factor exerts its effect at those stages in protein synthesis involved with chain initiation and not chain elongation (18, 19). It has been possible to demonstrate that in the presence of T4 factor, the 70S initiation complex composed of ribosomes, fMet-tRNA and mRNA readily forms with T4 and T5 mRNA but is restricted in formation with MS2 and E. coli template RNAs (Fig. 1).

The extensive alterations in the translational machinery of the host following bacteriophage infection suggest that these changes may be important for phage replication, but it is not clear exactly how these alterations contribute to the phage developmental process. It is well known, however, that during phage development strict control may be imposed over such functions as host macromolecular synthesis, the transcription of phage mRNA, and the production of early and late phage proteins. The control of some of these important functions could be regulated by the synthesis of specific phage proteins resulting from the translational changes imposed upon the host by the invading phage.

The necessity for certain bacteriophages to induce the synthesis of new tRNA species might be explained by their codon response. The tRNAs manufactured by a cell are those required for the recognition of code words in its own mRNA. If phage mRNA contained code words that were infrequently used or perhaps not present in host mRNA, then host mRNA species might be inadequate for the translation of phage messenger either because their concentrations were too low or because they were completely missing. It is possible that the synthesis of several key phage proteins could require the recognition of mRNAs

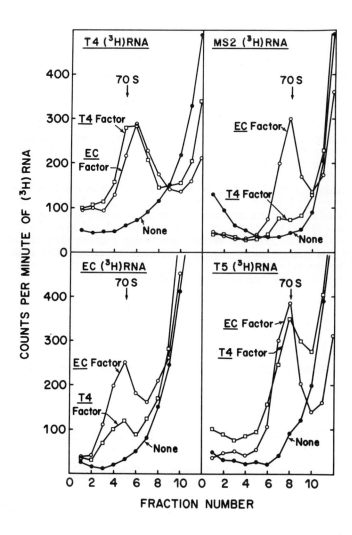

Fig. 1. Analysis of 70S initiation complex formation by sucrose gradient centrifugation. E. coli ribosomes were incubated with fMet-tRNA and (^{3}H) mRNA in the presence and absence of either E. coli (EC) or T4-infected (T4) factors which were extracted from normal and infected-cell ribosomes by 1 M NH_4Cl (19).

containing some "unique" code word (not present in host mRNA), thus the presence or absence of the proper tRNA could regulate the synthesis of these phage proteins.

The selective translation of phage messengers by modified host ribosomes represents another possible way in which the replicating phage could exert control over its own development. Control of messenger read-out by the modification of host ribosomes could be accomplished by the introduction of new initiation factors capable of recognizing only select initiation sites on phage mRNAs. The synthesis of inhibitors that would block the recognition of only specific initiation sites on mRNA molecules could accomplish a similar control over the selection of mRNA for translation.

Finally, control at the level of translation need not be restricted to bacteriophage development. Modifications in tRNA and translational control may occur in developing chick embryos (20, 21). Although the experimental evidence in favor of translational control mechanisms needs further documentation it has now become increasingly difficult to accept the idea that the protein synthetic apparatus of a cell is simply a workbench for the translation of mRNA.

REFERENCES

1. F. Jacob and J. Monod, J. Molec. Biol., 3 (1961) 318.

2. W. Gilbert and B. Mueller-Hill, Proc. Natl. Acad. Sci. U.S., 56 (1966) 1891.

3. M. Ptashne, Nature, 214 (1967) 232.

4. N. Sueoka and T. Kano-Sueoka, Proc. Natl. Acad. Sci. U.S., 52 (1964) 1535.

5. T. Kano-Sueoka and N. Sueoka, J. Molec. Biol., 20 (1966) 183.

6. L. C. Waters and G. D. Novelli, Proc. Natl. Acad. Sci. U.S., 57 (1967) 979.

7. E. Wainfan, P. R. Srinivason and E. Borek, Biochemistry, 4 (1965) 2845.

8. W.-T. Hsu, J. W. Foft and S. B. Weiss, Proc. Natl. Acad. Sci. U.S., 58 (1967) 2028.

9. S. B. Weiss, W.-T. Hsu, J. W. Foft and N. H. Scherberg, Proc. Natl. Acad. Sci. U.S., 61 (1968) 114.

10. V. Daniel, S. Sarid and U. Z. Littauer, FEBS Letters, 2 (1968) 39.

11. T. W. Tillak and D. E. Smith, Virology, 36 (1968) 212.

12. N. H. Scherberg and S. B. Weiss, Proc. Natl. Acad. Sci. U.S., 67 (1970) 1164.

13. N. D. Zinder, Perspectives Virol., 3 (1963) 58.

14. Z. Neubauer and V. Závoda, Biochem. Biophys. Res. Commun., 20 (1965) 1.

15. S. Hattman and P. H. Hofschneider, J. Molec. Biol., 29 (1967) 173.

16. S. Hattman and P. H. Hofschneider, J. Molec. Biol., 35 (1968) 513.

17. W.-T. Hsu and S. B. Weiss, Proc. Natl. Acad. Sci. U.S., 64 (1969) 345.

18. S. K. Dube and P. S. Rudland, Nature, 222 (1970) 820.

19. E. B. Klem, W.-T. Hsu and S. B. Weiss, Proc. Natl. Acad. Sci. U.S., 67 (1970) 696.

20. J. C. Lee and V. M. Ingram, Science, 158 (1967) 1330.

21. S. Sassa and S. Granick, Proc. Natl. Acad. Sci. U.S., 67 (1970) 517.

DISCUSSION

Dr. G. D. Novelli, Oak Ridge National Laboratory: I have a question and a comment. The question first: as you know with our unique chromatographic system, we were able to observe the appearance of 2 new leucine tRNAs following t2 and t4 infection. But in our experiments they appeared rather late in infection. I notice by your hybridization that you were able to pick these up rather early, within 2 and one half to 5 minutes. Were these leucine tRNAs or do you know?

Dr. S. B. Weiss: Well, we think that the charging experiments tell us that one of them is leucine. I believe that in experiments that you have done leucine tRNA was isolated some 20 minutes after t2 infection. Of course, you were working with a superinfecting system and I don't think anyone really knows what superinfection does with respect to the transcriptional process, whether in the presence of superinfection the same order or the same sequence of phage events appear as in the absence of superinfection. Either the new leucyl tRNAs which you have observed are different than the leucyl tRNAs which we see early after infection, or they are the same, or the superinfecting process has in some way shifted the time of transcription.

Dr. G. D. Novelli: I think the time is normal during a normal infection. Our chromatographic system is very sensitive to tertiary changes. That may be that the basic oligonucleotide is laid down early and the subsequent changes have to occur later before we can pick them up.

Dr. S. B. Weiss: Yes, I'm sorry, that's another possibility.

Dr. S. S. Cohen, University of Pennsylvania: Do I interpret your chloramphenicol experiment to suggest that the RNA polymerase is relatively unmodified at the time that it is transcribing the DNA to a new tRNA?

Dr. S. B. Weiss: I don't think our chloramphenicol experiment says that. I think that from other experiments it would seem that if we're going to get transcription of tRNA molecules as early as we have observed, within 2.5 minutes, then it is probable that we are using host RNA polymerase of an unmodified nature. What the chloramphenicol experiment is saying is that if there's any further modification of the transcribed tRNAs, as there should be, and if these modifications arise as a result of viral induced enzymes, chloramphenicol is blocking that type of modification. In spite of those blocks in the introduction of functional groups on tRNA molecules those viral tRNAs can still accept amino acids.

Dr. S. Greer, University of Miami: I'd like to know what group on transfer RNA is sulfated.

Dr. S. B. Weiss: Our information suggests that uridine is being sulfated and we don't know for sure but 4 thiouridine is the major thiolated product.

Dr. R. J. Mans, University of Florida: Does the isolated protein factor in anyway modify the fMET-tRNA or the messenger in your last series of experiments--that is where you incubated the tRNA and the messenger RNA in the presence and absence of the partially purified protein?

Dr. S. B. Weiss: We can't say for sure but it seems that that's not the case.

Dr. R. J. Mans: The reason I ask is that quite sometime ago an enzyme had been isolated by Klemperer from rat liver and this enzyme activity does add AMP moieties to ribosomal RNA. We have pursued this in corn seedlings and find that a similar enzyme isolated, although not from the ribosomal fraction, does add lots of AMP moieties onto tRNA than it does on ribosomal RNA.

Dr. S. B. Weiss: Well, whatever it's doing it's causing a selection in the type of messenger RNA. I think this is what we'd like to know, mechanistically, how is this type of selection in translation manifested by this particular factor? We're somewhat in a difficult position because the factor is not particularly stable, at least, not in our hands.

Dr. R. Armstrong, University of Michigan: Do you have any evidence whether this inhibitory factor is coming from the 50-s or the 30-s ribosome?

Dr. S. B. Weiss: We don't have any evidence, but I think there are other workers who have evidence that it's coming from the 30-s ribosomal sub unit.

Dr. G. D. Novelli: With regard to the possibility of a function for these new transfer RNAs I can only say we've looked at some amber mutant that are blocked late--they manufacture most of the parts but don't assemble them. The new leucine tRNAs that we're looking at are not present in these cells. John Abbleson at San Diego is looking at a mutant that's blocked still later; it makes complete phage but lacks some release function so that it doesn't come out of the cell. That mutant does have these 2 leucines, somewhere between the manufacture of the parts and the assembly. Maybe the region where these leucines are involved.

Dr. S. S. Cohen: I notice that your inhibition of the normal *E.coli* message was not really as great as that of MS2. Do you have any feeling as to whether this modification mechanism of the ribosomes can play some role in the shutting off of host translation?

Dr. S. B. Weiss: Of course, this is one of the first things we thought of, however, there are some problems here. If its going to shut off host synthesis, it has to do it very early in the game; it must do it within the first 2 to 3 minutes after infection. We have a very difficult time seeing this effect in the first 2 to 3 minutes after infection. However, this doesn't mean that the effect isn't there because we're using an *in vitro* assay which means that when we grind up cells and isolate the ribosomes we're getting a mixture of host ribosomes and any other ribosomes that might have been modified by the virus. If there were a new protein made, we would be reducing our chances of seeing it in our *in vitro* assay system. We really can't say at the present time whether this phenomenon is important in the host shut off mechanism.

Dr. F. C. Neidhardt, University of Michigan: I'm sympathetic with your inability to find a function for the new transfer RNAs that appear in the T4 infected cell. T4 also has a gene which apparently codes for a polypeptide factor which is made early after infection and which combines very tightly with the valyl tRNA synthetase. This apparently highly specific factor seems to be without function at least as one sees function in a laboratory experiment infecting *E. coli* cells and looking at the yield of viable phage particles which are produced because Dr. McLain working in our laboratory was able to isolate a phage mutant with an amber mutation an amber suppressible mutation in the phage gene such that there was no detectable production of this modifying factor and the valine activating enzyme of the host appeared to be completely unmodified and yet phage development proceeds normally with such a phage mutant.

Dr. S. B. Weiss: I think that there are, I know that there's one group that has a T4 phage mutant in viral tRNA and this may give some insight as to whether the tRNA is required for phage replication. The other thing is that although we look for tRNA function, I'm not all worried about it. I remember in our chairman's laboratory when hydroxymethylase was first found and there was no known function for this enzyme. Later on, it turned out to be very important in the production of viral DNA.

Dr. J. Schultz, Papanicolaou Cancer Research Institute: I have a very naive question, but maybe it will help clarify a point for others in the audience too. You introduced the idea of functional heteriogeneity of the ribosomes of the host and that rather than say that a virus infection results in new ribosomes being formed de novo you indicate a modification of existing ribosomes, can you explain why?

Dr. S. B. Weiss: Yes, because as far as we know, it was established a long time ago that in virus infection there is no new ribosome synthesis. There's no new synthesis of ribosomal RNA or ribosomal protein. In higher cell development, there are indications that during the developmental process there's sometimes a complete turnover of ribosomes. We feel there must be some important significance for cell development in this observation. We don't know whether it's tied up with translational control, but it certainly seems like a reasonable area to begin looking.

TRANSCRIPTION IN VITRO OF THE ESCHERICHIA COLI tRNA^{Tyr} GENE CARRIED BY THE TRANSDUCING BACTERIOPHAGE $\phi 80psu_3^+$

Uriel Z. Littauer, Violet Daniel, Jacques S. Beckmann and Sara Sarid
Department of Biochemistry
The Weizmann Institute of Science
Rehovot, Israel

Abstract: The Escherichia coli tRNA^{Tyr} gene carried by $\phi 80psu_3^+$ transducing phage was transcribed by E. coli RNA polymerase. The assay used to detect the formation of tRNA^{Tyr}-like polynucleotide chains was based on the specific ability of the product to compete with purified E. coli ^{32}P-tRNA^{Tyr} for hybridization with the light strand of $\phi 80psu_3^+$ DNA carrying the E. coli gene. Employing this procedure significant amounts of in vitro synthesized tRNA^{Tyr}-like chains were detected. When transcription was carried out with RNA polymerase lacking the σ factor, the percentage of tRNA^{Tyr}-like chains decreased significantly. The tRNA^{Tyr}-like chains synthesized in vitro possess a heterogenous distribution with respect to sedimentation coefficients, with a broad peak around 8 S. The presence of the termination factor, ρ, during transcription did not significantly reduce the average size of the in vitro synthesized tRNA^{Tyr}-like product.

INTRODUCTION

Transfer RNA molecules, like other species of RNA in the cell, are formed by the transcription of genetic information stored in DNA (1, 2). In spite of the extensive work that has been done to establish the nucleotide sequence and structure of the tRNA molecules, little is known about their synthesis, except that the modification leading to the appearance of the odd bases (such as methylation (3), thiolation (4) and formation of pseudouridylic acid (5, 6)) occur after the polynucleotide chain has been formed. Transfer-RNA synthesis, like that of all types of cellular RNA's is thought to be mediated by the DNA-dependent RNA polymerase. It is known that the fraction of the DNA genome which specifies the tRNA genes is quite low in all cells. Thus, 0.025% of the DNA of E. coli consists of the various tRNA genes (1, 2). If one considers only the gene of a single amino-acid specific tRNA, this fraction is decreased by a factor of 50 or more. Attempts to transcribe a single tRNA gene in vitro must, therefore, take into account the difficulty of detecting minute quantities of the particular tRNA synthesized in the presence of a large excess of other products (e.g., mRNA, rRNA and other specific tRNA molecules). There are two ways to overcome this problem: (1) to transcribe a fragment of DNA enriched for the specific tRNA gene (2) to transcribe the DNA of a transducing phage carrying the desired tRNA gene. We have chosen (7) to transcribe, in vitro, the E. coli tRNA^{Tyr} gene carried by $\emptyset 80psu_3^+$ transducing phage.

The su_3^+ gene carried by the $\emptyset 80$ phage is the structural gene which specifies a tRNA molecule (8, 9) that enables the amber codon, UAG, to be read as tyrosine (10, 11). There are two types of tyrosine tRNA's (I and II) in E. coli differing solely by two nucleotides in the variable loop. The main species in E. coli cells is tRNA^{Tyr} II; the location of its gene on the bacterial chromosome is unknown. The minor tRNA^{Tyr} I species is thought to be specified by two identical genes, one of which can undergo a mutation resulting in a single base change in the anticodon region of the tRNA (su_3^+). The two tRNA^{Tyr} I genes are located near the $\emptyset 80$ attachment site on the bacterial chromosome and can be transduced by $\emptyset 80$ bacteriophage (12). The $\emptyset 80$-derived transducing phages contain 1-2 (8, 9, 13) tRNA^{Tyr} genes per phage genome having a molecular weight of about 3×10^7 daltons. The DNA from these phage particles is approximately 100-fold enriched for the tRNA^{Tyr} gene as compared with the DNA from E. coli cells, which has a molecular weight of about 3×10^9 daltons.

RESULTS AND DISCUSSION

The non defective $\emptyset 80psu_3^+$ phage obtained from Dr. H. Ozeki (8) has been used as a source for the tRNA^{Tyr} gene. DNA isolated from this phage does not carry any other tRNA genes besides the tRNA^{Tyr} gene. This is demonstrated in the annealing experiments presented in Table I.

TABLE I
Specific hybridization of $\emptyset 80psu_3^+$ DNA with ^{32}P-tRNA^{Tyr}

Samples of tRNA in hybridization mixture	tRNA hybridized to $\emptyset 80psu_3^+$ DNA
	cpm
^{32}P-tRNA	620
^{32}P-tRNA^{Tyr}	600
^{32}P-tRNA(devoid of tRNA^{Tyr})	40

(Table I cont.)

Filters containing 5 μg Ø80psu$_3^+$ DNA were incubated for 16 hr at 68° with 0.03 μg of the ^{32}P-labeled tRNA samples (3.2x10^5 cpm/μg) in a volume of 0.4 ml 2xSSC. The filters were then removed, washed, digested with pancreatic RNase (20 μg RNase/ml in 2xSSC) for 60 minutes at room temperature, again washed and counted. ^{32}P-tRNA was prepared from E.coli B cells grown on Tris-glucose medium as described previously (14). ^{32}P-tRNA^{Tyr} was separated from other species of tRNA by acylation with tyrosine and fractionation on a benzoylated DEAE-cellulose column as described by Maxwell et al (15). The mixture of the other tRNA species was collected and used as ^{32}P-tRNA (devoid of tRNA^{Tyr}).

These experiments show the specific hybridization of purified ^{32}P-tRNA^{Tyr} to Ø80psu$_3^+$ DNA. A mixture of unfractionated ^{32}P-tRNA lacking tRNA^{Tyr} did not hybridize with this DNA.

We then determined which of the two DNA strands transcribes the tRNA^{Tyr} chains. The complementary strands of Ø80psu$_3^+$ DNA were separated according to the procedure of Hradecna and Szybalski (16). Alkali-denatured DNA was interacted with poly U,G and DNA strands separated by equilibrium centrifugation in a CsCl density gradient. Figure 1a shows the separation of the two complementary strands relative to the native DNA in an analytical CsCl gradient equilibrium centrifugation and Figure 1b shows their separation in the preparative ultracentrifuge. The main fractions of each band in figure 1b representing the separated heavy and light strands, were pooled and further purified through a second centrifugation to equilibrium in CsCl. The separated light and heavy strands of Ø80psu$_3^+$ DNA were then hybridized with E. coli ^{32}P-tRNA in order to identify the transcribed strand as well as to determine the purity of the separated strands. Table 2 shows that two successive equilibrium CsCl gradient centrifugations were sufficient to separate the light from the heavy strand. It is also evident that only the light strand (which binds a lesser amount of guanine-rich ribopolymers) hybridizes with tRNA. Similar results were obtained with a defective Ø80dsu$_3^+$ phage (17).

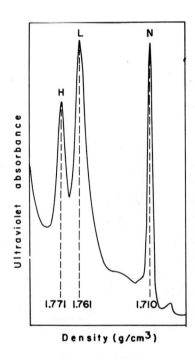

Fig. 1 Separation of the complementary DNA strands of phage $\emptyset 80psu_3^+$ by CsCl density gradient centrifugation.

Alkali denatured DNA was interacted with poly U, G, and the solution was adjusted with CsCl to a density of 1.75 g/ml.

(a) Analytical CsCl gradient equilibrium centrifugation at 44,700 rev/min for 20 hrs at 24°C. Native $\emptyset 80psu_3^+$ DNA was added as density marker (N).

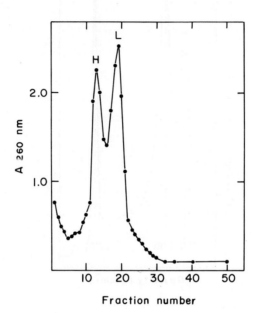

(b) Preparative CsCl gradient equilibrium centrifugation at 40,000 rpm for 48 hrs at 5°C in a type 50 fixed-angle rotor of a Spinco Model L ultracentrifuge (7).

Conditions for transcription of $\emptyset 80 psu_3^+$ phage DNA

RNA polymerase (containing σ factor) was purified from E. coli cells as described by Chamberlin and Berg (18). Transcription of $\overline{\emptyset 80 psu_3^+}$ DNA did not require the presence of KCl (Figure 2). This is in contrast to the transcription of T_4 DNA where both the rate and extent of the reaction are stimulated by 0.2 M KCl (Figure 3).

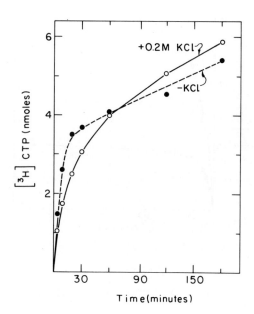

Fig. 2. The effect of KCl concentration on transcription of $\emptyset 80psu_3^+$ DNA.

Reaction mixtures of 0.1 ml contained 0.05 M Tris-HCl buffer, pH 7.9; 0.01 M $MgCl_2$; 0.5 mM ATP, GTP, UTP and 0.4 mM ^{3}H-CTP(7.1×10^5 cpm/µmole); 1 mM 2-mercaptoethanol, 5 µg $\emptyset 80psu_3^+$ DNA and 6 units of E. coli MRE 600 RNA polymerase (prepared by the method of Chamberlin and Berg (18)). Reaction mixtures were incubated at 37° and RNA synthesis was terminated at the specified times by cooling to 0° and the addition of 50 µg BSA and 5% TCA. The precipitates were collected on Millipore filters, washed with 5% TCA and counted in a liquid scintillation counter.

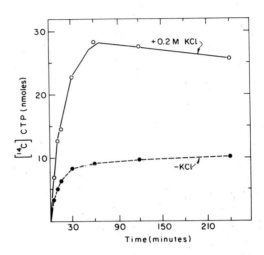

Fig. 3. The effect of KCl concentration on transcription of T4 DNA. Reaction mixture of 0.1 ml contained 0.05 M Tris-HCl buffer, pH 7.9; 0.01 M MgCl$_2$; 1 mM ATP, GTP, UTP and 0.45 mM ^{14}C-CTP (8.99x10^5 cpm/μmole); 1 mM 2-mercaptoethanol, 9 μg T4 DNA and 28 units E. coli MRE 600 RNA polymerase (prepared by the method of Chamberlin and Berg (18)). Reaction mixtures were incubated at 37° and RNA synthesis was terminated as in Fig. 2.

Competition between RNA transcribed in vitro on ∅80psu$_3^+$ DNA and E. coli ^{32}P-tRNA

The double-stranded DNA preparation from phages ∅80psu$_3^+$ and ∅80 were transcribed in vitro by DNA-dependent RNA polymerase from E. coli cells. The reaction mixture was treated with DNase and the RNA which was synthesized was

TABLE 2

Hybridization of E. coli ^{32}P-tRNA with the separated strands of $\emptyset 80psu_3^+$ DNA

Purification of strands on CsCl density gradient		Hybridization with the separated strands	
		L strand	H strand
		cpm	
First centrifugation	Expt. 1	2137	250
	Expt. 2	1967	385
Second centrifugation	Expt. 1	1850	0
	Expt. 2	1300	0

2 µg of DNA from the L or H strand of $\emptyset 80psu_3^+$ phage were hybridized with 0.9 µg of ^{32}P-tRNA (2x10^6 cpm/µg) in 0.3 ml of 2xSSC for 2 hr at 68°. The hybrids were loaded on filters, washed with 50 ml 2xSSC, digested with pancreatic RNase (25 µg/ml in 2xSSC), washed again and counted (7).

---

then extracted with 0.25% SDS and 70% phenol. The presence of tRNA^{Tyr}-like chains in the in vitro-synthesized-RNA was identified by their ability to compete with E. coli ^{32}P-tRNA for hybridization sites on the $\emptyset 80psu_3^+$ DNA. An annealing experiment was carried out using constant amounts of the light strand of $\emptyset 80psu_3^+$ DNA and E. coli ^{32}P-tRNA in the presence of increasing amounts of in vitro synthesized $\emptyset 80psu_3^+$ or $\emptyset 80$ RNA (Figure 4). The results clearly show that the $\emptyset 80psu_3^+$ RNA synthesized in vitro contains polynucleotide chains homologous to those found in the in vivo transfer RNA molecules. The competition is specific since RNA transcribed on $\emptyset 80$ phage does not decrease the hybridization with E. coli ^{32}P-tRNA. Similar results were obtained by competing the $\emptyset 80psu_3^+$ RNA synthesized in vitro with purified tyrosine-specific E. coli ^{32}P-tRNA (Figure 5). This demonstrates the homology of the RNA polymerase product to tRNA^{Tyr}.

The effects of σ and ρ factors on the transcription of $\emptyset 80psu_3^+$ DNA

Transcription of RNA requires that both initiation and termination of the reaction take place at their proper sites. RNA polymerase is composed of

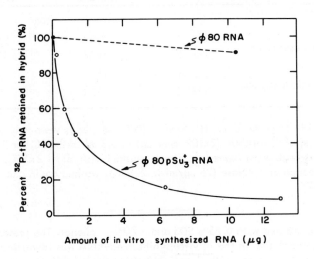

Fig. 4. Competition of RNA transcribed in vitro on ∅80psu$_3^+$ and ∅80 DNA's with E. coli ^{32}P-tRNA.

The hybridization mixture contained 2 μg of the light strand of ∅80psu$_3^+$ DNA, 0.9 μg of ^{32}P-tRNA (1.6 × 10^6 cpm/μg), and different amounts of in vitro synthesized ∅80psu$_3^+$ or ∅80 RNA in a total volume of 0.3 ml 2 × SSC. Mixtures were incubated for 2 hr at 68°C, loaded on filters, washed with 50 ml of 2 × SSC on each side, treated with pancreatic RNase (25 μg/ml in 2 × SSC), washed again and counted. The radioactivity remaining on the filter was expressed as percentage of control which contained no competing RNA (7).

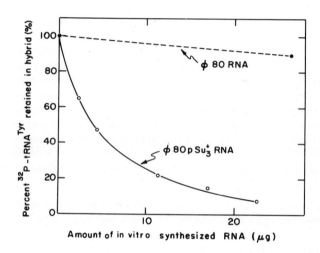

Fig. 5. Competition of RNA transcribed in vitro on ∅80psu$_3^+$ and ∅80 DNA's with purified tyrosine-specific E. coli ^{32}P-tRNA.

Filters containing 5 µg of alkali-denatured ∅80 psu$_3^+$ DNA were incubated with 0.017 µg E. coli ^{32}P-tRNA^{Tyr} (3.2 × 10^5 cpm/µg) in a total volume of 0.3 ml 2 × SSC at 68°C for 10 hr. Different amounts of unlabeled RNA, transcribed in vitro on ∅80psu$_3^+$ or ∅80 DNA's were added as specified. At the end of the hybridization reaction the filters were removed, washed with 2 × SSC, treated with pancreatic RNase (25 µg/ml in 2 × SSC), washed again and counted. The radioactivity remaining on the filters was expressed as percentage of control which contained no competing RNA (7).

several subunits: α_2, β, β' and σ. After chromatography on phosphocellulose, the σ subunit can be separated from the core enzyme (α_2, β, β'). Although the core enzyme is able to transcribe calf thymus DNA almost normally, it is much less active when assayed with T4 DNA as the template. Addition of the σ factor (which by itself has no catalytic activity) to the core enzyme restores its ability to transcribe T4 DNA. It is believed that the σ factor is required for initiation of RNA synthesis at the correct sites on the DNA template. Without the σ factor, RNA polymerase will initiate polymerization along the DNA molecule in a random fashion (19). A similar requirement for the σ factor in the transcription of $\emptyset 80 psu_3^+$ DNA has been found in our work. $\emptyset 80 psu_3^+$ DNA, is a poor template for the core enzyme lacking σ (not shown). It should be noted that centrifugation of RNA polymerase through a glycerol gradient followed by phosphocellulose chromatography not only separates the core enzyme from the σ factor but further purifies the enzyme preparation. Thus, RNA synthesis proceeds for more than 10 hours when the purified core enzyme & factor σ are used in the reaction mixture (Figure 6). The transcription product produced in the presence of σ shows, however, no difference in size distribution upon sedimentation in sucrose gradients. Figure 7 shows the pattern for the product with the complete enzyme. An identical pattern (not shown) was obtained when only the core enzyme was used.

An additional protein factor involved in the transcription reaction, has been recently reported by Roberts (22). This factor, ρ, enables the in vitro transcription of λ DNA to terminate at specific sites and produces RNA's of about the same size as that transcribed in vivo. As shown in Figure 6 both the rate and extent of RNA synthesis on $\emptyset 80 psu_3^+$ DNA are markedly depressed when the ρ factor was added to the reaction mixture. Furthermore, the average size distribution of the total RNA synthesized was considerably reduced, particularly with respect to molecules having sedimentation constants of 16 S or higher (Figure 7).

The size distribution of the tRNA^{Tyr} transcribed in vitro

RNA was synthesized using core enzyme alone, core enzyme with σ factor and core enzyme with σ and ρ factors in order to examine the effects of σ and ρ factors on the size distribution of tRNA^{Tyr}-like chains transcribed on $\emptyset 80 psu_3^+$ DNA. The in vitro synthesized ^{14}C-labeled $\emptyset 80 psu_3^+$ RNA preparations were fractionated by centrifugation through sucrose gradients. Each gradient was divided into four regions containing RNA of different sizes (in a fashion similar to that indicated in Figure 7). The location of the in-vitro-synthesized-tRNA^{Tyr}-chains on the gradient was then identified by their ability to compete with E. coli ^{32}P-tRNA for hybridization to the light strand $\emptyset 80 psu_3^+$ DNA. Use of the purified light DNA strand in the annealing reaction enabled us

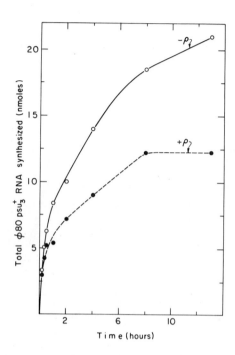

Fig. 6. The effect of factor ρ on the time course of RNA synthesis.

The reaction mixture, in a volume of 0.5 ml, contained: 0.05 M Tris-HCl buffer pH 7.9; 0.01 M $MgCl_2$; 1 mM DTT; 0.1 M KCl; 1 mM ATP, UTP, GTP and 0.3 mM ^{14}C-CTP (3.8 × 10^6 cpm/μmole); 25 μg $\phi 80 psu_3^+$ DNA; 120 μg core RNA polymerase; 9 μg σ factor and 12.5 μg ρ factor where indicated. The reaction mixture was incubated at 37° and samples of 50 μl removed at the specified times. Fifty μg of BSA was then added and the RNA precipitated by cold 5% TCA, was collected on Millipore filters, washed and counted in a liquid scintillation counter. RNA polymerase (prepared according to Chamberlin and Berg (18)) was subjected to low salt glycerol gradient centrifugation (20) followed by phosphocellulose chromatography (21).

---

to perform hybridization in liquid (23), thus, eliminating any danger of DNA-DNA annealing. The "liquid" procedure was also less time consuming than hybridization to DNA fixed on filters (24).

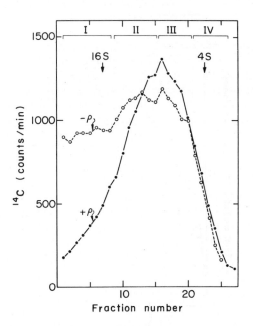

Fig. 7. Sucrose density gradient analysis of RNA synthesized in vitro.

One hundred μg $\emptyset 80 psu_3^{\pm}$ DNA were transcribed in vitro by RNA polymerase in the presence or absence of the ρ termination factor in the reaction mixture. Each preparation of synthesized RNA was dissolved in 0.1 ml of 1 x SSC, layered onto a 5 ml of 5-20% sucrose gradient in 1 x SSC, and centrifuged for 4 hr. at 50,000 rpm and 4°C in SW 50.1 rotor of Spinco model L centrifuge. Fractions of six drops each were collected. Sedimentation constants were estimated by centrifuging 16 S E. coli ribosomal RNA and 4 S E. coli tRNA in a separate sucrose gradient in the same rotor. Aliquots of 10 μl of each fraction were precipitated with trichloroacetic acid and passed through nitrocellulose filters. Distribution of the ^{14}C-labeled-synthesized RNA was determined by counting the filters in a scintillation spectrophotometer. For the competition-hybridization studies the individual fractions were combined into four samples (I-IV) representing different S regions of the sucrose gradient as indicated by the brackets (7).

Figure 8 shows that the tRNA^{Tyr}-like chains synthesized with an enzyme lacking σ have a wide size distribution since all four fractions show competing activity of comparable magnitude (with an indication of a broad peak around fraction II). The size of the tRNA^{Tyr}-like chains, synthesized in the presence of σ is somewhat lowered and the maximal amount is now found in fraction III. Figure 8 also shows that the relative amount of tRNA^{Tyr}-like chains in the total RNA transcribed on Ø80psu$_3^+$ DNA, is markedly increased when synthesis is conducted in the presence of σ (i.e.less RNA is required to achieve the same level of competition). Thus σ factor is required for correct initiation of tRNA^{Tyr} synthesis.

It should be noted that even in the presence of σ the synthesized tRNA^{Tyr}-like chains are heterogenous in their size distribution and that most of them have a sedimentation constant which is much larger than 4 S. In fact, the 4 S region contains the smallest amount of tRNA^{Tyr}-like molecules while the maximal amount is found in the 7-8 S regions. In an attempt to reduce the size of the in vitro synthesized tRNA^{Tyr}, the transcription of Ø80 psu$_3^+$ DNA was conducted in the presence of the termination factor (ρ). Under these conditions (as shown above), the average size distribution of the total RNA synthesized was considerably reduced (Fig. 7). However, when the same sucrose gradient fractions were assayed for their competing ability, it was found that the tRNA^{Tyr}-like chains were still concentrated in those fractions corresponding to sedimentation values of 7-8 S (fig. 8). From the amounts of in vitro-synthesized-Ø80psu$_3^+$ RNA showing 50% competition with a given amount of E. coli ^{32}P-tRNA (assuming that tyrosine-specific tRNA represents 5% of the whole tRNA mixture) the relative amounts of tRNA^{Tyr} in the in vitro synthesized RNA may be roughly calculated. It is found that tRNA^{Tyr}-like polynucleotide chains amount to 3.6-13.5% of the total RNA synthesized.

From the above results it can be concluded that the transcription system of RNA polymerase still lacks a factor necessary for the synthesis of a 4 S molecule. Under the conditions of our assay, ρ factor does not seem to fulfill this function. Whether the control of the size of the tRNA would be mediated by ψ_r factor (25), CAP factor (26) or requires a new factor remains to be determined. An alternative (or additional) hypothesis would be to postulate the existence of a specific nuclease which cleaves large RNA molecules to a smaller size. In fact, precursor tRNA molecules, which seem to be somewhat longer than the normal tRNA molecules, have been observed in mammalian (27-32) and possibly in E. coli cells (33).

It is hoped that further studies of the in vitro transcription of a specific tRNA gene by RNA polymerase may elucidate the basic molecular mechanisms involved in the formation of a tRNA polynucleotide and its transformation into a biologically active tRNA molecule. One can envisage many

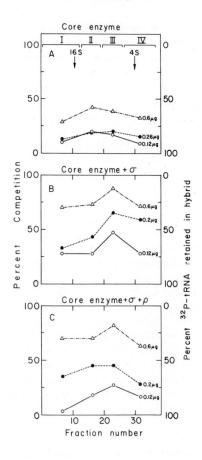

Fig. 8. Competition of in vitro-synthesized-RNA from different regions of a sucrose gradient with E. coli ^{32}P-tRNA for hybridization to the light strand of Ø80psu$_3^+$ DNA.

Fractions from each sucrose gradient were combined into four samples (I-IV) in a fashion similar to that described in the legend to Fig. 7. Equal amounts from each of the four fractions were allowed to compete with E. coli ^{32}P-tRNA for hybridization to the light strand of Ø80psu$_3^+$ DNA. The hybridization mixture contained 2 μg of the light strand of Ø80psu$_3^+$ DNA, 0.27 μg of ^{32}P-tRNA (1.1 x 10^6 cpm/μg) and 0.12 μg, 0.2 μg or 0.6 μg of one of the four (I-IV) samples of in vitro-synthesized-RNA. Five μl of phenol were added to the mixture containing 2 x SSC in a final volume of 0.3 ml. Hybridization was performed at 68°C for 2 hr. The hybrids were then loaded on filters, washed by 50 ml of 2 x SSC, treated with pancreatic RNase (25

(Fig. 8 cont.)
μg/ml in 2 x SSC), washed again and counted in a scintillation spectrophotometer. The radioactivity remaining on the filters from annealing reactions carried out in the absence of competing RNA was taken as the 0 percent competition level.
A - The competing $\emptyset 80psu_3^+$ RNA was synthesized by core RNA polymerase
B - The competing $\emptyset 80psu_3^+$ RNA was synthesized by core RNA polymerase + sigma (σ) factor.
C - The competing $\emptyset 80psu_3^+$ RNA was synthesized by core RNA polymerase + sigma (σ) factor in the presence of termination factor (ρ).

topics for further studies; we shall consider only a few here. It should now be possible to determine whether the 3'-terminal CpCpA sequence of tRNA is transcribed by RNA polymerase or inserted by the enzyme, tRNA..pCpCpA pyrophosphorylase (34). If the latter possibility is correct, it would be interesting to examine whether the CpCpA sequence is added while the tRNA is still attached to the DNA or only after it has been freed from the template. The same question may be asked regarding the various tRNA-modifying enzymes. Will the nucleoside modification take place while transcription is still in progress? Do these enzymes act independently of each other or in concert? Is the putative precursor RNA the natural substrate for the modifying enzyme or is its conversion to tRNA a prior first requirement? Does the conversion involve methylation and what is the recognition site for the cleaving nuclease? Are the extra nucleotides located at the 3'- or 5'-termini or on both ends of the chain? Finally, it would be of importance to learn whether the extra nucleotides on the precursor tRNA molecules are transcribed from the promoter site; if so, base sequence studies could be attempted.

REFERENCES

1. D. Giacomoni and S. Spiegelman, Science, 138 (1962) 1328.
2. H.M. Goodman and A. Rich, Proc. Natl. Acad. Sci. U.S., 48 (1962) 2101.
3. E. Fleissner and E. Borek, Proc. Natl. Acad. Sci. U.S., 48 (1962) 1199.
4. R.S. Hayward and S.B. Weiss, Proc. Natl. Acad. Sci. U.S., 55 (1966) 1161.
5. S.B. Weiss and J. Legault-Demare, Science, 149 (1965) 429.
6. R.L. Heinrikson and E. Goldwasser, J. Biol. Chem., 239 (1964) 1177.
7. V. Daniel, S. Sarid, J.S. Beckmann and U.Z. Littauer, Proc. Natl. Acad. Sci. U.S., 66 (1970) 1260.
8. T. Andoh and H. Ozeki, Proc. Natl. Acad. Sci. U.S., 59 (1968) 792.
9. A. Landy, J. Abelson, H.M. Goodman and J.D. Smith, J. Mol. Biol., 29 (1967) 457.
10. S. Kaplan, A.O.W. Stretton and S. Brenner, J. Mol. Biol., 14 (1965) 528.

11. M.G. Weigert, E. Lanka and A. Garen, J. Mol. Biol., 14 (1965) 522.
12. H.M. Goodman, J.N. Abelson, A. Landy, S. Zadrazil and J.D. Smith, Eur. J. Biochem., 13 (1970) 461.
13. R.L. Russell, J.N. Abelson, A. Landy, M.L. Gefter, S. Brenner and J.D. Smith, J. Mol. Biol., 47 (1970) 1.
14. V. Daniel, S. Lavi and U.Z. Littauer, Biochim. Biophys. Acta, 182 (1969) 76.
15. I.H. Maxwell, E. Wimmer and G.M. Tener, Biochemistry, 7 (1968) 2629.
16. Z. Hradecna and W. Szybalski, Virology, 32 (1967) 633.
17. H.A. Lozeron, W. Szybalski, A. Landy, J. Abelson and J.D. Smith, J. Mol. Biol., 39 (1969) 239.
18. M. Chamberlin and P. Berg, Proc. Natl. Acad. Sci. U.S., 48 (1962) 81.
19. E.K.F. Bautz, F.A. Bautz and J.J. Dunn, Nature, 223 (1969) 1022.
20. R.R. Burgess, J. Biol. Chem., 244 (1969) 6160.
21. D. Berg, K. Barett and M. Chamberlin, in: Methods in Enzymology, eds. L. Grossman and K. Moldave (Academic Press Inc., 1971) in press.
22. J.W. Roberts, Nature, 224 (1969) 1168.
23. A.P. Nygaard and B.D. Hall, Biochem. Biophys. Res. Commun., 12 (1963) 98.
24. D. Gillespie and S. Spiegelman, J. Mol. Biol., 12 (1965) 829.
25. A.A. Travers, R.I. Kamen and R.F. Schleif, Nature, 228 (1970) 748.
26. G. Zubay, D. Schwartz and J. Beckwith, Proc. Natl. Acad. Sci. U.S., 66 (1970) 104.
27. B.M. Lal and R.H. Burdon, Nature, 213 (1967) 1134.
28. R.H. Burdon, B.T. Martin and B.M. Lal, J. Mol. Biol., 28 (1967) 357.
29. R.H. Burdon and A.E. Clason, J. Mol. Biol., 39 (1969) 113.
30. D. Bernhardt and J.E. Darnell, Jr., J. Mol. Biol., 42 (1969) 43.
31. D. Bernhardt Mowshowitz, J. Mol. Biol., 50 (1970) 143.
32. E.J. Smillie and R.H. Burdon, Biochim. Biophys. Acta, 213 (1970) 248.
33. B. Pace, R.L. Peterson and N.R. Pace, Proc. Natl. Acad. Sci. U.S., 65 (1970) 1097.
34. V. Daniel, S. Sarid and U.Z. Littauer, Science, 167 (1970) 1682.

Acknowledgement: We thank Mr. Y. Tichauer for his valuable assistance. This work was supported in part by USPHS research grant RO5-TW-00260 and by Agreement 45514.

DISCUSSION

Dr. S. S. Cohen, University of Pennsylvania: Perhaps I can ask Dr. Littauer whether he has observed a conversion of his larger tRNA fraction to a 4S component on an addition of an appropriate extract.

Dr. U. Z. Littauer: Well, it's obvious that I have listed so many possibilities that we're attempting to do some of them but we're still attempting.

Dr. S. B. Weiss, University of Chicago: Have you done the converse? Have you taken RNA transcribed from $\phi80$ DNA not containing the su_3^+ gene and will it compete with your isolated fractions for hybridization sites on $\phi80psu_3^+$ DNA?

Dr. Littauer: Do you mean taking a tRNA which is devoid of tyrosine and trying to hybridize? It does not.

Dr. Weiss: This suggests that the RNA that you're transcribing is only tyrosyl tRNA?

Dr. Littauer: No, the RNA we transcribe contains a large excess of messenger RNA, there is no question about that. However, the control we are using here is the RNA transcribed on $\phi80$ phage DNA which does not contain the tRNA^{Tyr} gene. The product of the $\phi80$ DNA transcription does not compete with our annealing mixture, as I've shown in two slides.

Dr. G. D. Novelli, Oak Ridge National Laboratory: Dr. Littauer, do you have any notion of the nucleotides at either end, the five prime or three prime?

Dr. Littauer: No, we don't have yet any notion of that.

POLYAMINES AND THE MULTIPLICATION OF T-EVEN PHAGES*

Seymour S. Cohen and Arnold Dion
Department of Therapeutic Research
University of Pennsylvania School of Medicine
Philadelphia

Abstract: T-even phages contain large amounts of polyamines, whose biosynthesis for considerable periods parallels that of phage DNA. At the present time this phage system is unique for the study of the interrelations between the production of DNA and these organic cations. As previously demonstrated in the study of the biosynthesis of the polyamines and RNA in uninfected cells, synthesis of putrescine and spermidine in bacteria infected by normal T4 or by ultraviolet-irradiated or various DO amber mutant stocks is essentially independent of DNA synthesis. On the other hand in a putrescine- or spermidine-requiring mutant of E. coli the initiation and rate of DNA synthesis are markedly stimulated by addition of either polyamine.

INTRODUCTION

In the past decade, it has become evident that the polyamines, putrescine and spermidine, are closely tied to the synthesis and accumulation of RNA in bacteria and animal cells (1). These substances account for about 1/5 to 1/4 of the cation necessary to neutralize the RNA of these cells. Not only are these polyamines present in the RNA-containing structures, e.g. ribosomes, in these cells but can facilitate the organization of these organelles (2). They have also been found on various isolated RNA fractions, such as the tRNA of bacteria (3) and of mouse fibroblasts (4) under conditions suggesting that spermidine, for example, is significant in organizing the structure of the particular RNA (3). Spermidine has also been found in many plant viruses, and in several such systems, e.g. cowpea chlorotic mottle virus, the polyamines can replace Mg^{++} in the reaggregation of disaggregated nucleoprotein to active, apparently normal virus particles (5).

Furthermore spermidine has been found to participate in many reactions of transcription and translation at concentrations found in the entire cell (1). In the first group of reactions it has been shown that spermidine can participate in strand selection,

---

* This investigation has been supported by a USPHS grant 7005 from the National Institute of Allergy and Infectious Diseases.

initiation of strand transcription and in extension of RNA chains. The latter effects appear to relate to the ability of spermidine to facilitate the release of nascent RNA chains from the complex of template DNA, RNA polymerase and newly synthesized RNA. Among reactions of translation it has been found that spermidine assists association of ribosomal subunits, formation of an active configuration of tRNA, aminoacid transfer from an activating enzyme-aminoacid-adenylate complex to tRNA and attachment of the acylated tRNA to the ribosome in the presence of message. It appears also that spermidine stimulates the methylation of tRNA. In many of these activities spermidine is significantly more effective than is Mg^{++}.

It is not clear to what extent spermidine fulfills these roles within the cells. A teleonomist would argue that compounds present in cells in such large amounts and capable of fulfilling the activities indicated above undoubtedly have been conserved to fulfill significant functions. Our problem at present is not that of convincing anyone that the compounds are important but of finding out precisely how they work in cells, as well as in the reactions studied in vitro.

As noted earlier, a possible mechanism is known for the organization of helical structure of nucleic acids by spermidine and spermine. This was first proposed by Tsuboi (6) as a result of studies of DNA and its interaction with polyamines. It was suggested that both primary and secondary amino groups of spermine for example interact with 4 DNA phosphates, with the secondary amino groups making a bridge across the shallow groove in the double helix, while the primary amino groups were bound to the adjacent nucleoside phosphate along a single polynucleotide chain. Tsuboi supposed that, in forming an oblique bridge across the shallow groove between the two polynucleotide helices, the polyamine may function as a clamp for binding two chains together, i.e. to stabilize an ordered double-stranded structure in solution. Essentially this model has been adopted by Liquori (7) who has presented some X-ray evidence and additional binding data to indicate that spermine and spermidine can fit in DNA as proposed (Fig. 1). Suwalsky et al (8) have interpreted their recent X-ray study of a spermine-DNA complex to suggest that the polyamine can stabilize the naturally occurring B form of DNA in solution. The data on their complex were best interpreted as a combination of binding of spermine across the small DNA groove in one molecule plus intermolecular crosslinks from DNA to DNA.

Although the Tsuboi-Liquori model was developed on DNA-polyamine complexes, we have adopted it as a working hypothesis to account for the association of the polyamines with RNA. The organization of polyribonucleotides by polyamines is structurally and functionally demonstrable (1) and it seems possible that the

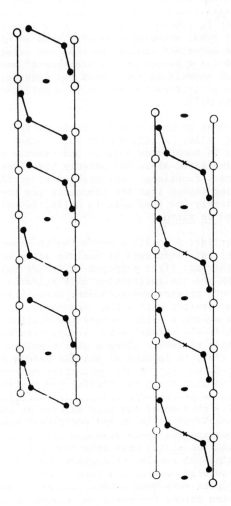

Figure 1 Models showing spermidine and spermine molecules in the narrow groove of DNA (7).

presence of spermidine in RNA can help to organize relatively unstable sequences of complementary base pairs such as exist in tRNA, rRNA and even in some polycistronic RNA into more stable double stranded regions which are not only more resistant to enzymatic degradation but also thereby rendered non-translatable in these regions. For these reasons the physiology of the polyamines might be expected to relate to the metabolism and utilization of RNA more significantly than to that of DNA, whose double-stranded structure would seem to be sufficiently stable not to need polyamines.

In these terms therefore we note somewhat wryly that we have explored the polyamines in connection with the multiplication of a DNA phage and that no one has yet studied these substances in relation to multiplication of RNA phages. Nevertheless the physiology of DNA, i.e. replication, recombination, repair, and even many aspects of transcription are functions of the development of single strandedness. The organization of the essential qualities of helicity, rigidity and suppleness in DNA necessary to restrict or to facilitate these physiological events must itself have sets of physiological controls, among which the polyamines may participate.

Histones in eucaryotic cells and protamines in gametes must fulfill two as yet mysterious roles, presumably of these types, in the physiology of eucaryotic chromosomes. Although polyamines are known to be present in nuclei (with little if any in sperm) their concentration does not appear to relate specifically to the presence of DNA. Bacteria have not been shown as yet to have evolved histones or protamines of comparable functions, nor is it known that bacterial DNA is associated with, or neutralized by, polyamines.

Michaels and Tchen (9) have reported that the polyamine content of nucleated and enucleated ("minicell") _E. coli_ were approximately the same when calculated on the basis of RNA content. Since bacteria may have a great excess of RNA over DNA and the analysis of polyamine is tricky on occasion, these data do not exclude the possible presence of spermidine on bacterial DNA. On the other hand, Johnson and Bach (10) have reported that radioactive spermine taken up in a non-exchangeable form by spheroplasts of _E. coli_ is associated with both the DNA and RNA isolable from the cells. Nevertheless spermine is not synthesized by _E. coli_ and the experiment of Johnson and Bach does not prove the presence of normal polyamines on the DNA of untreated growing cells. Indeed no one has attempted to isolate bacterial DNA under conditions which might permit the detection of bound spermidine. Even if this were done, i.e. if bound polyamine were found on isolated DNA, it would have to be proven that the bound polyamine did not associate with the DNA after cell disruption.

Recently Inouye and Pardee (11) have described a marked effect of the polyamines on the course of bacterial division and have suggested that the molar ratio of putrescine to spermidine is a significant control on this event. It is not known if this effect relates in any way to the control of DNA replication. On the other hand, polyamines are reported to be antimutagenic (12, 13) and to affect zygotic induction of lysogens (14), events suggested to relate more directly to DNA structure and metabolism.

POLYAMINES AND PHAGE

Following this rather thin evidence relating polyamines to DNA physiology, we turn to somewhat firmer data which tie polyamines to the T-even phages. In 1957 Hershey (15) discovered the existence of two ninhydrin-positive non-amino acid components in the T-even phages; he was unaware of the chemical identity of his substances. Nevertheless, he developed a sound chromatographic method for their isolation, demonstrated that arginine was a precursor, and showed not only that they were injected into the host with virus DNA, but that similar substances were present in the uninfected host. Hershey recognized that their origin in arginine probably conferred basic properties to the substances and thought that they might neutralize charges on phage DNA. Ames and Dubin (16, 17) identified the substances in phage as putrescine and spermidine. They analysed T4 phage and in accounting for almost 80 to 90% of the total cation of the virus showed that the polyamines accounted for about 40-50% of the identifiable cation, confirming Hershey's hypothesis.

The polyamines in phage were transmitted unchanged from parent to offspring (15); in this transmission the substances were presumably associated with virus DNA. However a mixture of the host-derived polyamines added to the medium competed with the transmission of this material from parent to progeny. Noting also their separability from DNA and protein during dialysis of disrupted phage against buffer, Hershey concluded that these substances were probably neither genetically nor structurally important constituents of the phage DNA.

Arginine, added to cultures of infected bacteria, is converted to polyamines, subsequently isolable from the phage (15). Exogenous ^{14}C-putrescine is converted to ^{14}C-spermidine isolable from phage with only moderate dilution of the latter (16). We have shown that a considerable net synthesis of putrescine and spermidine does occur during infection (18). Immediately after infection small amounts of the compounds leak from the cell. Concomitant with DNA synthesis, the substances are taken back into the cell, and a net increment occurs thereafter. We have reasoned that since these substances are largely made after infection and packaged into phage, their synthesis is probably important in some measure to phage multiplication.

Ames and Dubin (17) showed that almost all of the polyamines were displaced from the permeable 0 mutants of T4 by washing with buffer containing 10 mM Mg^{++}. Although this experiment is often described as suggesting that polyamine can be largely displaced without loss of phage activity, the low infectivity of the control stock calls for additional data to prove this point unequivocally. In any case, it should be remembered that polyamine-free DNA is injected into a bacterium rich in polyamine, i.e. containing 1-2 mM spermidine and 10-20 mM putrescine. It may be mentioned that T4 can be grown in E. coli in minimal media containing spermine, and the spermidine of this phage will be almost totally replaced by spermine (17).

That the divalent cations or polyamines are essential to phage biosynthesis can be inferred from several types of experiments. For example, the mutant strain T4rII is unable to multiply in E. coli K12 (λ) although it can multiply in E. coli B. Garen showed that high Mg^{++} permits multiplication of T4rII in E. coli K12 (λ) when this cation is added at the 8th minute (19). Ferroluzzi-Ames and Ames (20) have shown that spermidine alone will restore phage synthesis. Since infection leads to leakage of putrescine, they used the intrabacterial concentration of both putrescine and spermidine for exogenous addition and almost completely restored virus multiplication. It is thought by some workers (21) that these substances preserve a weakened membrane rather than facilitate DNA synthesis directly.

We have examined the course of DNA synthesis in bacteria infected by the phage mutant, am N116 (22). Infection by am N116 is characterized by a marked delay in DNA synthesis in the presence of low concentration of Mg^{++}. It was suggested that this mutant is unable to direct the development of an essential Mg^{++} pump in EDTA-treated cells. As shown in Figure 2, spermidine improves initiation of DNA synthesis in the restrictive cell. Putrescine is ineffective in this activity.

We have analysed the polyamines in these systems by the sensitive dansyl method (23) applied to small aliquots (0.2 ml) of acidified cultures or extracts of filtered cells, infected at 2-3 X 10^8 per ml. Free polyamines have been analysed directly; total polyamines or media have been determined after hydrolysis in 6 N HCl at 110° for 18 hours. Unlike the rII infection, in which putrescine is lost from the cell, this diamine is not lost during infection of the "restrictive" cell, strain B by am N116. This virus permits an accumulation of free spermidine to slightly greater than normal levels (Fig. 3). Actually cells infected by the mutant appear to make more putrescine than in infection by the wild type, T4D, although most of this appears in the medium,

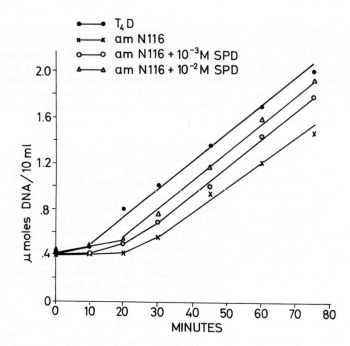

Figure 2 The effect of spermidine on DNA synthesis in E. coli B infected by T4r+ am N116. Exponentially growing cultures of E. coli B at 2×10^8/ml were infected with T4D or am N116 at an m.o.i. of 5. At the indicated times 10 ml aliquots were removed from each culture and analysed for DNA. Spermidine, where indicated, was added 5 minutes after infection.

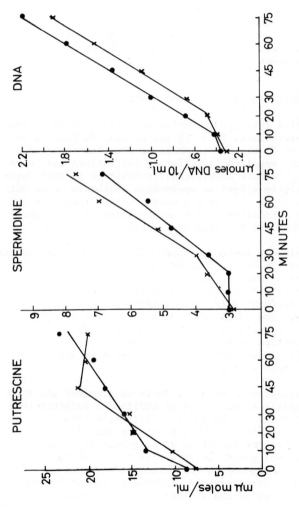

Figure 3 Intracellular free polyamine levels and DNA synthesis in T4D- and am N116-infected E. coli B. Exponentially growing cultures were infected with T4D or am N116 at an m.o.i. of 5. For polyamine analyses 2 ml samples were millipore-filtered and the filter was immediately transferred to a vial containing 0.2 N perchloric acid. X, am N116; ●, T4D.

partly in bound form. In the "permissive" cell, E. coli strain
CR63, there is still a slight delay in the initiation of DNA synthesis induced by the phage mutant but very minor differences in
polyamine syntheses are seen in this system.

Although the lesion in infection by this phage mutant does
not appear to be the gross membrane defect seen in T4rII infections, it appears that the lesion can be alleviated to a reproducibly significant extent by spermidine. The nature of the
lesion itself is unclear, although manifested in a delay of DNA
synthesis. Of interest is the point that in the restrictive
system spermidine seems to increase even earlier than in normal
infections.

DOES POLYAMINE SYNTHESIS DEPEND ON DNA SYNTHESIS?

We have studied this problem under conditions of infection
with ultraviolet-irradiated T4D. As can be seen in Fig. 4, infection at a moi of 5 with normal phage or phage suffering 3.4
or 6.5 lethal hits, resulted in different initiations and rates
of DNA synthesis. Despite considerable inhibition of this function, there was little effect on spermidine synthesis or on putrescine synthesis (Fig. 5). In the most severely inhibited system, there was an inhibition in the accumulation of polyamines
within the cells but not in total synthesis.

We then examined a system in which infection was carried out
with the DO mutant, am N122. This mutant, which is blocked in
the dCMP hydroxymethylase, prevents DNA synthesis almost totally.
As can be seen in Fig. 6, there is essentially no difference in
the synthesis of spermidine in cells infected by T4D or am N122.
The latter even appears to evoke a somewhat greater synthesis of
putrescine in this experiment. Thus as in the study of polyamine
synthesis as a function of RNA synthesis (24) the net synthesis
of these materials is unaffected by the presence or absence of
DNA synthesis in these systems.

In E. coli infected by am N122, there is scarcely any accumulation of any nucleic acid. It is of additional interest that
the course of polyamine synthesis is nevertheless comparable in
the two systems, characterized by a brief lag and linear production, suggesting that the inception of polyamine synthesis after
interruption of normal growth by infection is not in fact set off
by DNA synthesis but depends on quite independent factors. Unlike some other uninfected systems, spermidine is not converted
largely to a bound form in the absence of nucleic acid synthesis.

DOES DNA SYNTHESIS DEPEND ON POLYAMINE SYNTHESIS?

As has been described earlier (18), it was not possible to

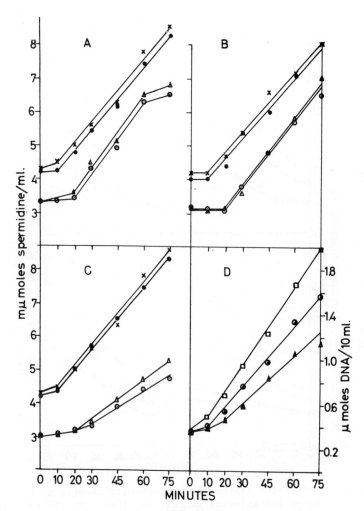

Figure 4 Spermidine and DNA synthesis in UV-irradiated T4D-infected E. coli B. Exponentially growing cultures of E. coli B were infected at an m.o.i. of 5 with unirradiated T4D (A) and T4D containing 3.4 (B) and 6.5 (C) lethal hits per phage. At the indicated times samples were taken for DNA and polyamine analyses.
A-C: Spermidine levels in cells (Δ, total; O, free) and cells plus mediun (X, total; ●, free).
D: DNA: ☐, unirradiated; ◐, 3.4 lethal hits/phage; ▲, 6.5 lethal hits/phage.

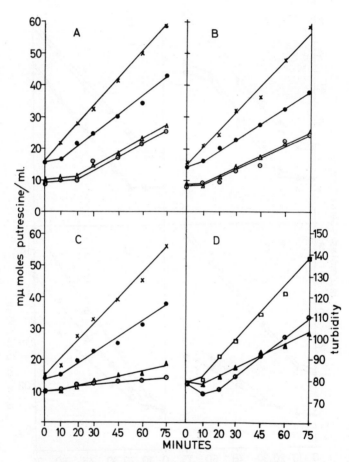

Figure 5 Putrescine synthesis and turbidity changes in UV-irradiated T4D-infected E. coli B. These are the same samples described in Figure 4.
A-C: Putrescine levels in cells (Δ, total; O, free) and cells plus medium (X, total; ●, free).
D: Turbidities: □, unirradiated; ◐, 3.4 lethal hits/phage; ▲, 6.5 lethal hits/phage.

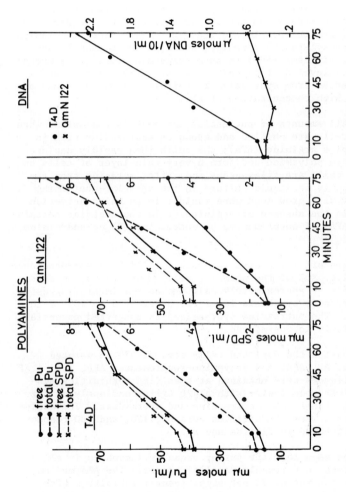

Figure 6 Polyamine and DNA synthesis in T4D- and am N122-infected E. coli B. Exponentially growing cultures were infected at an m.o.i. of 5 with T4D or am N122. Samples for DNA and polyamines were removed at the designated times; polyamine levels are expressed as concentration/ml of cells plus medium.

explore a relation of DNA synthesis to spermidine synthesis with
a methionine-requiring E. coli depleted in methionine and therefore arrested in spermidine synthesis. Since those studies, Maas
and his colleagues have described a mutant of E. coli extremely
limited in its ability to make putrescine (and hence spermidine)
under defined nutritional conditions (25). We are grateful to
Dr. Maas for making the organism available to us.

E. coli can make putrescine by one of two routes shown in
Fig. 7. In many strains putrescine is derived from the decar-
boxylation of ornithine. In the presence of arginine, this amino-
acid is decarboxylated to agmatine, which is hydrolysed by an
agmatine ureohydrolase (AUH) to putrescine and urea. This mutant
has a very low level of this enzyme (4% of normal) and in addition is prevented from synthesizing ornithine by the presence of
a relatively high concentration of arginine.

Very small amounts of putrescine are made in the mutant when
broth-grown cells are chilled and grown on agar medium rich in
amino acids plus arginine (AFA); the cells then rapidly deplete
their available polyamine and form a very thin layer of cells on
agar. These cells are filamentous and, after scraping from the
agar, grow slowly in liquid culture in the arginine-containing
medium. They will grow much more rapidly in an amino acid-rich
medium (AF) in the absence of arginine or in the arginine-containing medium (AFA) if supplemented by putrescine (P) or spermidine
(S) (25).

When taken off the plates the mutant cells have no detectable
putrescine, i.e. <5% of the normal level. However they do contain spermidine at approximately half the normal level, suggesting that putrescine synthesis is of the order of 5 to 10% of the
normal level. The putrescine synthesized is converted essentially quantitatively to spermidine.

On incubation the depleted cells grow at 37° in aerated
liquid medium, AF, AFA, and polyamine-supplemented AFA, i.e. AFAP
and AFAS. Aliquots were obtained at specified turbidities measured in a Klett colorimeter with a 420 filter and analysed for
their content of RNA, DNA, putrescine and spermidine. The analyses for cultures whose turbidities were 60, 120, and 180 respectively are presented in Fig. 8A and B.

It can be seen that the parent organism increases in all of
these parameters in a roughly parallel manner. The AFA medium,
inhibitory to the mutant is not significantly inhibitory with
this organism. On the other hand, the AFA medium is inhibitory
to the mutant which is quite slow in DNA, RNA and spermidine
synthesis, as compared to these syntheses in the AF or AFAP media.

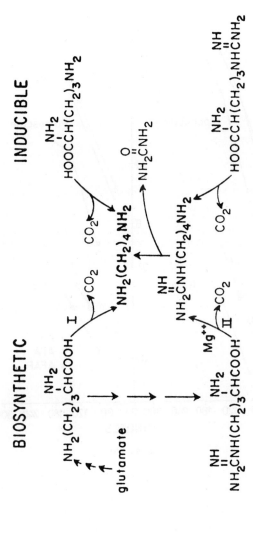

Figure 7 Biosynthetic and inducible pathways of putrescine synthesis in E. coli.

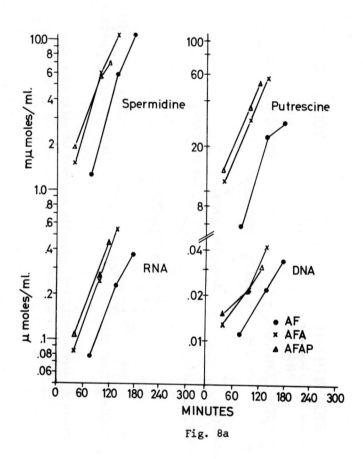

Fig. 8a

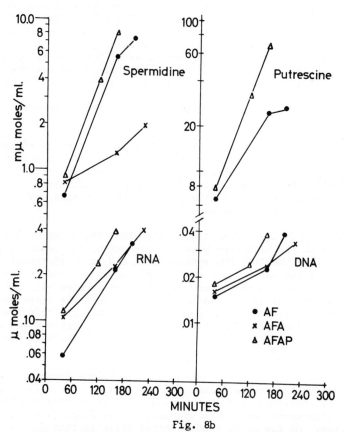

Fig. 8b

Figure 8 Relationship between nucleic acid synthesis and polyamine levels in E. coli K12 (wild) and E. coli K12 (AUH⁻). Experimental conditions for polyamine depletion in the mutant strain consisted of chilling overnight broth cultures of the two strains at 4° C for 5 hours, followed by a 1 to 40 dilution of each culture in AFA medium and overnight incubation at 37° C. Depleted cultures were washed free of excess arginine by centrifugation and resuspension in Davis-Mingioli medium, omitting glucose (25). Aliquots of the washed cell suspensions were added to various AF media, as indicated, at an initial turbidity of 33 and 45 Klett units for wild and mutant strains, respectively. Samples for the assay of intracellular polyamines and nucleic acids were removed at turbidities of 60, 120, and 180 Klett units.
A: Wild
B: AUH⁻

Nevertheless spermidine can be made at a significant rate in putrescine-deficient conditions; at no time can putrescine be detected in these cells. In all media, RNA and DNA content at constant turbidity (T = 120) were fairly similar; however it can be seen that the spermidine content of the AFA culture is markedly lower than the others, although the curve of increase of spermidine does resemble that of RNA. A problem remaining for the apparent unbalance between the RNA and spermidine content of this organism in the AFA medium turns on the possible use of agmatine as an organic cation to neutralize neucleic acid. These analyses are being done.

Despite the fact then that the organism is capable of making small amounts of spermidine, and larger amounts of the physiological unknown agmatine, we have gone on to infect the mutant (and the parent). After growth (1 division) of the agar-grown depleted cells in AFA medium, the appropriate polyamine was added 15 minutes prior to infection.

Under conditions of multiple infection, the rates of synthesis of DNA in the wild type were independent of the medium (Fig. 9). In the AFA medium, DNA synthesis in the infected mutant started late and was about a quarter of the rate found in the infected parent in the AF medium. In AFA supplemented with putrescine or spermidine (0.5 µmole per ml), DNA synthesis began earlier as in AF medium and was approximately double the rate found in the unsupplemented AFA medium. Thus the polyamines added at a physiological level actively and reproducibly stimulate synthesis of virus DNA in a putrescine-depleted cell.

We have found that the infected cells in the inhibitory AFA medium do synthesize small amounts of spermidine, as well as DNA. It appears therefore that this system does not provide the baseline of rigorously controlled spermidine deficiency that one would like. On the other hand infected cells in this medium do not contain any detectable putrescine, even as in growth conditions. It will obviously be useful to have additional nonleaky mutants blocked in either putrescine or spermidine synthesis or in both to extend these studies.

Despite these less than optimal qualities of our biological system, we have begun to seek possible correlations between polyamine deficiency and phage morphogenesis. We have been surprised to find that under conditions of single infection in the amino acid-rich media there is a loss of about 90% of the infectious centers, which do not liberate phage for many hours. Nevertheless if $CHCl_3$ is added to the dilution tubes, phage is eventually liberated from these infected cells, as shown in Fig. 10. It appears that complete phage is produced quite late but that such phage are held in these bacteria in an inactive state until freed by

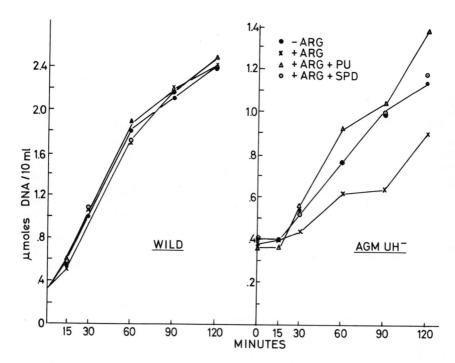

Figure 9 The effect of polyamines on DNA synthesis in T4D-infected $\underline{E.\ coli}$ K12 (wild) and $\underline{E.\ coli}$ (AUH$^-$). Experimental conditions for polyamine depletion are described in the text. Cultures, at an initial turbidity of 40 Klett units were allowed to increase to a turbidity of 80; putrescine or spermidine was then added as designated, and infection with T4D ensued 15 minutes later at an m.o.i. of 5. Final concentrations of arginine, putrescine, and spermidine were 100 µg, 1 µmole, and 0.5 µmoles per ml, respectively.

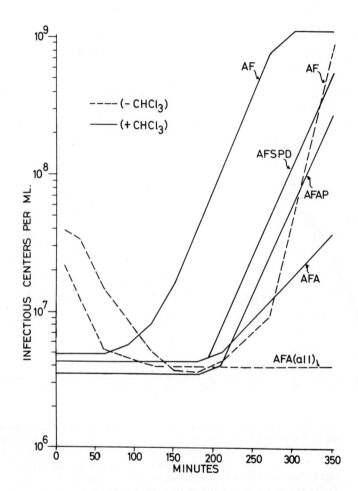

Figure 10 Effect of polyamines on one step growth of T4D-infected E. coli K12 (AUH⁻). Conditions for polyamine depletion and growth prior to infection were exactly as described in Figure 9. After infection with T4D at an m.o.i. of 0.1, each culture was diluted 10^{-4}. Assays of infectious centers and chloroform lysis were performed according to Adams (26).

this agent from an inactivating association, perhaps with a membrane component, in lysed cells. In any case the formation and liberation of intact phage in a one-stage growth experiment occurs long after the inception of DNA synthesis observed at high multiplicities. Electron microscopic studies currently under way should tell us if there are significant effects of the polyamines in facilitating morphogenetic changes in phage reproduction as well as in stimulating synthesis of phage DNA.

CONCLUSION

An increasing body of data suggest many direct and indirect effects of the polyamines, particularly spermidine, on nucleic acid metabolism, structure and function. This physiological role of the compounds is being actively investigated in relation to many areas of the biochemistry of RNA. The T-even phages provide an unusual opportunity to explore the interactions of viral DNA with these substances. Although the syntheses of DNA and of polyamines occur concurrently, and both newly synthesized substances are eventually found together in phage, it has been shown that the synthesis of the polyamines is not controlled by that of the nucleic acids. On the other hand, a cell inhibited to a considerable extent in the synthesis of the polyamines synthesizes DNA and phage quite slowly. Further these syntheses are stimulated quite significantly by the presence of physiological concentrations of putrescine which is convertible to spermidine of of spermidine which does not generate putrescine. The latter result suggests that, as in studies of RNA synthesis, spermidine is the most active polyamine in evoking these stimulations. Nevertheless it is not yet demonstrated unequivocally that the effects noted are directly on the synthesis or packaging of DNA rather than indirectly in protecting cellular structure and metabolism. Studies to determine the mechanism of these effects are being continued.

REFERENCES

1. S. S. Cohen, Introduction to the Polyamines (Prentice Hall, Englewood Cliffs, 1971).

2. S. S. Cohen and J. Lichtenstein, J. Biol. Chem. 235 (1960) 2112.

3. S. S. Cohen, S. Morgan and E. Streibel, Proc. Nat. Acad. Sci. 64 (1969) 669.

4. S. S. Cohen, Ann. N. Y. Acad. Sci. 171 (1970) 169.

5. E. Hiebert and J. Bancroft, Virology 39 (1969) 296.

6. M. Tsuboi, Bull. Chem. Soc. Japan 37 (1964) 1514.

7. A. M. Liquori, 'L. Constantino, V. Crescenzi, V. Elia, E. Giglio, R. Puliti, M. DeSantis Savino, and V. Vitagliano, J. Mol. Biol. 24 (1967) 113.

8. M. Suwalsky, W. Traub, U. Shmual and J. A. Subirana, J. Mol. Biol. 42 (1969) 363.

9. R. Michaels, and T. T. Tchen, J. Bact. 95 (1968) 1966.

10. H. G. Johnson and M. K. Bach, Arch. Biochem. Biophys. 128 (1968) 113.

11. M. Inouye and A. B. Pardee, J. Bact. 101 (1970) 770.

12. M. G. Sevag and W. T. Drabble, Biochem. Biophys. Res. Commun. 8 (1962) 446.

13. H. G. Johnson and M. K. Bach, Proc. Nat. Acad. Sci. 55 (1966) 1453.

14. Z. Kwiatkowski and H. Heleszko, Acta Microbiol. Polonica 1 (1969) 101.

15. A. D. Hershey, Virology 4 (1957) 237.

16. B. N. Ames, D. T. Dubin and S. M. Rosenthal, Science 127 (1958) 814.

17. B. N. Ames and D. T. Dubin, J. Biol. Chem. 235 (1960) 769.

18. S. S. Cohen and A. Raina in: Organizational Biosynthesis, eds. H. J. Vogel, J. O. Lampen and V. Bryson (Academic Press, New York, 1967) p. 157.

19. A. Garen, Virology 14 (1961) 151.

20. G. Ferroluzzi-Ames and B. N. Ames, Biochem. Biophys. Res. Commun. 18 (1965) 639.

21. C. S. Buller and L. Astrachan, J. Virology 2 (1968) 298.

22. B. S. Guttman and L. Begley, Virology 36 (1969) 687.

23. E. J. Herbst and A. S. Dion, Fed. Proc. 29 (1970) 1563.

24. S. S. Cohen, N. Hoffner, M. Jansen, M. Moore and A. Raina, Proc. Nat. Acad. Sci. 57 (1967) 721.

25. I. N. Hirschfield, H. J. Rosenfeld, Z. Leifer and W. K. Maas, J. Bact. 101 (1970) 725.

26. M. H. Adams, Bacteriophages (Interscience Publishers, Inc., New York, 1959).

DISCUSSION

Dr. S. Greer, University of Miami: It would be of interest to relate the polyamines to Albert's 32 gene product to see whether it is antagonistic or cooperative; has anything been done in this area?

Dr. S. S. Cohen: No, it has not. Dr. Alberts has the 32 product; we don't have that. Perhaps he'll tell us.

Dr. B. Alberts, Princeton University: I did only one experiment with spermidine and that was to see if, first of all, since there is a magnesium requirement in this system for renaturation when the single strand DNA is fully covered with 32 protein, that is, magnesium greatly speeds the rate and I did look to see if one millimolar of spermidine would replace the magnesium and it clearly did not. I also looked to see if the one millimolar of spermidine plus magnesium would affect the rate and make the rate either faster or slower than the rate with just magnesium alone, and there again there was no effect. So the only experiments that have been done don't show any effect of spermidine, but that doesn't mean that there isn't one and some other way of testing this might show that up.

Dr. Cohen: May I remark that in this last experiment in which we are stimulating DNA synthesis in this depleted, infected cell, we're stimulating with half millimolar exogenous spermidine. It should be mentioned that the depleted cell still contains half of the normal level of spermidine.

Dr. U. Z. Littauer, Weizmann Institute of Science: In your last experiment, when depleting the cells of polyamines, can you reverse this effect by increasing the magnesium concentration of your medium?

Dr. Cohen: We haven't tried it. There are a good many messy things about the particular system we have. These cells, in order to be depleted of polyamine, are grown on agar plates in the AFA condition, and are scraped off these plates in order to get enough to do these kinds of experiments. We do not have conditions as yet for depleting polyamines from the cell in liquid culture.

Dr. F. C. Neidhardt, University of Michigan Medical School: In these last experiments, Seymour, were you measuring accumulation or synthesis? I didn't understand the technique.

Dr. Cohen: I am measuring synthesis of DNA and, I may say, that in the system slowly synthesizing DNA in the absence of exogenous polyamine, one can show, as we have shown, that there really is a slow synthesis of polyamine at the same time. Now, we need really good mutants and no one has these yet and that's where we are.

Dr. Neidhardt: What about the stability of the DNA though? Do you think there is any degradation under the conditions of low intracellular spermidine?

Dr. Cohen: Well, it's a good point. We haven't looked.

Dr. M. Rabinovitz, NIH: You indicated that putrescine might stabilize two chains by fitting between them. Has an homologous series been prepared with different numbers of methylenes between the amino groups to see whether there is a sharp cutoff in such stabilization.

Dr. Cohen: Yes, there is. There is a series of experiments by Stevens. He did a number of experiments on Tm for DNA as a function of lengthening the C4 central piece. He found that when you went up to C5, the analogue was incapable of increasing the Tm as well as the natural compound.

Dr. R. Werner, University of Miami: You mentioned that you get about a two- or three-fold increase in the rate of phage DNA synthesis after adding spermidine. This could have several reasons. Either the number of growing points is increased, or the growing points are moving at a faster rate. I would suggest a third possibility - namely that the frequency of recombination is reduced and thus leaving more DNA available for replication. You could test this very easily.

Dr. Cohen: There have been no studies on recombination of phage. There have been some studies on recombination with Hfr strains of *E. coli*. Some Polish workers have reported that the polyamines will affect recombination.

Dr. J. Schultz, Papanicolaou Cancer Research Institute: Inasmuch as the relationship of magnesium to spermidine has been raised, under what circumstances in the experiments you have carried out can magnesium replace spermine and spermidine?

Dr. Cohen: I am almost invariably confronted with this question. Why worry about an organic cation if magnesium is going to be doing the same thing? There are a number of systems in which magnesium doesn't do the same thing. For example, in the methylation of tRNA by a liver methylase, according to Leboy, you methylate bases in the presence of spermidine that you do not methylate in the presence of magnesium. And furthermore, the extent of methylation and its rate is about four times greater with spermidine. That is one sort of effect. Tanner has demonstrated that the binding of message to a protein synthesizing system and of tRNA is much better in spermidine than in optimal magnesium, and so forth. One can indicate a number of these more specific effects. But it seems to me that even if we have the situation where Mg^{++} and polyamine are quite identical, a cell which has worked out a mechanism for using half of two very important amino acids, arginine and methionine, is using this for something. That's a weak argument, I suppose.

Dr. M. M. Sigel, University of Miami: In view of the current interest in synthetic polynucleotides in connection with interferon production and cancer chemotherapy, etc., is anything known about the effect of polyamines on the function of the synthetic polynucleotides?

Dr. Cohen: The Merck group published a paper last year which says that it's possible to stimulate interferon production with poly IC if you make polyamine salts of poly IC. This was the title, but the actual paper was concerned mainly with the formation of neomycin salts, rather than with spermidine or spermine salts. But apparently they have done some work on it.

Dr. Schultz: In your charts you use as the parameter the synthesis of DNA. What relationship is this DNA to the host DNA, or is it host DNA?

Dr. Cohen: Oh, no, after infection you don't make host DNA. There are no ribosomal genes on the virus DNA. But that isn't the answer to your question. The answer really relates to the problem of how you manage to shut off host syntheses generally and this is something which has been mysterious for twenty-five years.

DNA POLYMERASES AND OTHER ENZYMES OF RNA TUMOR VIRUSES

HOWARD M. TEMIN, SATOSHI MIZUTANI and JOHN COFFIN
McArdle Laboratory
University of Wisconsin
Madison, Wisconsin 53706

Abstract: RNA-dependent and double-stranded nucleic acid dependent DNA polymerases, DNA exo- and endonucleases, and ligase activities are found in purified virions of Rous Sarcoma Virus. Cells infected with Rous Sarcoma Virus contain several different types of particulate DNA polymerases.

INTRODUCTION

RNA Tumor Viruses have been of interest since the first decade of 1900 because of their connection with cancer in animals. They have been of interest to academic virologists: the class of viruses which first was found to have an infection in which virus production and cell division were not mutually exclusive. Now, the RNA Tumor Viruses are of interest to molecular biologists because they appear to use the previously unknown mode of RNA → DNA information transfer.

In this paper, we shall review the history of the discovery of this mode of information transfer, the evidence for several kinds of nucleic acid metabolizing enzymes in the virion, and discuss some of the properties of and significance of DNA polymerases in RNA Tumor Virus-infected and in uninfected cells.

DNA PROVIRUS: Chicken cells infected with Rous Sarcoma Virus (RSV) are able to continue multiplying while they are producing virus (1). Other viruses, for example SV5 in monkey kidney cells (2), seem to have a somewhat similar virus-cell state. The Rous Sarcoma Virus differs from SV5 in that RSV appears to enter into some special relationship with the chicken cell, and in that there is regular inheritance of the information for virus production and for cell transformation (3). This inherited information is called the provirus. The first indication that the provirus might be DNA came from experiments using actinomycin D. This compound was very useful for the study of other RNA viruses, because it inhibited cellular RNA synthesis without affecting viral RNA synthesis. However, production

Table 1. HOMOPOLYMERS AS TEMPLATES FOR VIRION POLYMERASE

Substrate	Incorporation (pmoles)	None	Poly dA·dT	Poly rA·dT	Poly dA	Poly dT
α-^{32}P-dATP	dAMP	0.03	1.8	0.25	0.6	0.075
^{3}H-dTTP	dTMP	0.045	2.2	1.65	0.35	0.06
α-^{32}P-dATP and ^{3}H-TTP	dTMP	0.065	5.95	1.85	4.9	0.45
	dAMP	0.03	5.4	0.4	4.3	0.47

Disrupted pronase-treated virions were incubated with 50 μg/ml ribonuclease A for 60 min at 0°C. 0.4 μg of the indicated templates (kindly prepared by Dr. R. Wells) and 0.2 nmoles of α-^{32}P-dATP (550 c.p.m./pmole) and/or 0.2 nmoles of ^{3}H-dTTP (2.65 x 10^4 c.p.m./pmole) were added and a standard polymerase assay was carried out for 1 h. Samples were counted for ^{3}H and ^{32}P, and the results calculated as pmoles of substrate incorporated, assuming a counting efficiency of 20 per cent for ^{3}H. With 10 μg poly rA as template, no incorporation was found (from 18).

of Rous Sarcoma Virus and its RNA appeared to be sensitive to
actinomycin (4,5). This observation, in light of the known
specificity of action of actinomycin D, was interpreted to
mean that the provirus was DNA (6). Further experiments using
inhibitors of DNA synthesis and nucleic acid hybridization (7)
were compatible with this hypothesis. The hybridization experiments have been extended by Baluda. The
experiments with inhibitors of DNA synthesis have more recently
been carried out with stationary cultures to exclude the effects
on cell division, which were present in most earlier experiments
(8-10).

At present the most convincing experiments to demonstrate
that there is new proviral DNA synthesized soon after infection,
comes from experiments (11,12) using 5-bromodeoxyuridine
labeling of stationary cells soon after infection, and specific
inactivation of the provirus DNA, but not of the cellular DNA,
by visible light treatment. Especially important is the experiment of Boettiger (12) which showed that the target for
inactivation of the provirus was dependent upon the multiplicity
of infection.

The availability of the stationary cell system enboldened us
to look again for the enzyme system responsible for formation
of the provirus. The insensitivity of provirus formation to
treatment by cycloheximide (Mizutani cited in 13,14) suggested
that the polymerase could be in the virion. We found it there
(15), and at the same time David Baltimore of MIT did too (16).

VIRION ENZYMES: The RNA-dependent DNA polymerase of the
virion seems to be in the virion core (Fig. 1), explaining the
necessity for treatment with detergent to find enzyme activity.
The enzyme is easily separated from its endogenous template by
salt or spontaneously, and is then found to use DNA as well as
RNA as template. Many synthetic homopolymers will also act as
templates (Table 1).

Our study of the product of the endogenous SRV virion polymerase showed that small pieces of nicked double-stranded DNA
were produced (18). This result suggested to us that in
addition to the RNA- and double-stranded nucleic acid-dependent
DNA polymerase(s) there might be an endonuclease in the virion.
Such a nuclease could not only explain our results, but could
also be a component of the machinery needed to integrate the
provirus DNA within the host cell chromosome.

When we incubated RSV virions with labeled DNA from bacteriophage T7, we found that the S-value of the DNA decreased (18).
We also found that poly d(IC)·d(IC) was more rapidly degraded

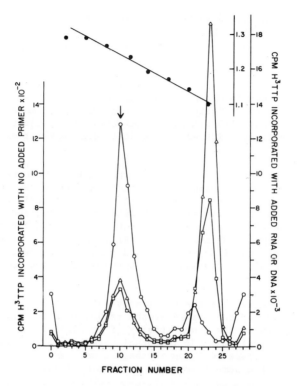

Figure 1. Sucrose density centrifugation of disrupted virions of RSV. A preparation of Schmidt-Ruppin Virus was disrupted with Nonidet and centrifuged in a 30-70% sucrose-D_2O gradient for 2.5 hours at 45000 RPM in a Spinco SW50.1 rotor. Fractions were collected from the bottom of the gradient and assayed in a standard polymerase assay with nothing (o), 5 µg yeast RNA (□), or 0.5 µg calf thymus DNA (▲) added. Selected fractions were weighed to determine density in g/cc (●). The arrow marks the location of the peak of H^3uridine labelled SRV disrupted and centrifuged on the same gradient (from 17).

by exonuclease I after incubation with virions. This endonuclease activity requires the presence of Mg^{2+}. It has been found with several strains of avian RNA Tumor Viruses, when sufficient concentrations of virions were assayed. At present we are not sure whether the endonuclease is located in the same portion of the virion as the polymerase. Although the endonuclease is resistant to preincubation of the virions with pro-

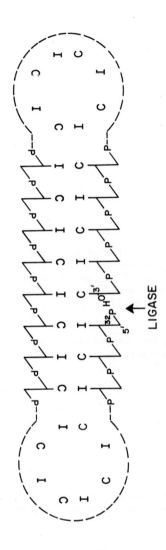

Figure 2. Substrate for ligase assay -- ^{32}P-poly d(IC)·d(IC). Chain length was about 150 nucleotides.

Table 2. REQUIREMENTS FOR NUCLEASE ACTIVITY OF VIRIONS OF RSV

Complete system		TCA insoluble 32P c.p.m.	Per cent
with no virions		12,600	(100)
with non-disrupted virions		9,080	72
with disrupted virions		4,280	34
with heated disrupted virions		15,000	119
with disrupted	+ EDTA (10mM)	12,500	99
"	− Mg^{++}, + Mn^{++} (6.6 mM)	9,820	78
"	− ATP	4,660	37
"	+ NaF (0.5 M)	9,700	77
"	+ pCMB (2 × 10^{-4} M)	4,150	33

Complete system consisted of 0.014 mM ^{32}P-poly d(IC)·d(IC), 6.6 mM MgCl$_2$, 0.066 mM ATP in 6.6 mM tris-HCl pH 7.6 containing 10 mM dithiotreitol in 60 µl, and disrupted virions of Schmidt-Ruppin strain of RSV. Heating was 100°C for 2 min. The mixture was incubated at 37°C for 60 min.

nase, it often is somewhat active without detergent treatment of the virion (Boettiger, personal communication). We have not yet studied isolated virion cores.

Study of other conditions required for endonuclease activity led us to suspect ligase activity in the virion, which in turn led us to an exonuclease as well as a ligase.

To assay for ligase we used a ^{32}P labeled alternating poly d(IC)·d(IC) (19) prepared with Dr. Kodama (Fig. 2). Wells and Kodama (personal communication) have shown that transfer of the 5'^{32}P to a form resistant to digestion with bacterial alkaline phosphatase is a good assay for ligase. However, when disrupted RSV virions were incubated with ^{32}P poly d(IC)·d(IC), a decrease in the amount of acid insoluble ^{32}P was found without any treatment with bacterial alkaline phosphatase (Fig. 3). This decrease was not prevented by 5 mM phosphate, which inhibits bacterial alkaline phosphatase.

After incubation of ^{32}P poly d(IC)·d(IC) with RSV virions, the mixture was chromatographed on Whatman No. 1 paper in a phosphate buffered amonium sulfate: propanol system. About twenty per cent of the counts were found in 5' dIMP and 5' dCMP. An equal fraction was found in very small oligonucleotides. Therefore, the virion contains an exonuclease as well as the

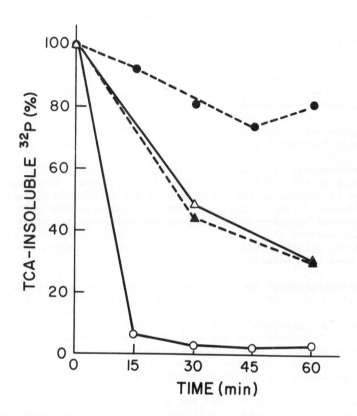

Figure 3. Kinetics of nuclease reaction. ^{32}P-poly d(IC)·d(IC) was incubated at 40°C in the presence (dotted line, filled symbols) or absence (straight line, open symbols) of 5 mM phosphate. Tubes contained in 60 µl either 64 µg protein of disrupted RSV (triangles) or 10 µg of bacterial alkaline phosphatase (circles).

endonuclease. Previously, we did not find any exonuclease activity with T7 or chicken DNA as substrates. The difference in results probably occured because of the greater sensitivity of the present assay. The exonuclease activity required virion disruption and was inactiviated by heat (Table 2). Activity required a divalent cation, Mg^{2+} preferred to Mn^{2+}, and had a pH optimum between 7.0 and 7.5.

When disrupted RSV virions were incubated with ^{32}P poly d(IC)·d(IC), an increase in bacterial alkaline phosphate resistant ^{32}P was found (Fig. 4). This reaction was proportional to the amount of disrupted virions in the incubation mixture. No inhibition of bacterial alkaline phosphatase activity by disrupted RSV virions was found. Therefore, the reaction probably represents a joining of the poly d(IC)·d(IC).

Considerable difficulty was found in degrading the final products of the joining-reaction to 3'nucleotides. The large number of enzymes in the virion and in the incubation mixture contributed to this. However, after partial digestion with micrococcal nuclease and speen phosphodiesterase, the major product behaved like 3' dIMP in paper chromatography using two different solvent systems. Therefore, a phosphodiester bond must have been formed.

The requirements for ligase activity are seen in Table 3. Disruption of virions was required for ligase activity. The activity was destroyed by heating, and required a divalent cation, Mn^{2+} preferred to Mg^{2+}. The optimum pH was 8.5. However, no requirement for ATP could be demonstrated. Addition of large amounts of ATP or dATP partially inhibited the reaction. NAD did not affect the reaction. At present we do not know how to explain the lack of ATP requirement.

These results suggest that the RSV virion contains RNA-directed DNA polymerase, double-stranded nucleic acid-dependent DNA polymerase, DNA endonuclease, DNA exonuclease, and DNA ligase activities. If all of these activities are real, and if they are all active inside cells, the virion would have all of the machinery necessary to transfer information from RNA into DNA integrated with host cell DNA. However, the fact that the machinery is present does not prove that it has the postulated function. The presence and specificities of these enzyme activities may help to explain the differences between leukemia viruses, sarcoma viruses and protoviruses. They may also help to explain the differences between infection of homologous and heterologous cells.

<u>DNA POLYMERASES IN CELLS</u>: Because of our interest in these

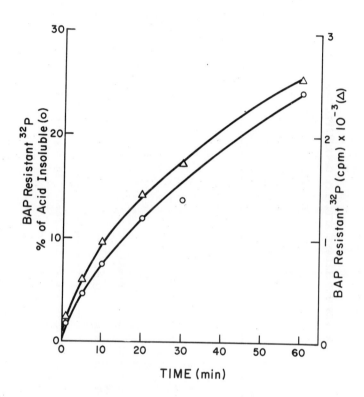

Figure 4. Kinetics of ligase reaction. Disrupted RSV was incubated at 37°C with ^{32}P-poly d(IC)·d(IC). At intervals portions were taken, and TCA precipitable counts were determined, either before or after heating in a boiling water bath for 2 min and double incubation with bacterial phosphatase.

Table 3. REQUIREMENTS FOR LIGASE ACTIVITY OF RSV VIRIONS

Complete system	% BAP resistant ^{32}P (% of acid insoluble)	
	Pronase treated virions Mg^{++} (6.6 mM)	Non-pronase treated virions Mn^{++} (10 mM)
– Virions	13	13
with non-disrupted virions	18	8
with disrupted virions	40	59
with disrupted heated virions	5	4
with disrupted virions, – Mg^{++} or – Mn^{++}		
+ EDTA (4 mM)	9	15
– ATP	30	58
" , – ATP, +dATP (0.1 mM)	–	38
" , – ATP, + NAD^+ (1.7 μmoles)	40	41
" , ATP (0.132 mM)	–	38
" , + dITP, dCTP (each 10 μmoles)	41	–

Complete system as in legend to Table 2 except as indicated. Incubation was at 37°C for 30 min.

enzymes as markers for virus-specified production in non-virus producing cells, and the interest of others in these enzymes as markers for human cancers, we looked for particulate DNA polymerases in Nonidet extracts of infected cells. Uninfected chicken cells did not contain significant quantities of particulate endogenous or DNA-dependent DNA polymerase (Fig. 5). Chicken cells producing RSV seemed to contain two particulate endogenous RNA-directed DNA polymerases (Fig. 6). One appeared to have the density of virion cores. The other was much denser. Since no DNA polymerase activities were found in parallel cultures of multiplying uninfected chicken cells, this denser polymerase also must be virus-related. Perhaps it is an earlier precursor of the core. The 1.15 g/cc DNA-dependent DNA polymerase also must not have been present in the parallel uninfected cells. At present, we cannot say if it is related to virus production, to neoplastic conversion, or something else.

Rat cells converted by infection with RSV are interesting because although they contain the full information of RSV, they do not produce virus or virions (17,20). Nonidet extracts of these cells also contained endogenous RNase sensitive DNA polymerase. This polymerase had a different density (1.14 g/cm^3) from either the virion cores (1.22 g/cm^3) or the dense polymerase (1.3 g/cm^3) in infected chicken cells (Fig. 7). However, this endogenous polymerase activity in the rat-Rous cells had the same detergent sensitivity and divalent cation requirements as the virion enzyme. The DNA-dependent DNA polymerase in rat-Rous cells is also in a ribonucleoprotein particle. The major endogenous DNA polymerase can be separated from the other particulate polymerase by sedimentation (Fig. 8).

The major question in interpreting these particulate DNA polymerases is their relationships to DNA polymerases in uninfected rat cells. With different batches of uninfected rat cells, we have had different amounts of particulate endogenous DNA polymerase activity. Since it takes several transfers before a completely transformed culture of rat-Rous cells is secured, it is not possible to have a strictly parallel control as with uninfected and infected chicken cells. However, when endogenous DNA polymerase was found in a particulate fraction from uninfected rat cells, it always had a different response to cations than did the polymerase from rat-Rous cells. Mn^{2+} did not substitute as well for Mg^{2+}.

Since with the parallel chicken fibroblasts we did not find any particulate DNA polymerases (Fig. 5), we looked at soluble enzymes from uninfected chicken cells and embryos for their DNA polymerase activity with natural RNA, and single-stranded or double-stranded DNA templates. We have separated by DEAE-cell-

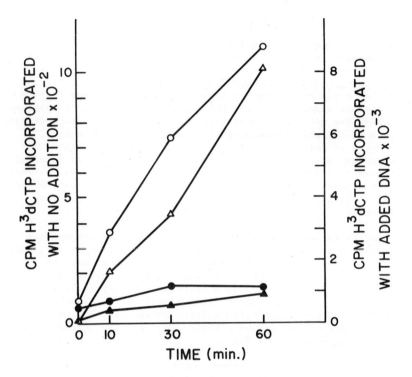

Figure 5. DNA polymerase activity in extracts of uninfected and RSV-infected chicken cells. Parallel cultures of uninfected (closed symbols) and B77-virus infected (open symbols) chicken cells were treated with Nonidet and high speed pellet fractions prepared. These were assayed for DNA polymerase activity with no additions (o,●) or with added DNA (∆,▲), and TCA insoluble radioactivity determined at the indicated times (from 17).

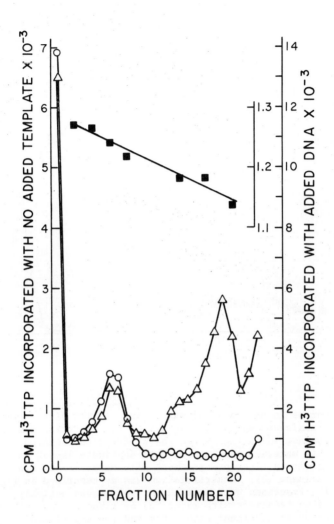

Figure 6. Particulate DNA polymerase activity in RSV-infected chicken cells. A high speed pellet fraction of Nonidet-treated RSV-infected chicken cells was layered on a 30-70% sucrose D_2O gradient and centrifuged as described in Fig. 1. All fractions were assayed for DNA polymerase activity with no addition (o) or with added DNA (Δ). Density in g/cc (■) was determined for selected fractions (from 17).

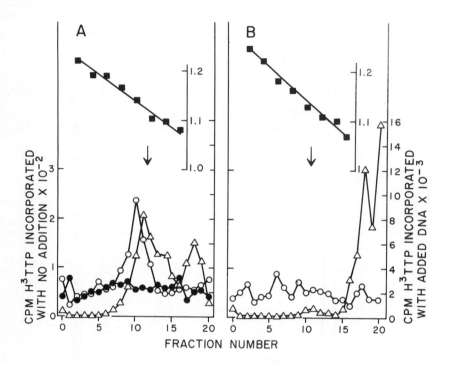

Figure 7. Endogenous and DNA-dependent DNA polymerase in RSV-infected rat cells. A line of rat cells transformed by the B77 strain of RSV was treated with Nonidet and a high speed pellet fraction prepared. This material was layered onto duplicate 15-65% sucrose gradients after pretreatment with water (A) or ribonuclease (B). Centrifugation was as described in Fig. 1. Fractions were assayed for polymerase activity after treatment with water (o) or ribonuclease (●) or after treatment with water and the addition of 5μg of calf thymus DNA (▲). Density of selected fractions in g/cc was also determined (■).

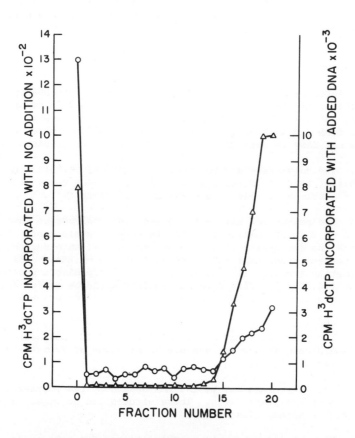

Figure 8. Sucrose velocity gradient centrifugation of DNA polymerase from RSV-infected rat cells. A high speed pellet fraction of Nonidet disrupted B77V-infected rat cells was layered on a 5-20% sucrose gradient and centrifuged at 20,000 RPM for 1 hour in a Spinco SW50.1 rotor. Fractions were collected from the bottom of the tube and assayed for DNA polymerase activity with no addition (o) or with 5 µg calf thymus DNA (△).

ulose chromatography at least two soluble DNA polymerases from chicken embryos. These chicken DNA polymerases differ from the RSV virion DNA polymerase in having a lower Mg^{2+} optimum and in preferring single-stranded DNA to double-stranded DNA as a template. So far experiments using RNA as a template have not given consistent results.

DISCUSSION

In interpreting the significance of RNA-dependent DNA polymerases or new DNA polymerases in normal cells and in tumor cells, the following possibilities must be kept in mind. It is clear from our work with rat-Rous and chicken-Rous cells, that it is not easy to separate these possibilities experimentally.

1. The polymerase may be involved in normal cellular DNA synthesis; that is in replication and repair, but under cell-free conditions be able to use RNA as a template (21).

2. The polymerase may be an unusual cell polymerase involved in RNA → DNA information transfer in normal development. In considering the origin of RNA Tumor Viruses, we have suggested a protovirus hypothesis which states that DNA → RNA → DNA information transfer is a normal part of an organism's genetic system (14,22,23). This type of transfer would be useful for amplifying regions of DNA, and possibly for differentiating cells.

3. The polymerase may be an RNA-dependent DNA polymerase of a non-oncogenic virus, for example Visna Virus or Primate Syncytial Virus (23,24).

4. The polymerase may be the RNA-dependent DNA polymerase of a passenger RNA Tumor Virus. These viruses can be widely distributed in normal tissues (25).

5. The polymerase may be an RNA-directed DNA polymerase of an RNA Tumor Virus activated by the neoplastic transformation, with the RNA Tumor Virus neither causing nor maintaining the neoplastic transformation. This possibility is similar to 4, but the RNA Tumor Virus would come from a provirus in the cells of the tumor.

6. The polymerase may be the RNA-dependent DNA polymerase of the RNA Tumor Virus which caused the neoplastic transformation, but the polymerase may not be essential to the persistence of the neoplastic transformation. We imagine this is the explanation for most of the polymerases in the rat-Rous and chicken-Rous cells discussed in this paper.

7. The polymerase may be the RNA-dependent DNA polymerase of the RNA Tumor Virus which caused the neoplastic transformation. The polymerase may be essential to the persistence of the transformation. This possibility would be most hopeful for therapy. However, the present experiments with RNA Tumor Viruses have so far given no evidence supporting the existence of this possibility.

In each case when a new DNA polymerase is reported, especially an RNA-directed one, we shall have to do further experiments to determine the origin and biological role of such polymerases.

REFERENCES

1.) H. M. Temin and H. Rubin. Virology 8 (1959) 209.

2.) K. V. Holmes and P. W. Choppin. J. Exp. Med. 124 (1966) 501.

3.) H. M. Temin. Virology 13 (1961) 158.

4.) H. M. Temin. Virology 20 (1963) 577.

5.) H. M. Temin. Virology 23 (1964) 486.

6.) H. M. Temin. Nat. Cancer Inst. Monog. 17 (1964) 557.

7.) H. M. Temin. Proc. Nat. Acad. Sci. 52 (1964) 323.

8.) H. M. Temin. J. Cell. Physiol. 69 (1967) 53.

9.) H. M. Temin. Cancer Res. 28 (1968) 1835.

10.) R. Murray and H. M. Temin. Int. J. Cancer 5 (1970) 320.

11.) P. Balduzzi and H. R. Morgan. J. Virol. 5 (1970) 470.

12.) D. Boettiger and H. M. Temin. Nature 228 (1970) 662.

13.) H. M. Temin. Proc. Xth Int. Cancer Conf. (1970).

14.) H. M. Temin. in: Proc. 2nd Lepetit Coll. (North-Holland, Amsterdam, 1971).

15.) H. M. Temin and S. Mizutani. Nature 226 (1970) 1211.

16.) D. Baltimore. Nature 226 (1970) 1209.

17.) J. Coffin and H. M. Temin. J. Virology (1971).

18.) S. Mizutani, D. Boettiger and H. M. Temin. Nature 228 (1970) 424.

19.) Y. Mitsui, R. Langridge, B. E. Shortle, C. R. Cantor, R. C. Grant, M. Kodama and R. D. Wells. Nature 228 (1970) 1166.

20.) C. Altaner and H. M. Temin. Virology 40 (1970) 118.

21.) L. F. Cavaliere and E. Carroll. Biochem. Biophys. Res. Com. 41 (1970) 1055.

22.) H. M. Temin. Pers. Biol. Med. 14 (1970) 11.

23.) F. H. Lin and H. Thormar. J. Virol. 6 (1970) 702.

24.) E. Skolnik, E. Rands, S. A. Aaronson and G. J. Todaro. Proc. Nat. Acad. Sci. 67 (1970) 1789.

25.) R. M. Dougherty and H. DiStefano. Prog. Med. Virol. 11 (1969) 154.

This research was supported by Public Health Service Research Grant CA 07175 from the National Cancer Institute. H.M.T. holds Research Career Development Award 10K 3-CA8182 from the National Cancer Institute. J.C. is a predoctoral trainee of the National Cancer Institute.

DISCUSSION

Dr. P. Duesberg, University of California: You mentioned that you can do this protein inhibitory experiment with stationary cultures; does that imply that there is no mitosis in these cultures? A previous biological paper of yours in Journal of Cell Physiology linked virus replication with mitosis.

Dr. H. M. Temin: This is the difficulty in communicating between biologists and biochemists. I have been trying to explain this point to Peter for three years. Could I have slide four again? Now as Peter has said and I said and I will try and say again -- in the replication of this virus there are the initial infection and formation of the virus, the requirements for two DNA strands. The requirements are for synthesis of two DNA in order to get conversion. So if one takes these stationary cells and exposes them to virus, nothing will happen. These cells will not change in morphology and will not produce virus unless at a later time when serum is added which stimulates DNA synthesis. Then the cells divide; then they increase in number and they make virus antigen and become converted. Now this is a different process which we've called activation, to give it another name, and we think at the present time that it might have something to do with transport or processing of the product RNA of the virus so it can get out and make the specific protein but we're not sure of that. That's a completely different requirement from everything that's talked about here. All of these things, if they have a biological role, will be concerned right in the first twelve hours where there is another DNA synthesis period which we can only tell by inhibitor experiments in the BUDR labeling and, as I emphasized, we can not yet isolate this new DNA and this is the one where we think all of these enzymes are operative to go from RNA into the DNA provirus in the cell. Then it appears that once that is present, in presumably integrated state, with the chromosome, it is not active, at least to the extent of making the cell a tumor cell and making virion antigens and making virus. Something further must happen and this is correlated with cell division--this second step and perhaps relates to the processing of the product RNA. It's of course intriguing to wonder about how the information is made from the DNA and what kind of polymerases are active there. But this is the second step which perhaps seems simpler, yet we don't have any idea such as we do about the first DNA synthesis.

Dr. D. Baltimore, M.I.T.: In relation to what you were saying, do you think it's possible that the requirements for cell division is a requirement for getting the DNA product through the nuclear membrane.

Dr. Temin: Well, no, not the DNA product but the RNA product. We've done experiments which perhaps are preliminary to mention but if one takes these kind of cells and then looks with fluorescent antibodies when they are stationary, one sees no virion products and when they go into mitosis before the new interphase has begun one already sees the antigen. So it suggests that it's kind of blocked starting protein synthesis. But this clearly has to be worked on a lot more. And there are a lot of experiments to do about that. My present idea is that it has to do with the transport of the product RNA, the provirus DNA specified RNA out of the nuclear membrane. I think that's the most likely thing now. But not much is known about that either.

Dr. S. Greer, University of Miami: My laboratory is interested in the selective killing of viruses in cells using BUDR and radiation. I'm interested, therefore, to know how much cell death you are getting with the visible light BUDR combination as you are inactivating the virus?

Dr. Temin: We are getting no detectable cell death. The importance of this experiment is it's carried out with a stationary cell. Therefore, there is very little BUDR incorporated into DNA just because there is very little DNA synthesis. This paper with the data was published in Nature in, I forget, September or something. Betiger is the senior author if you want to look at the data but by our tests there is no detectable cell killing while we are killing the infection.

DNA POLYMERASE OF AVIAN TUMOR VIRUSES

P. DUESBERG, E. CANAANI and K. v. d. HELM
Department of Molecular Biology and Virus Laboratory
University of California, Berkeley, California 94720

Abstract: The unpurified DNA polymerase associated with avian tumor viruses was found to transcribe more than 90% of the sequences of endogenous 60-70S tumor virus RNA into small pieces of DNA.

The DNA polymerase of Prague Rous sarcoma virus of subgroup C (PR RSV-C) and Schmidt-Ruppin RSV of subgroup A (SR RSV-A) was solubilized with Triton X-100. The soluble enzyme represented <2% of the total soluble protein of the virus. It is probably different from the viral group-specific antigen and it was separated from 90% of the viral glycoprotein. The sedimentation coefficient ($S_{20,w}$) of the soluble DNA polymerase was 8S before and 6S after incubation with pancreatic RNase. The molecular weight (MW) of the 8S DNA polymerase was estimated from its $S_{20,w}$ to be around 170,000 daltons and that of the 6S DNA polymerase to be around 110,000 daltons.

Purified DNA polymerase had a high activity using 60-70S viral RNA as well as salmon DNA as templates, but had low activities if primed by influenza virus RNA and the RNA of tobacco mosaic virus (TMV). Both the 8S and the 6S DNA polymerase had no endogenous template activities. The DNA-dependent and the RNA-dependent DNA polymerase activities of PR RSV-C coincided in sucrose gradients, both in the 8S and the 6S forms. It is concluded that the RNA-dependent and the DNA-dependent DNA polymerase activities of the avian tumor viruses are probably the same basic enzyme.

INTRODUCTION

Only nine months ago an RNA-dependent DNA polymerase was discovered in avian and murine RNA tumor viruses by Temin and Baltimore (1,2). Since this discovery, RNA tumor viruses have "infected and completely transformed" a great many scientists to mine the secrets of RNA tumor virology. The excitement about this enzyme derives from assumptions which for the time being are all based on rather indirect evidence--that the enzyme transcribes in vivo viral RNA to DNA and that this DNA serves as a template for the replication of viral RNA (3,4). In addition this DNA is assumed to be integrated into the cellular chromosome to cause stable transformation of the cell (3,4).

The present report deals with the questions whether in vitro the enzyme has properties which are compatible with those expected from a virus-specific RNA-dependent DNA replicase. The enzyme was originally found to be RNA-dependent using endogenous viral RNA as template (1,2,5,6). Later it was found that both native and denatured double-stranded DNAs from various sources (7-11), as well as synthetic nucleic acids (7,12), could serve as templates. It has not been determined, however, whether the enzyme exhibits template-specificity for natural RNAs, in particular whether it prefers 60-70S tumor virus RNA over the RNA from other sources as might be expected for a specific replicase. Further, it has not been determined previously whether the DNA polymerase activities primed by endogenous viral RNA, by native or denatured DNAs or synthetic nucleic acids are carried out by a single enzyme, modifications of one enzyme or different enzymes. It is also not known whether the viral DNA polymerase activity corresponds to known structural proteins of the virus. Such questions may be answered if the viral DNA polymerase becomes available in a soluble and purified form. The present report extends our previous findings (5) that the virus-associated enzyme transcribes in vitro most of the sequences of 60-70S viral RNA into small pieces of DNA. In addition, we have solubilized the viral DNA polymerase to determine what it is and to examine how it works.

HYBRIDIZATION OF SR RSV ^{32}P-RNA WITH EXCESS OF *IN VITRO* SYNTHESIZED ^{3}H-DNA AND DNA FROM OTHER SOURCES

To determine how much of the endogenous 60-70S RNA of SR RSV-A was transcribed in vitro to small pieces of DNA (5) by the virus-associated enzyme, 60-70S SR RSV-A ^{32}P-RNA was annealed with various concentrations of in vitro synthesized viral ^{3}H-DNA and other DNAs. The extent of hybridization of ^{32}P-RNA with ^{3}H-DNA was measured by comparison of the RNase-resistance of the ^{32}P-RNA before and after hybridization (5). As seen in Table 1, up to 100% of the 60-70S ^{32}P-RNA was rendered RNase-resistant by hybridization with excess of ^{3}H-DNA made during an 18-hour incubation period by the virus-associated enzyme. By contrast, annealing with E. coli DNA, chick cell DNA or DNA of SR RSV-infected chick cells did not significantly increase the RNase-resistance of SR RSV ^{32}P-RNA over the 7-10% background obtained under these conditions. It may be concluded from these experiments that viral DNA polymerase is able to transcribe probably all viral RNA to DNA in vitro.

TABLE 1

Hybridization[a] of 60-70S SR RSV ^{32}P-RNA[b] with in vitro-synthesized ^{3}H-DNA[c] and DNA from other sources

Experiment	Added RNA		Added DNA (µg)					SR RSV ^{32}P-RNA resistant to RNase[e]	
	SR RSV-^{32}P (cpm)	TMV (µg)	In vitro SR-^{3}H	E. coli	chick[d]	SR-chick[d]		(cpm) av. 2 ± range	%
1	2950 (= 0.02 µg)	10	1.8	40	–	–		1980 ± 150	69
	2950 (= 0.02 µg)	10	–	30	–	–		216 ± 25	7.4
2	3000 (= 0.01 µg)	10	3.75	12	–	–		2570 ± 200	85.5
	3000 (= 0.01 µg)	10	1.25	12	–	–		2135 ± 160	71
	3000 (= 0.01 µg)	10	–	30	–	–		300 ± 20	10
	3000 (= 0.01 µg)	10	–	–	30	–		300 ± 30	10
	3000 (= 0.01 µg)	10	–	–	–	35		335 ± 25	11.5
3	2320 (= 0.01 µg)	10	4.6	–	–	–		2320 ± 130	100
	2320 (= 0.01 µg)	10	1.4	–	–	–		1820 ± 40	78.5
	2320 (= 0.01 µg)	10	–	30	–	–		170 ± 10	7.3
	2320 (= 0.01 µg)	10	–	–	30	–		200 ± 10	8.5
	2320 (= 0.01 µg)	10	–	–	–	30		230 ± 30	10

[a]*As described by Duesberg and Canaani (5).*

[b]*Prepared as described (13).*

c_Prepared as described by Duesberg and Canaani (5) by incubating 50-100 A_{260} units (determined in 0.1% SDS) of purified SR RSV in a 20-40 ml reaction mixture, containing 10^{-4} $\underline{M}$ ^{3}H-dTTP, for 18 hours. ^{3}H-DNA treated for 20 hours at room temperature with 0.2 $\underline{N}$ KOH usually contained some more material absorbing at A_{260} than could be accounted for by ^{3}H-DNA and carrier DNA. This material may be viral or cellular DNA._

d_Cellular DNAs were prepared by incubating cells in 1% SDS, 0.05 $\underline{M}$ EDTA pH 7.4, 0.2 $\underline{M}$ NaCl, 1% mercaptoethanol and 100 µg pronase per ml at 37°C for 30 minutes. After 3 phenol extractions and ethanol-precipitation the pellet was incubated in 0.2 $\underline{N}$ KOH for 20 hours at room temperature and once reprecipitated._

e_After hybridization the reaction mixture was diluted 10-fold with low salt buffer and the nucleic acids were precipitated with 2 vol ethanol. The average recovery of TCA-precipitable SR RSV ^{32}P-RNA determined from suitable aliquots at this step was taken as 100% input ^{32}P-RNA for calculation of percentages of hybridized, RNase-resistant ^{32}P-RNA. This was justified because between 90 and 100% of the 60-70S SR RSV ^{32}P-RNA and ^{3}H-DNA added to a reaction mixture remained TCA-precipitable after incubation under annealing conditions. RNase-digestion was in 1 ml 2 x SSC for 45 minutes at 30°C with 20 µg pancreatic RNase (heated 100°C for 5 minutes). TCA-precipitable radioactivity was determined as described previously (5). Experiments 1 and 2 of this Table are from Duesberg and Canaani (5); experiment 3 is from Duesberg, Vogt and Canaani (6)._

HOW THE VIRAL DNA POLYMERASE WAS SOLUBILIZED

The information, however, that virus-associated DNA polymerase transcribes besides endogenous viral RNA all kinds of other natural and synthetic nucleic acids, with the possible exception of non-viral natural RNAs, raised the question about the identity or identities of the viral RNA- and DNA-dependent polymerase activities. To answer these questions the virus-associated enzyme was solubilized with Triton X-100 and analyzed by chromatography and sucrose gradient sedimentation (14). Approximately 75% of the virus-associated DNA polymerase was rendered soluble by incubation with Triton X-100 (14). The solubilized DNA polymerase was assayed in the following experiments using KOH-denatured (0.2 $\underline{N}$ KOH, 24 hours, 20°) salmon DNA as template. Salmon DNA was found to increase the resident DNA polymerase activity of Triton X-100 disrupted virus, using presumably only endogenous viral RNA template, by a factor of 5-10.

CHROMATOGRAPHY OF THE SOLUBLE DNA POLYMERASE ON DEAE-CELLULOSE

Chromatography of the soluble DNA polymerase of ^{14}C-amino acid-labeled PR RSV-C is shown in Fig. 1A. About 30% of the starting enzymatic activity eluted at a KCl concentration between 0.15 - 0.2 M together with about 2% of the total ^{14}C-protein recovered from the column. Some activity eluted at higher KCl concentration. Due to the loss of 70% of enzymatic activity this corresponds only to a 15-fold enrichment of the enzyme. Since 80% of the ^{14}C-protein which was applied to the column was recovered, it may be concluded that the enzyme represented at most 2% of the soluble ^{14}C-proteins of the virus.

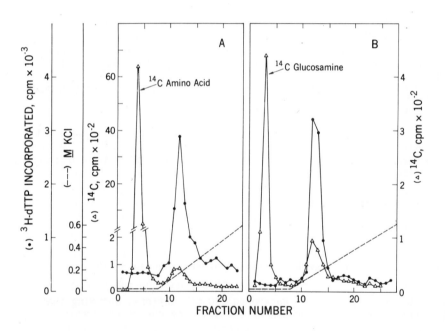

Fig. 1. *DEAE-cellulose chromatography of the soluble DNA polymerase derived from ^{14}C-amino acid-labeled (A) or ^{14}C-glucosamine-labeled (B) PR RSV-C. A) 30 µl PR RSV-C (18 A_{260}/ml in 0.1% SDS, 0.1 M NaCl, 0.01 M Tris, pH 7.4, 1 mM EDTA) and 20 µl ^{14}C-amino acid-labeled PR RSV-C (~ 100,000 cpm), 0.5 ml buffer containing 0.02 M KCl, 0.01 M Tris, pH 8.0, 2 mM EDTA, 1 mM DTT and 25 µl Triton X-100 (10% v/v) were processed and centrifuged at 150,000g as described for Fig. 2A. The following was at 5°C. The*

*supernatant was applied to a DEAE-cellulose column
(3 x 0.6 cm) equilibrated in buffer containing 0.05 $\underline{M}$
Tris, pH 7.4, 0.01 $\underline{M}$ MgCl$_2$, 0.02 $\underline{M}$ KCl, 0.2 m$\underline{M}$ EDTA,
1 m$\underline{M}$ DTT and 5% glycerol. After washing with several ml of this buffer a linear KCl gradient in the
same buffer was applied and 10-min (~ 1 ml) fractions
were collected. Analysis of the gradient fractions
was as described for Fig. 2A.   B) Same as for A
but 20 μl ^{14}C-glucosamine-labeled PR RSV-C (~ 100,000
cpm) was used. The data are from Duesberg, v. d. Helm
and Canaani (14).*

Fig. 1B illustrates that about 30% of the radioactive glycoprotein (15) of solubilized virus which eluted from the column cochromatographed with the viral DNA polymerase. Elution from DEAE-cellulose at 0.15 - 0.2 M KCl indicates that the enzyme has an isoelectric point below 6 (16), which is similar to that of other nucleic polymerases (17) of similar size (see below).

SEDIMENTATION COEFFICIENT ($S_{20,w}$) OF THE SOLUBILIZED VIRAL DNA POLYMERASE BEFORE AND AFTER INCUBATION WITH RNase

The soluble DNA polymerase sedimented as a single component with an estimated (18) $S_{20,w}$ of 8S based on the 4.3S ^{14}C-bovine serum albumin (BSA) (19) marker (Fig. 2A). After incubation with pancreatic RNase the $S_{20,w}$ of the enzyme was reduced to 6S (Fig. 2D). The recovery of DNA polymerase activity after sucrose gradient sedimentation was 80 ± 5% (av. 4 ± S. E.) without RNase treatment and varied between 40 and 80% after RNase treatment.

Using a ^{14}C-BSA standard (MW = 67,000) (19) the MW of the 8S DNA polymerase can be estimated by the formula S_{w_1}/S_{w_2} = $(MW_1/MW_2)^{2/3}$ (20) to be about 170,000 daltons and that of the 6S DNA polymerase to be around 110,000 daltons, assuming that the relationship between MW and $S_{20,w}$ of the enzyme resembles that of BSA. The difference between the estimated MWs of the 8S and the 6S form of the DNA polymerase may be due to the removal by RNase of small pieces of enzyme-associated RNA but may have other explanations.

HYDRODYNAMIC AND PHEROGRAPHIC ANALYSES OF THE 8S AND THE 6S
VIRAL DNA POLYMERASE AND OTHER SOLUBLE COMPONENTS OF THE VIRUS

It is shown in Fig. 2B that the 8S DNA polymerase did not correspond to a distinct peak of soluble ^{14}C-protein of the virus.

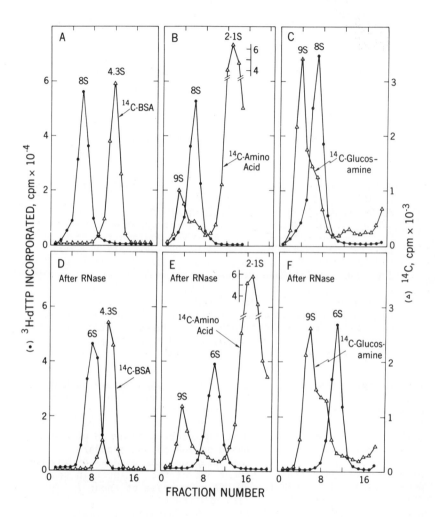

Fig. 2. *Sucrose gradient sedimentation of the soluble DNA polymerase of PR RSV-C before (A,B,C) and after (D,E,F) RNase treatment. A) 35 μl PR RSV-C in 50% buffered glycerol at a concentration of 18 A_{260}/ml (cf., Fig. 1), 200 μl 0.1 M KCl, 0.01 M Tris, pH 8, 2 mM EDTA, 1 mM DTT, and 15 μl Triton X-100 (10% v/v) were incubated at 38°C for 15 min. Subsequently this solution was centrifuged in a Spinco 40 rotor at 40,000 rpm, 20°C for 15 min in a polycarbonate tube. The supernatant was mixed with ^{14}C-BSA*

and layered on a linear sucrose gradient (15 - 30% v/v) containing 0.1 $\underline{M}$ KCl, 0.01 $\underline{M}$ Tris, pH 8.0, 1 m$\underline{M}$ EDTA, 0.2 m$\underline{M}$ DTT, 0.2% Triton X-100. The gradient was poured over a 0.3-0.4 ml cushion of 65% sucrose in the same buffer. Centrifugation was in a polyallomer tube in a Spinco SW 65 rotor, 20°C for 9.5 hr at 65,000 rpm. Twelve-drop fractions were collected. 10-μl aliquots were assayed for the polymerase activity during a 90-min incubation period in a 50-μl reaction mixture (5) using 5 μg denatured salmon DNA as template. ^{14}C-radioactivity was determined by placing appropriate aliquots in toluene-based scintillation fluid containing 20% NCS (Nuclear Chicago) and counting in a TriCarb scintillation counter. B) ^{14}C-amino acid-labeled PR RSV-C containing about 750,000 cpm and 30 μl PR RSV-C at a concentration of 18 A_{260}/ml were disrupted and centrifuged for 9 hr as described for A. C) ^{14}C-glucosamine-labeled PR RSV-C containing about 250,000 cpm were mixed with 30 μl PR RSV-C (18 A_{260}/ml) and processed as described for B. D) Same as A but 1 μg pancreatic RNase (heated 100°C, 5 min) was included in the disruption mixture. E) Same as described for B but disruption was in presence of pancreatic RNase as described for D. F) Same as C but disruption was in presence of pancreatic RNase as described for D and centrifugation was for only 7 hr. The data are from Duesberg, v. d. Helm and Canaani (14).

The enzyme overlapped with a ^{14}C-protein component which had a slightly higher $S_{20,w}$ than the enzymatic activity and consisted of about 12% of the radioactive protein in the gradient. This component had been identified previously (15) as an "8S" glycoprotein complex and it will here be referred to as 9S glycoprotein component (Fig. 2E,F), based on a 4.3S BSA standard. [The previous estimate (15) was based on tRNA used as 4S instead of a 4.5S standard (18)]. About 85% of the ^{14}C-protein sedimented more slowly than the 8S DNA polymerase at about 1 - 2S (Fig. 2B,E). Both the $S_{20,w}$ (15) and the electrophoretic properties (not shown, ref. 21) of this material indicated that it was the viral group-specific antigen. The three peak fractions (80%) of the 8S DNA polymerase in Fig. 2B cosedimented with about 5% of the soluble ^{14}C-protein, representing a 15-fold purification of the enzyme.

If the $S_{20,w}$ of the DNA polymerase was reduced to 6S by treatment with RNase (Fig. 2E) the sedimentation distribution of the major viral protein components was not significantly altered but the three peak fractions (80%) of the enzymatic activity cosedimented with only 2% of the viral ^{14}C-proteins (Fig. 1E).

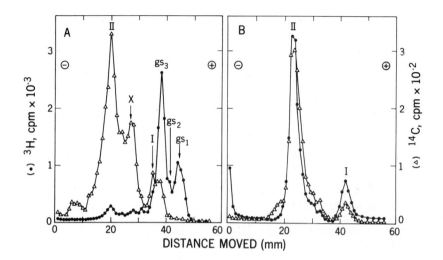

Fig. 3. *Electrophoretic analysis of the ^{14}C-amino acid- or ^{14}C-glucosamine-labeled material of solubilized virus which cosediments with the viral DNA polymerase. A) Aliquots of fractions 9, 10, 11 (Fig. 2E) of ^{14}C-amino acid-labeled virus were precipitated with 5 vol ethanol and 50 µg BSA. The precipitate was mixed with an appropriate amount of unfractionated ^3H-amino acid-labeled virus to provide electrophoretic markers and analyzed by electrophoresis in 5% SDS-polyacrylamide gels as described previously (15). B) Aliquots of fractions 10, 11, 12 (Fig. 2F) of ^{14}C-glucosamine-labeled virus were precipitated as described for A. Prior to electrophoresis unfractionated ^3H-glucosamine-labeled PR RSV-C was added to provide glycoprotein reference markers (15). The data are from Duesberg, v. d. Helm and Canaani (14).*

This represents a 40-fold purification of the enzyme. Fig. 3A shows an electrophoretic analysis of the ^{14}C-protein which cosedimented with the 6S DNA polymerase shown in Fig. 1E. Unfractionated ^3H-amino acid-labeled virus was added to provide electrophoretic markers for the major structural proteins of the virus, which include the proteins of the group-specific antigen gs_1, gs_2 and gs_3 and the two viral glycoproteins I (37,000 daltons) and II (105,000 daltons) (15). The majority of the ^{14}C-protein had the same electrophoretic mobility as the two glycoproteins I and II (cf., Fig. 3B) (15); very little coelectrophoresed with the components of group-specific antigen. In addition, at least one new ^{14}C-peak, labeled X (Fig. 3A), appeared in the gel, which had not been identified previously (15,21). A very similar electrophoretic pattern was obtained when the ^{14}C-protein, which cochromatographed with the enzyme on DEAE-cellulose, was analyzed under these conditions (not shown). It can only be deduced from these experiments that the DNA polymerase is probably not identical with the group-specific antigen of the virus.

Cosedimentation of the 8S DNA polymerase with the soluble ^{14}C-glycoprotein of the virus is shown in Fig. 2C. The 8S polymerase overlapped with 25% of the 9S glycoprotein complex whereas the 6S DNA polymerase overlapped with only 10% of the soluble ^{14}C-glycoprotein (Fig. 2F). Resedimentation of the enzymatic activity after overnight precipitation in 60% ammonium sulfate in the presence of 50 μg BSA did not change the relative sedimentation distributions of the enzyme and the glycoprotein (not shown).

Electrophoresis of the ^{14}C-glycoprotein which cosedimented with the 6S DNA polymerase (fractions 10-12, Fig. 2F) is shown in Fig. 3B. More than 90% of this ^{14}C-glycoprotein coelectrophoresed with the two known glycoproteins I and II of unfractionated virus (15) and no new distinct glycoprotein component appeared. The same electrophoretic pattern was obtained when the peak fractions of the 9S ^{14}C-glycoprotein component (fractions 4-7, Fig. 2F), which did not sediment with enzymatic activity, were analyzed (not shown) or when the ^{14}C-glycoprotein which cochromatographed with the enzyme on DEAE-cellulose (Fig. 2B) was electrophoresed (not shown). Results very similar to those described for PR RSV-C were obtained with SR RSV-A.

It may be concluded that the 6S DNA polymerase is different from 90% of the viral glycoprotein. The 10% of viral glycoprotein which could not be separated from the 6S DNA polymerase by our methods had the same electrophoretic distribution as the two known glycoproteins of the virus.

TEMPLATE SPECIFICITY OF THE SOLUBLE DNA POLYMERASE

The soluble DNA polymerase presented an opportunity to examine whether the enzyme was able to use added RNA templates, and whether it preferred 60-70S PR RSV RNA over other RNAs. This question could not be answered previously because the enzyme had not been solubilized, nor was it experimentally possible to eliminate the endogenous RNA template without at the same time affecting added natural RNA templates.

The 60-70S RNA of PR RSV-C was found to have per unit weight 20 - 40% of the template activity of salmon DNA for the 8S and the 6S DNA polymerase (Fig. 4A,B). The kinetics on the peak fraction of RSV RNA- and salmon DNA-primed enzymatic activities of the 8S DNA polymerase were nearly linear and almost indistinguishable (insert in Fig. 4A). TMV RNA had only 2% and influenza virus RNA only 5% of the template activity of 60-70S PR RSV RNA for the 8S DNA polymerase (Fig. 4A). Both the RNA- and the DNA-primed polymerase activities of the 8S and the 6S form of the enzyme coincided in the gradients shown in Fig. 4.

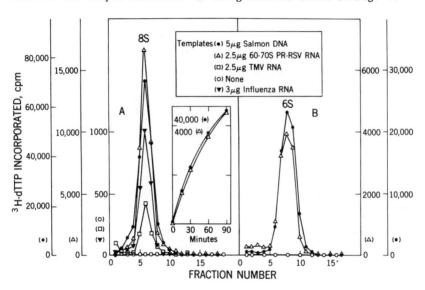

Fig. 4. *Analysis of the template activities of various nucleic acids for the 8S (A) and the 6S (B) viral DNA polymerase activity. A) The sucrose gradient used for assay of the 8S DNA polymerase is the same as in Fig. 2A. 10-µl aliquots of each fraction were incubated for 90 min with either 5 µg salmon DNA or 2.5 µg 60-70S PR RSV-C RNA, or 2.5 µg TMV RNA, or 3 µg influenza virus RNA or no added nucleic acid in the standard DNA polymerase assay (Fig. 2, ref. 14). Preparation of 60-70S PR RSV RNA: PR RSV-C RNA (13,22) was incubated in 250 µl, 0.01 M Tris, pH 7.4, 0.15 M NaCl, 2 mM $MgCl_2$ containing 10 µg DNase (RNase free) for 15 min at 38°C and 1 hr at room temperature to degrade associated DNA (23). Subsequently it was diluted with 1 ml standard buffer containing 0.2% SDS, once extracted*

with phenol and after ethanol-precipitation the 60-70S PR RSV-C RNA was prepared by sucrose gradient sedimentation as described previously (13). The insert shows the kinetics of ^{3}H-dTTP-incorporation on 10-µl aliquots of fraction 6 in the presence of 60-70S PR RSV RNA or salmon DNA. B) The sucrose gradient of the 6S DNA polymerase is the same as in Fig. 2D. The respective fractions were tested as described for A. The data are from Duesberg, v. d. Helm and Canaani (14).

Moreover both the 8S and the 6S DNA polymerase had no detectable endogenous template activities. Further it was found when different fractions were mixed that the RNA-primed enzymatic activities of all fractions of the 8S DNA polymerase peak tested were 70-100% additive; the same was true when DNA-primed activities were measured. This suggests that no enhancing nor inhibiting factor was in this region of the gradient.

DISCUSSION

The experiments reported here indicate that the small pieces of DNA made by the unpurified polymerase activities in Rous sarcoma virus in vitro include at least 90% of the sequences in the 60-70S viral RNA. These results are compatible with the hypothesis of Temin (3) that the RNA of avian tumor viruses is replicated via a DNA intermediate and that the virus-associated enzyme may play the role of an RNA-DNA replicase in the infected cell.

It has been found that the $S_{20,w}$ of the soluble DNA polymerase can be reduced by treatment with RNase from 8S to 6S without affecting the $S_{20,w}$'s of the components of the viral glycoprotein or group-specific antigen. This distinctive RNase-sensitivity of its $S_{20,w}$ permitted a 40-fold purification of the enzyme based on the total soluble radioactive protein of the virus. Since the particle weight of RSV is about 5×10^8 daltons and its protein content 60% (24), at most 1.5% of the mass of the virion can be DNA polymerase. This calculation assumes that the specific radioactivities of the enzyme and the structural proteins of the virus are the same. It follows that a maximum of 70 6S DNA polymerase molecules (MW = 110,000) may be present per virion. However, this is considered an upper estimate, because the 40-fold purified enzyme was found to be contaminated by other structural proteins of the virus. The purified enzyme contained at least one protein component, which had not previously been identified as a structural component of the virus and it contained 10% of the soluble viral glycoprotein.

However, the glycoprotein of the purified enzyme may not be an essential part of it because it consisted mainly of the two components of the 9S glycoprotein complex of the virus which had no onzymatic activity.

Both the 8S and the 6S DNA polymerase were found to have no endogenous template activities. This leaves open the question whether the reduction of the $S_{20,w}$ of the enzyme from 8S to 6S by RNase treatment is due to removal of short pieces of viral RNA or perhaps of cellular RNA without template activity. It is conceivable that RNase removes from the 8S DNA polymerase very short pieces of viral RNA, which are unable to serve as a template but are capable of affecting its $S_{20,w}$. Such short pieces of viral RNA presumably arise by degradation of the 60-70S viral RNA (22) by virus-associated RNase (25), as tumor virus RNA is known to become RNase-sensitive in the presence of nonionic detergents (22). Short pieces of enzyme-associated RNA might enhance the activity of templates added to the purified enzyme. This is compatible with the decrease of enzymatic activity observed after RNase treatment.

The coincidence in sucrose gradients of both the RSV RNA and the DNA-primed DNA polymerase activities in both the 8S and the 6S form makes it likely that the two activities reside in the same basic enzyme. This evidence does not exclude, however, that different sites of the enzyme or different factors might specify the RNA- or DNA-dependent activities. It would appear then that this enzyme, which is known to have rather nonspecific template requirements if primed by natural DNAs (7-11) or synthetic nucleic acids (7,12), has specific template requirements for natural RNAs, i.e., a preferential affinity for 60-70S RSV RNA.

REFERENCES

1. H. M. Temin and S. Mizutani, Nature, 226 (1970) 1211.

2. D. Baltimore, Nature, 226 (1970) 1209.

3. H. M. Temin, Virology, 23 (1964) 486.

4. News and Views, Nature, 227 (1970) 998.

5. P. H. Duesberg and E. Canaani, Virology, 42 (1970) 783.

6. P. H. Duesberg, P. K. Vogt and E. Canaani, II Lepetit Colloquium, Paris, Nov. 6-8, 1970 (North-Holland, Amsterdam, in press).

7. S. Mizutani, D. Boettiger and H. M. Temin, Nature, 228 (1970) 424.

8. S. Spiegelman, A. Burny, M. R. Das, J. Keydar, J. Schlom, M. Travnicek and K. Watson, Nature, 227 (1970) 1029.

9. J. Riman and G. S. Beaudreau, Nature, 228 (1970) 427.

10. J. P. McDonnell, A-C Garapin, W. E. Levinson and N. Quintrell, Nature, 228 (1970) 433.

11. K. Fujinaga, J. T. Parsons, J. W. Beard, D. Beard and M. Green, Proc. Nat. Acad. Sci. U.S.A., 67 (1970) 1432.

12. S. Spiegelman, A. Burny, M. R. Das, J. Keydar, J. Schlom, M. Travnicek and K. Watson, Nature, 228 (1970) 430.

13. P. H. Duesberg and P. K. Vogt, Proc. Nat. Acad. Sci. U.S.A., 67 (1970) 1673.

14. P. H. Duesberg, K. v. d. Helm and E. Canaani, Proc. Nat. Acad. Sci. U.S.A., in press.

15. P. H. Duesberg, G. S. Martin and P. K. Vogt, Virology, 41 (1970) 631.

16. W. Zillig, E. Fuchs and R. Millette, in: Procedures in Nucleic Acid Research, eds. G. L. Cantoni and D. R. Davies (Harper and Row, New York, 1966) p. 323.

17. M. Chamberlin, J. McGrath and L. Waskell, Nature, 228 (1970) 227; R. Knippers, Nature, 228 (1970) 1050.

18. R. G. Martin and B. N. Ames, J. Biol. Chem., 236 (1961) 1372.

19. In: Handbook of Biochemistry, ed. H. A. Sober (Chemical Rubber Company, Cleveland, Ohio, 1968) p. C-17.

20. H. K. Schachman, Ultracentrifugation in Biochemistry (Academic Press, New York, 1959) 269 pp.

21. P. H. Duesberg, H. L. Robinson, W. S. Robinson, R. J. Huebner and H. C. Turner, Virology, 36 (1968) 73.

22. P. H. Duesberg, in: Current Topics in Microbiology and Immunology, Vol. 51 (Springer-Verlag, Berlin, 1970) p. 79.

23. W. Levinson, J. M. Bishop, N. Quintrell and J. Jackson, Nature, 227 (1970) 1023.

24. P. K. Vogt, in: Advances in Virus Research, Vol. 11, eds. K. M. Smith and M. A. Lauffer (Academic Press, New York, 1965) p. 293.

25. M. Rosenbergova, F. Lacour and J. Huppert, C. R. Acad. Sci. Paris, 260 (1965) 5145.

This work was supported by U.S. Public Health Service Research Grant CA 11426 from the National Cancer Institute and by Cancer Research funds of the University of California.

DISCUSSION

Dr. D. Baltimore, Massachusetts Institute of Technology: I'd like to ask just 2 short questions. Have you done the sucrose gradient and column chromatography purifications serially?

Dr. P. Duesberg: Yes, I've done that too and as you saw with the DAE column it didn't help very much separating the glycoprotein from the enzyme. The gradient was the best method of function reaction so far. We also precipitated the enzyme with ammonium sulphate and resedimentated it and didn't separate it any further from glycoprotein. So, I guess, a different method altogether has to be tried to test whether the glycoprotein is associated with the enzyme, a part of it, or contaminating it.

Dr. Baltimore: In the experiments in which you look at the covering of the RNA by the DNA product, you're putting in a large excess of DNA. Is this necessary kinetically or is it necessary in order to get to saturation?

Dr. Duesberg: It was necessary in order to get it 100% resistant to RNA-ase. It might suggest now random transcription of the viral RNA to small segments of DNA. It could be, say that 30% of the RNA's are transcribed much better than the remaining 70%. In order to get complete protection of viral RNA, you might have to use a large excess of DNA in order to cover all sequences. But that's only a speculation. Mike Bishop told me, that he has done some Cot experiments, you know, Cot experiments like Britten and Cohn. He says also that the populations of DNA molecules made by non-purified enzyme using endogenous RNA template are heterogeneous; there is one fraction which seems to self-anneal much faster than another fraction.

Dr. H. M. Temin, University of Wisconsin: In the experiments where you were making all that DNA, how much DNA was made for the amount of template that was present in the virus. In other words, just what was the efficiency of this stuff you used in that reaction. You can't tell from your paper and I'm really curious.

Dr. Duesberg: Well I should have done it then, but I was more qualitatively minded than now, and I haven't calculated it. Okay, I'll do it.

Dr. J. Hurwitz, Albert Einstein College of Medicine:
I just have 2 brief questions. Number 1, what is the amount of ribonuclease which is used for the conversion of the 8S material to the 6S?

Dr. Duesberg: I varied the concentration; the lowest concentration which I used here was 1 microgram in 250 microliters.

Dr. Hurwitz: And the second question is with regards to the product of the RNA dependent deoxynucleotide incorporation.

Dr. Duesberg: Yes....

Dr. Hurwitz: Have you looked at the product at all? Is the DNA attached to the RNA covalently?

Dr. Duesberg: No, that I haven't tested, no. We have done some preliminary sedimentations and it may be a little bit bigger than it used to be which is maybe what you really wanted to know.

TEMPLATE AND PRIMER REQUIREMENTS FOR THE AVIAN MYELOBLASTOSIS DNA POLYMERASE

DAVID BALTIMORE and DONNA SMOLER
Department of Biology
Massachusetts Institute of Technology
Cambridge, Massachusetts
02139

INTRODUCTION

Virions of the RNA tumor viruses contain DNA polymerase activity (1, 2). The enzyme (or enzymes) responsible for this activity can copy the viral RNA (the "endogenous" reaction) or can copy added templates of various types including homopolymers (the "exogenous" reaction) (3, 4, 5). In order to systematically study the template requirements of the exogenous reaction we have developed an assay which employs a long homopolymer template and a short oligodeoxyribonucleotide primer. With this assay we have compared the template activity of ribohomopolymers with that of their deoxyribohomopolymer analogues. This comparison shows that the enzymes in tumor viruses prefer ribohomopolymer templates while E. coli DNA polymerase I (the "Kornberg enzyme") prefers a deoxyribohomopolymer template.

It has been shown by Spiegelman et al (5) that the duplex poly A·poly dT will stimulate poly dT synthesis by the RNA tumor virus DNA polymerase; poly A alone, however, is inactive as a template. This result suggests that the function of the poly dT in the duplex may be to act as a primer for the copying of poly A. To test this possibility small amounts of dT oligomers were added to poly A in the presence of avian myeloblastosis virus (AMV); ^{3}H-TTP; the non-ionic detergent, NP40; and the appropriate co-factors for DNA synthesis (Table 1). Without any added dT oligomer as primer there was no detectable poly dT synthesis but oligomers from $(dT)_4$ to $(dT)_{12-16}$ supported extensive synthesis of poly dT on the poly rA template. The oligomer, $(dT)_{12-16}$, by itself was unable to stimulate synthesis, indicating the need for both a template and a primer. Much less primer than template is required for stimulation of a maximal rate of poly dT synthesis.

If the reaction was allowed to proceed for 2 hours under an atmosphere of N_2 (to avoid oxidation in the reaction mixture which inhibits enzyme activity after an hour of synthesis), then the amount of poly dT synthesis was equal to 80% or more of the amount of added poly A. However, synthesis of further poly dT did not occur indicating that a true poly A·poly dT duplex is inactive as a template for poly dT synthesis (Baltimore and Smoler, unpublished results).

TABLE 1
Poly A as a Template for the AMV DNA Polymerase

Additions	Counts/min ^{3}H-TMP Incorporated
None	198
Poly A, 5 µg/ml	605
Poly A, 5 µg/ml; (dT)$_4$, 5 µg/ml	18,196
Poly A, 5 µg/ml; (dT)$_6$, 5 µg/ml	23,182
Poly A, 5 µg/ml; (dT)$_8$, 5 µg/ml	56,776
Poly A, 5 µg/ml; (dT)$_{12-16}$, 0.5 µg/ml	82,734
Poly A, 5 µg/ml; (dT)$_{12-16}$, 0.02 µg/ml	40,949
(dT)$_{12-16}$, 5 µg/ml	331

Reaction mixtures of 0.1 ml contained the following: 0.05M Tris-HCl, pH 8.3; 0.006M Mg acetate; 0.02M dithiothreitol; 0.06M NaCl; 0.2% Nonidet P-40; 202 pmoles of ^{3}H-TTP (6,940 counts /min/pmole) and 0.8 µg of avian myeloblastosis virus protein. Samples were incubated for 60 minutes at 37°C and then prepared for counting as described previously (1). The virus had been purified by differential centrifugation and isopycnic banding in a sucrose gradient from plasma of infected chickens kindly supplied by Dr. J. Beard. The ribohomopolymers were products of Miles Laboratories. The oligomeric deoxyribonucleotide primers were products of Collaborative Research, Inc., Waltham, Mass.

Poly C and Poly I as Templates: Similar experiments with poly C and poly I using the complementary oligodeoxyribonucleotide primer have shown that both of these ribohomopolymers are excellent templates for the AMV DNA polymerase (Table 2). In both cases, a primer and template are necessary and no reaction occurs when either is added separately. The oligomer, (dT)$_{12-16}$, cannot prime synthesis on a poly C template indicating that the primer and template must be able to form hydrogen-bonded base pairs.

Poly U has consistently been inactive as a template even in the presence of (dA)$_{12-16}$.

Poly A as a Template for E. coli DNA Polymerase I: Lee-Huang and Cavalieri (6) showed that poly A·poly U is a template for E. coli DNA polymerase I and Riley et al (7) have confirmed this result. Poly A plus (dT)$_{12-16}$ will also support poly dT synthesis using this enzyme (Table 3). The poly A is completely covered by poly dT and then the reaction ceases. Poly dA with

$(dT)_{12-16}$ is also covered by poly dT. Poly dA supports synthesis at a rate at least 10-fold greater than poly A under comparable conditions (unpublished results).

TABLE 2

Poly I and Poly C as Templates for the AMV DNA Polymerase

Additions	Counts/min/Incorporated ^{3}H-dCMP	^{3}H-dGMP
None	<200	
Poly I, 5 µg/ml	<200	
Poly I, 5 µg/ml; $(dC)_{12-16}$, 5 µg/ml	23,648	
$(dC)_{12-16}$, 5 µg/ml	<200	
None		<200
Poly C, 5 µg/ml		<200
Poly C, 5 µg/ml; $(dG)_{12-16}$, 0.2 µg/ml		441,875
Poly C, 5 µg/ml; $(dG)_{12-16}$, 5 µg/ml		420,766
Poly C, 5 µg/ml; $(dT)_{12-16}$, 1 µg/ml		<200
$(dG)_{12-16}$, 5 µg/ml		<200

Conditions were as described for Table 1 except that the nucleotides provided in the reaction mixtures were 94 pmoles of ^{3}H-dCTP (14,800 counts/min/pmole) in Experiment 1 and 274 pmoles of ^{3}H-dGTP (5,100 counts/min/pmole) in Experiment 2.

Neither poly C, poly I or poly U will act as a template for E. coli DNA polymerase I even in the presence of the corresponding oligomer. However, poly dI, poly dC and poly dT are excellent templates under the same conditions.

E. coli DNA polymerase I, therefore, shows a decided preference for deoxyribohomopolymers over ribohomopolymers. Even with poly A, where synthesis is evident, poly dA is the preferred template in experiments where rates of synthesis are measured rather than yields.

Deoxyribohomopolymers Templates for the AMV DNA Polymerase: Of the four deoxyribohomopolymers, only poly dC will act as a template for the AMV DNA polymerase. When rates of synthesis are measured, poly C supports a rate about 2-fold faster than poly dC.

TABLE 3

Poly A and Poly dA as Templates for E. coli DNA Polymerase I

Additions	pmoles ^{3}H-TMP Incorporated	% of Input Polymer
None	< 1	-
Poly A, 380 pmoles	< 1	< 0.1
Poly A, 380 pmoles, $(dT)_{12-16}$, 560 pmoles	344	91
None	< 1	< 0.1
Poly dA, 270 pmoles	< 1	< 0.1
Poly dA, 270 pmoles, $(dT)_{12-16}$, 560 pmoles	272	100

Reaction mixtures of 0.1 ml contained the following: 0.05M potassium phosphate buffer, pH 7.4, 0.006M Mg acetate, 18 nmoles of ^{3}H-TTP (3.8 x 10^4 counts/min/nmole) and 0.0076 μg of purified E. coli DNA polymerase I (kindly supplied by Dr. A. Kornberg) (8). The deoxyribohomopolymers were a kind gift from Dr. F. Bollum.

The AMV DNA polymerase, therefore, shows a strong preference for ribohomopolymer templates. The one deoxyribohomopolymer that is active (poly dC) supports a somewhat slower rate of synthesis than its ribohomopolymer analogue (poly C).

CONCLUSION

Although the AMV DNA polymerase can use both ribo- and deoxyribohomopolymers as templates, the enzyme shows a clear preference for ribohomopolymers in the assay described here. The uniqueness of this behavior is evident when AMV DNA polymerase is compared to E. coli DNA polymerase I. Mouse leukemia virus DNA polymerase also prefers ribohomopolymers over deoxyribohomopolymers although its spectrum of specific preferences is different from AMV (unpublished results). The tumor virus enzymes therefore appear to be truly RNA-dependent DNA polymerases.

The methods described here should allow any DNA polymerase to be screened for its relative activity on DNA and RNA templates. In searching for RNA-dependent DNA polymerases in cells it would appear to be best to compare the activity of paired polymers (such as poly dA and poly A) in order to differentiate an RNA-dependent polymerase from one which is DNA-dependent.

REFERENCES

1. D. Baltimore, *Nature*, 226 (1970) 1209.

2. H. Temin and S. Mizutani, *Nature*, 226 (1970) 1211.

3. S. Spiegelman, A. Burny, M.R. Das, J. Keydar, J. Schlom, M. Travnicek and K. Watson, *Nature*, 227 (1970) 1029.

4. S. Mizutani, D. Boettiger and H. M. Temin, *Nature*, 228 (1970) 424.

5. S. Spiegelman, A. Burny, M. R. Das, J. Keydar, J. Schlom, M. Travnicek and K. Watson, *Nature*, 228 (1970) 430.

6. S. Lee-Huang and L. F. Cavalieri, *Proc. Natl. Acad. Sci. U. S.*, 50 (1963) 1116.

7. M. Riley, B. Maling and M. J. Chamberlin, *J. Mol. Biol.*, 20, (1966) 359.

8. T. M. Jovin, P. T. Englund and L. L. Bertsch, *J. Biol. Chem.*, 244 (1969) 2996.

We wish to thank Drs. F. Bollum and O. Friedman for providing polymers used in these studies. This work was supported by grants AI-08388 from the U.S. Public Health Service and E-512 from the American Cancer Society. D.B. is a Faculty Research Awardee of the American Cancer Society.

REPLICATION OF REOVIRUS

A. F. Graham and S. Millward
Department of Biochemistry
McGill University School of Medicine
Montreal, P.Q. Canada

During the past few years at least half a dozen viruses have been discovered that have genomes of double-stranded RNA (dsRNA) (1,2,3). While this feature alone would serve to distinguish a new group of viruses perhaps the most unusual characteristic of the genomes is that they appear to be discontinuous structures. At least, none can be isolated as an unbroken length of double-helical RNA. Invariably the genomes are obtained as a number of fragments on removal from the virions.

Of these viruses only reovirus has yet been studied intensively and we shall consider it in detail to illustrate a number of interesting consequences that have arisen from this discontinuous type of genomic structure. We shall start by summarizing the major observations and then discuss each one in some detail. First, the reovirus genome is covalently discontinuous and contains ten segments of dsRNA. Second, all the information in six of the ten genomic segments could be used to code for the known viral capsid proteins. Third, transcription of the viral genome is regulated in the infected cell but late in the viral growth cycle all ten segments of the genome are copied into mRNA's of corresponding sizes. Fourth, reovirus contains an RNA polymerase as an integral constituent of the virion and this enzyme can transcribe the ten genomic segments <u>in vitro</u>.

Commencing with the structure of reovirus, the proteins of the virion are arranged in the form of a double capsid. The inner capsid, or core, is approximately 450 Å in diameter and contains the viral genome. An outer layer of capsomeres, probably 92 in number, is arranged on the surface of this core to give an icosahedral form and an overall diameter of 550 to 600 Å. The outer capsomeres can be stripped off most simply by brief digestion with chymotrypsin, and the inner core thereby released.

The total molecular weight of RNA in the virion is about 20×10^6 daltons. Approximately 5×10^6 daltons of this are represented by a heterogenous collection of single-stranded oligonucleotides whose maximum chain length is probably twleve nucleotides (4). No function is yet known for these oligonucleotides. Approximately 15×10^6 daltons of RNA are consigned to the double-stranded structure of the genome (5, 6). No one has yet been able to isolate an unbroken piece of dsRNA with this molecular weight in spite of many attempts. Instead, on extraction of the dsRNA either from virions or infected cells, it breaks up into smaller fragments. This fragmentation can be visualized most simply by sedimenting the extracted RNA through a sucrose gradient (7,8). Three size-classes of double-stranded RNA are seen in the gradient which will be termed L (large), M (medium) and S (small). The same trimodal distribution of sizes was also seen by several workers when they examined the RNA in the electron microscope. Average molecular weights can be calculated from the observed lengths (9) for each size-class and are shown in Fig. 1 (6). Clearly, from these estimates and the total weight of the genome of approximately 15×10^6 daltons there must be more than one piece of RNA in each size-class.

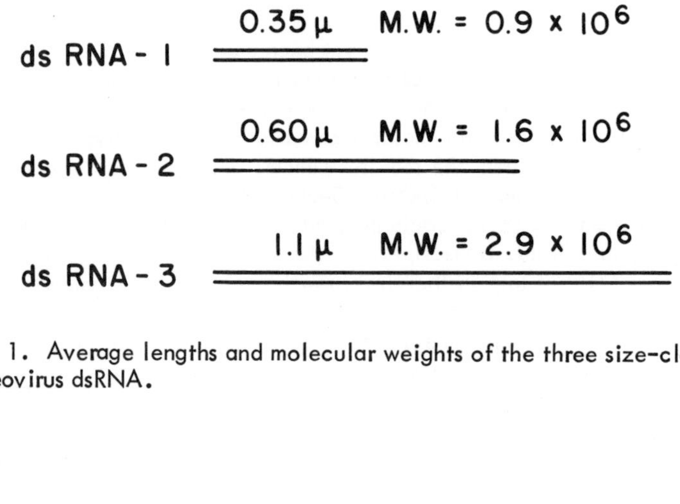

ds RNA - 1 $0.35\,\mu$ M.W. = 0.9×10^6

ds RNA - 2 $0.60\,\mu$ M.W. = 1.6×10^6

ds RNA - 3 $1.1\,\mu$ M.W. = 2.9×10^6

Fig. 1. Average lengths and molecular weights of the three size-classes of reovirus dsRNA.

The number of pieces in the genome has been defined by analyzing ^{3}H-labeled dsRNA by electrophoresis on polyacrylamide gels as shown in Fig. 2 (10). The three size-classes of dsRNA are apparent but each is resolved into several components. It can be calculated from the amount of radioactivity in the various peaks and from the corresponding molecular weight that each peak represents a single component. Thus, we conclude (5, 10) that the genome of reovirus contains ten pieces of double-stranded RNA. The L class has three components, the M class has three and the S class has four components. The approximate molecular weights of the individual segments can be found readily. When logarithims of the average molecular weight of the three size classes of dsRNA are plotted against distance migrated in a polyacrylamide gel they fall on a straight

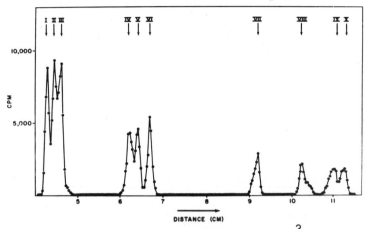

Fig. 2 Electrophoretic analysis on polyacrylamide gel of ^{3}H-labeled dsRNA (10). The ten segments fall into three size-classes L (I, II, III), M (IV, V, VI) and S (VII, VIII, IX, X).

line (5, 1). From this line one can determine the molecular weight of each segment from the distance migrated in the gel. These estimates are shown in Table 1.

Table 1
Correlation between classes of genomic segments, messenger RNAs and capsid polypeptides of reovirus

Genome (1)				Messenger RNA		Capsid polypeptides (13)			
Size Class	Segment	Mol. wt Daltons $\times 10^{-6}$	Function	Size Class	Component	Size Class	Component	Mol. wt Daltons $\times 10^{-4}$	Origin
L	I	2.7	Early	I	l_1	λ	λ_1	15.5	Core
	II	2.6	Late?		l_2		λ_2	14.0	Core
	III	2.5	Late		l_3				
M	IV	1.8	Late	m	m_1	μ	μ_1	8.0	Capsomere
	V	1.7	Late		m_2				
	VI	1.6	Early		m_3		μ_2	7.2	Capsomere
S	VII	1.1	Late	s	s_1	σ	σ_1	4.2	Capsomere
	VIII	0.85	Late		s_2				
	IX	0.76	Early		s_3		σ_2	3.8	Core
	X	0.71	Early		s_4		σ_3	3.4	Capsomere

An obvious question to ask now is how do these pieces of dsRNA arise. Studies of hybridization between the three size-classes (7, 11) showed that there was no homology between them. The segments could not have arisen as a result of random breakage in a continuous double-helical genome. There must therefore be specific weak points in the genome.

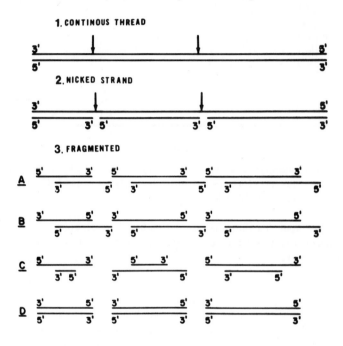

Fig. 3. Possible models for the reovirus genome. Arrows on Models 1 and 2 indicate potential points of breakage

Several types of structure can be postulated for the genome and these are shown in Fig. 3. In trying to decide between these models a technique was employed for specifically labeling 3'-OH terminals of ribonucleic acids. Essentially, the 3'-OH is oxidized with periodate; the resulting dialdehyde is then reduced with tritiated borohydride which introduces two tritium atoms per 3'-terminal. This procedure was applied to intact, purified virions and the RNA was extracted. The same procedure was also carried out on viral RNA after extraction and the two samples of oxidized and reduced RNA were compared by gel electrophoresis. The distribution of ^3H under the various peaks was the same in both analyses (6). This result means that when the dsRNA is in situ in the virion there are just as many 3'-OH ends free as there are after the RNA has been extracted from the virion. Thus, model 1 can be eliminated. Model 1 predicts that, within the virion, only the 3'-terminals at the two extreme ends of the genome would be open to the reaction. In this event, only two of the ten segments found in the gel analysis should have been labeled. On the other hand, Model 2 predicts that the left hand segment has two 3'-terminals available for oxidation-reduction. All other segments have only one 3'-terminal open. If this were so, one segment should contain twice as much ^{3}H as all others. In fact, all segments were equally labeled and model 2 can be eliminated. As a result of these experiments we are left with a model of the genome in which both strands of the structure are covalently discontinuous such as Model 3.

Models 3, A, B, C suggest that there are overlapping single-stranded tails on the ends of each segment by which the segments could be held together by hydrogen bonding through complementary regions. Model 3D suggests that the genome is comprised of discrete segments with no single-stranded tails. A choice between these models could be made by specifically labeling either the 3'- or 5'-terminals and then determining the resistance of the label to the action of various nucleases. For example the 3'-terminals of A should be resistant to attack by either pancreatic or T_1 ribonuclease because of their involvement in secondary structure. Conversely, the 5'-terminals of A should be completely hydrolyzed by treatment with the appropriate nuclease. Similar arguments can be applied to models B, C and D to set up a number of predictions that can be tested.

First, to determine whether any single-stranded tails exist, heavily ^{32}P-labeled RNA was obtained from purified virus (10). This material was passed through a sephadex G-100 column to remove the low-molecular weight polynucleotide fraction and the resulting dsRNA was subjected to the action of venom phosphodiesterase or spleen phosphodiesterase and phosphomonoesterase. Each reaction mixture was analyzed to determine the amount of P^{32} released. The results were consistent with having single-stranded tails on the 5'-terminals of the dsRNA segments. From the amount of ^{32}P released an average length of six to eight nucleotides was estimated for each of the twenty 5'-terminals. To further investigate this proposition a technique was used to specifically label the 5'-terminals with ^{32}P (10).

For this purpose the enzyme polynucleotide kinase was used. This enzyme transfers the γ-phosphate group of ATP to the 5'-OH ends of polynucleotides. The viral dsRNA labeled in this manner was then subjected to the action of pancreatic and T_1 ribonucleases. As a result of an extensive series of experiments it was concluded that the ^{32}P labeled,

5'-ends could be removed by T_1 ribonuclease but not by pancreatic ribonuclease. The results of these experiments are compatible with model 3 A of Fig. 3. In view of the specificities of T_1 and pancreatic ribonucleases, it is tentatively concluded that each genomic segment is terminated at the 5'-terminals in a run of six to eight purine nucleotides. The exact number of purines and their sequences are not yet known but with this type of 5'-terminal structure it is apparent that the segments cannot be held together by interactions between complementary base sequences in the tails. One important implication of this type of structure involves the synthesis of 5'-tails during replication of the genome. It is difficult to visualize how an enzyme that replicates the double-stranded structure can also replicate the single-stranded tails if replication proceeds by a semi-conservative mechanism involving hydrogen-bonding between basepairs. Perhaps a terminal addition enzyme activity is also necessary to complete the segments. At present, then, we conceive that the reovirus genome exists in situ as a segmented and not as a continuous polynucleotide structure (10). How the segments are held together in the specific alignment that must be essential is a question still to be answered.

Having come to some general conclusions about the structure of the genome let us summarize our present information on the other major structural components of the virus, the capsid proteins. We would expect the viral genome to provide the information for synthesis of these proteins. The general approach to this problem, by Loh and Shatkin (12) and by Joklik and his coworkers, (13, 14) has been through the controlled dissociation of the capsid into its fundamental polypeptide units and analysis of these products by gel electrophoresis. Reovirus is labeled with ^{3}H-amino acids during infection, the virus is purified, dissociated with sodium dodecyl sulfate and mercaptoethanol and analyzed by electrophoresis through a polyacrylamide gel. This type of analysis separates the proteins according to their relative sizes, the smallest proteins migrating the fastest.

The viral proteins fall into three major size-classes termed λ, μ and σ classes (12,13), Table 1. There are two components in the largest class (λ_1, and λ_2), two in the medium class (μ_1 and μ_2) and three in the small class (σ_1, σ_2, σ_3). Molecular weights can be obtained for these proteins in much the same way as for the genomic RNA segments. Markers of known molecular weights are used in the gel analysis and by plotting the distances these markers migrate a straight line relating distance to logarithim of molecular weight can be constructed. Molecular weights of the viral proteins are then read off this line. The average molecular weights of the various classes of proteins are entered in Table 1. A very interesting point now emerges. Knowing the molecular weights of the L, M and S classes of the viral genome and that the molecular weight ratio of dsRNA to protein is approximately 18, one can compute the approximate molecular weights of the proteins then could code for (6, 13). It turns out that there is just about enough information in the L class of viral segments to code the λ class of proteins, enough information in the M class of RNA to specify the μ proteins, and the S class of RNA could specify the σ proteins. Continuing the correlation, two L segments could code for the λ proteins, two M segments might code for the two μ proteins and three S segments of RNA could specify the three σ proteins. Probably protein μ_2 is not a primary gene product since it seems to be formed from μ_1 during intracellular synthesis of the virus (14). Thus, we might say that two L segments of RNA, one M segment of RNA and three S segments of RNA are entirely occupied in coding for capsid proteins. This would leave only four of the ten genomic segments with functions still to be specified.

It should be emphasized that this is still only a correlation, although a very interesting one. As yet no single capsid protein can be assigned unambiguously to a specific segment of the genome. Until this can be done the correlation will remain in the realm of speculation. One other pertinent point might be made here, however. We have stated that it is rather a simple matter to strip the capsomeres from the inner core of the virion. This enables one to analyze separately for the proteins of the core and the capsomere. Of the several capsid proteins three are known to be constituents of the core λ_1, λ_2, and σ_2 (12, 13).

Having discussed some of the structural features of the virus and how a good deal of the information in the genome might be used in specifying the capsid proteins, let us now consider some of the problems concerned with transcription of the genome in infected cells. Two kinds of virus-specific RNA are synthesized in cells infected with reovirus (15, 16). One species is dsRNA and is incorporated into progeny virus particles. The second species is single-stranded RNA. It is synthesized in infected cells in presence of sufficient actinomycin D to block the synthesis of cellular RNA: it hybridizes specifically and very efficiently to viral double-stranded RNA after the latter is denatured: it does not hybridize with itself. Moreover, the single-stranded RNA is associated with polyribosomes. This species of RNA is virus-induced mRNA transcribed from one strand of the viral genome (16, 17).

When virus-induced mRNA is isolated from infected cells and centrifuged through a sucrose gradient three size-classes of mRNA are separated according to sedimentation rate (7, 11). Molecular weights calculated from the sedimentation rates are approximately half those found for the three size-class of dsRNA in the viral genome. Hybridization studies then showed that the large class of mRNA hybridized only with the L-class of dsRNA and not at all with the M and S classes, the medium class of mRNA hybridized only with the M class of dsRNA and the small mRNA only with the S class of dsRNA (11, 18). These results mean that each class of dsRNA is uniquely transcribed into mRNA's that have approximately the same lengths as the genomic segments in that class.

The question of how many of the genomic segments are transcribed has only recently been answered (10). Late in the infectious cycle all segments are transcribed while at very early times probably only four mRNA's are synthesized. The experimental evidence was obtained in the following way. To determine the late transcription pattern cells were infected with reovirus and labeled with ^{3}H-uridine between 10 and 11 hours after infection. At this time virus is growing at its maximum rate. The ^{3}H-labeled mRNA was isolated from the cells at 11 hours after infection, hybridized with ^{14}C-labeled dsRNA and the hybrids were analyzed by gel electrophoresis (Fig. 4) (10). All ten segments of the genome were resolved as shown by the ^{14}C-labeled RNA. Each segment has hybridized with ^{3}H-labeled mRNA showing that it had been transcribed during the ten to eleven hour period after infection.

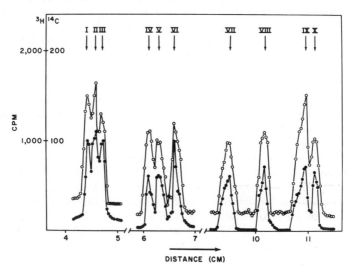

Fig. 4. Electrophoretic analysis on polyacrylamide gel of the hybrids formed between viral dsRNA and late mRNA (10). Closed circles, ^{14}C-labeled dsRNA. Open circles, ^{3}H-labeled mRNA hybridized with dsRNA.

We can derive a further piece of information from this analysis. Inspection of the profiles in Fig. 4 shows that the hybrids migrate at essentially the same rates as the corresponding dsRNA marker segments. From this observation we can make an estimate for the fraction of the genome that is transcribed during infection. The distance each genomic segment migrates in a polyacrylamide gel is proportional to the logarithim of its molecular weight, as was shown earlier. Thus, we can calculate that the length of one gel fraction is equivalent to a molecular weight of 15,000 daltons. If, in the analysis shown in Fig. 4, hybrids differed from their marker segments by more than a gel fraction in their rates of migration the shift would have been seen, and there were no such shifts. Thus, we can say that the hybrids do not differ from their corresponding marker segments by more than 15,000 daltons in molecular weight. This difference is approximately 46 nucleotides, 0.6% of the molecular weight of the largest genomic segment and 1.7% of that of the smallest. We can therefore place a lower limit on the fractions of the various segments that are transcribed, for example 99.4% of the largest and 98.7% of the smallest. We conclude that on the average at least 99% of the total genome is transcribed into mRNA late in the infectious cycle. Virtually all the information in the genome is thus available for protein synthesis at this stage.

It is well known for many viral infections that transcription of the viral genome is regulated during the infectious cycle. That is, the pattern of transcription changes considerably at different times during replication of the virus and our next question is whether this is the case also with reovirus. Putting the question in its most simple terms, can we distinguish an early pattern of mRNA synthesis as distinct from the late pattern we have just described in Fig. 4? The experimental approach to this problem was to infect cells with reovirus in the presence of an inhibitor of protein synthesis, cycloheximide. It has been shown that the virus will enter the cell, be uncoated and that the genome will then be transcribed at least in part (29, 30). Since protein synthesis is blocked during infection any mRNA synthesis must be carried out by a pre-existing RNA polymerase. Moreover, the genome of the parental virus only can be copied since no replication is permitted by the inhibitor. Continuous protein synthesis seems to be an absolute requirement for synthesis of dsRNA throughout the infectious cycle (19). Under these experimental conditions, ^{3}H-uridine was added to the infected cells for 8 hours to label the mRNA. The mRNA was then isolated, hybridized with viral dsRNA and the hybrids were analyzed by electrophoresis on a polyacrylamide gel. Fig. 5 shows such an analysis (10). The marker double-stranded RNA had previously been labeled with ^{14}C and is shown by the closed circles. Hybrids with mRNA are shown by the open circles. The pattern of transcription is quite

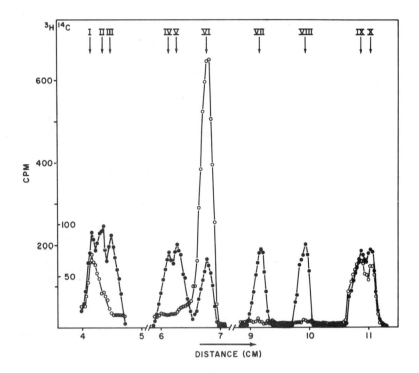

Fig. 5. Electrophoretic analysis on polyacrylamide gel of the hybrids formed between viral dsRNA and early mRNA (10). Cycloheximide was added at the time of infection to inhibit protein synthesis. Closed circles, ^{14}C-labeled dsRNA. Open circles, ^{3}H-labeled mRNA hybridized with dsRNA.

different from that seen in Fig. 4. Viral mRNA is hybridized only with genomic segments 1, VI, IX and X. Possibly segment II also has some hybrid but the resolution is not good enough to be definite about this. Thus, 1, VI, IX and X are the segments transcribed under the experimental conditions and by definition these four segments of the genome are concerned with early viral functions. Some, if not all, of these functions must be expressed before the genome can be replicated.

If this is the case one should be able to see a transition in the mRNA patterns from the early type shown in Fig. 5 to the late type shown in Fig. 4. Such a transition can indeed be seen as shown in Fig. 6. In this experiment cycloheximide was added at 2 hours after infection to block

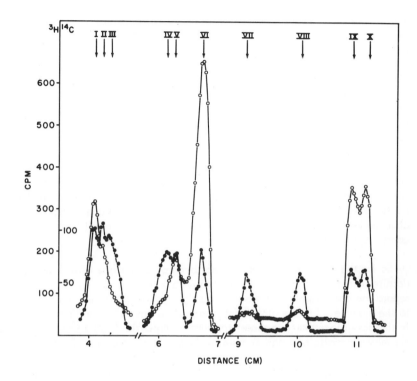

Fig. 6. Electrophoretic analysis on polyacrylamide gel of the hybrids formed between viral dsRNA and two-hour mRNA. Cycloheximide was added at 2 hours after infection to block synthesis of protein and dsRNA. Closed circles, ^{14}C-labeled dsRNA. Open circles, ^{3}H-labeled mRNA hybridized with dsRNA (M. Nonoyama, unpublished results).

synthesis of dsRNA and of protein (19). Under these conditions, however, mRNA synthesis proceeds (19). The mRNA was labeled with ^{3}H-uridine for 8 hours after the addition of cycloheximide, isolated, hybridized with dsRNA and the hybrids were analyzed. In addition to the early segments I, VI, IX and X, two other segments II and V are starting to be transcribed. Apparently, there is a sequential "turning on" of the various segments as the infection progresses. Cycloheximide has had the effect of "freezing" the pattern of transcription from the time it was added.

If certain segments of the genome are transcribed in infected cells when protein synthesis is blocked from the beginning, the synthesis of the mRNA's must be carried out by a pre-existing RNA polymerase. This must either be a cellular enzyme or an integral component of the parental virion. In fact, about two years ago a very active RNA polymerase was shown to be associated with purified reovirus (20, 21). This enzyme has been shown to be a component of viral cores: it is inactive in purified virions but when the capsomeres are removed the polymerase activity can be easily demonstrated (22, 23). The enzyme transcribes all ten segments of the genome into mRNA's of the same length as the genomic segments (22, 23). Obviously this must be the enzyme that transcribes the early segments during infection. The major unsolved question is why it only transcribes selected segments under these conditions because, in vitro, the polymerase copies all ten segments. No one has yet found how to control the in vitro action of the enzyme and restrict its action to a limited number of segments. Perhaps the cell supplies certain factors that control the polymerase action or perhaps certain limited conformational changes in the genome occur in the infected cell that do not happen when cores are formed in vitro (10,24).

So far the virion polymerase has not been obtained free of its template despite numerous attempts. It would be very useful to accomplish the separation and be able to transcribe a single genomic segment added back to the enzyme. On the other hand, no protein has yet been discovered in the core other than the three structural polypeptides λ_1, λ_2 and σ_2. We may specualte that one or more of the structural proteins of the viral core comprise the polymerase protein. In this regard we might point out as an interesting correlation that the RNA polymerase of E. coli is comprised of ß and ß' subunits with molecular weights of approximately 150×10^3 daltons, an α subunit of molecular weight 40×10^3 and a σ subunit of molecular weight 80×10^3 daltons. Thus, the reovirus core has its λ_1 and λ_2 proteins corresponding approximately in size to the ß and ß' subunits of the bacterial RNA polymerase and the σ_2 protein corresponding to the α subunit of the bacterial enzyme. The reovirus core might be considered to be analagous to the E. coli holoenzyme. The problem of initiation of transcription which is taken care of by the sigma subunit of the bacterial polymerase could be managed in another way by the viral polymerase. A good deal of further work is needed to find whether this speculation has any foundation.

Summarizing the situation to this point we see that the reovirus genome is comprised of ten segments of dsRNA with a total molecular weight of approximately 15×10^6 daltons. The segments must be arranged in a specific order in the genome but how they are held together is unknown. The linkage between them is a very labile one since any attempt to isolate the genome results in separation of the ten segments. One possibility is that the segments are held to the core proteins by interaction with the short 5'-tails on each segment, but if this is so it makes the problem of specific arrangement difficult to understand. In any event, the genome is transcribed into ten mRNA's of lengths almost exactly those of the segments from which they are transcribed. The process of transcription does not present the conceptual difficulty it did a year or so ago. Since it is now known that the genome is segmented <u>in situ</u>, each segment must have initiation and termination signals at the extreme ends. The segments can be transcribed independently of each other in the sense that some of them seem to be individually regulated. Four segments (perhaps five) are transcribed in absence of protein synthesis and these are related to early viral functions. Other segments are then turned on until late in infection the whole genome is transcribed. If we assume the capsid proteins to be all virus-specified, six of these proteins could be primary products of viral genes. The molecular weights of these proteins are such that they could each have been coded entirely by a corresponding genomic segment. If this is the case, we arrive at the conclusion that the corresponding mRNA's are monocistronic and thus that six of the ten mRNA's copied off the genome are monocistronic. This line of reasoning would leave the information of only four segments unaccounted for, one large, two medium and one small. Whether the corresponding RNA's are monocistronic or not is unknown since the functions involved have not yet been found.

Superficially, the whole story hangs together reasonably well at the present time but perhaps this should encourage one to be skeptical. For example, the impression one gets from this virus is that of great complexity, both in structure and mechanism of replication. Yet we are suggesting that nearly two-thirds of the information in the genome is assigned to code for structural protein and are implying that perhaps only four more functions remain to be discovered. To some extent, one can get round this paradox by postulating that one or more of the three known core proteins might be RNA polymerase proteins and that they are early proteins. This enzyme could transcribe the genome. One might also suggest that one or more of the core proteins could also behave as a replicase and, if the proper factor were supplied, would replicate the genome. Some of the factors necessary for transcription or replication of the genome might be supplied by the cell itself as for the phage Qß replicase, for example (31,32). There is no evidence for these statements but they are not unreasonable and can be tested. Intuitively, it would seem that several of the structural proteins must have some further function to reconcile such a large proportion of the genetic information of the virus being devoted to their synthesis.

So far we have said little about replication of the viral genome and, indeed, very little is known about this process. A crude enzyme has been isolated from infected cells that will synthesize dsRNA in vitro (25). The enzyme has not been purified or isolated free from its dsRNA template. Synthesis in vitro proceeds for a few minutes and then stops. All the segments seem to be made but probably the enzyme is merely finishing off the synthesis of the segments that had already commenced in the infected cell. The problem here is to find how to reinitiate the synthesis of dsRNA in vitro.

Another approach that will undoubtedly contribute much in the future is through study of the genetics of the virus and such work has commenced in several laboratories (26, 33). It is not difficult to isolate conditional lethal mutants of reovirus and Fields and Joklik (26) have published a preliminary genetic analysis. They studied recombination between 35 temperature-sensitive mutants and found that they fell into five groups based on recombination frequency. Interpretation of these data is still tentative. One might hypothesize that those mutants that show zero recombination frequency are in the same linkage group, or in physical terms, have their genetic defects in the same genomic segment. Mutants that do recombine have defects in different segments. Some of the high recombination frequencies (30-50%) between mutants could then be explained on the basis of reassortment of whole segments during replication, rather than by the classical breakage and rejoining type of mechanism. This interpretation may well turn out to be correct. Unfortunately, at present, there is a difficulty in it which may be posed in the following way. Twenty-eight mutants have been placed in group A since they recombine with each other with zero frequency. These mutants recombine with mutants of group B with frequencies ranging from 3% to 50%. Our hypothesis places group A mutants on one segment and group B mutants on another. If recombination occurs by shuffling of the two segments during replication, then all recombination frequencies between the two groups should be the same. Clearly, the explanation of recombination data is not a simple one. Equally clearly the genetic analysis of this virus will be extremely important because it is likely to be the means through which we discover the order in which the segments are aligned in the genome.

One recently discovered phenomenon may help in this analysis. Nonoyama (27,28) found that during replication of reovirus, defective particles frequently arose. These defective particles were found when the capsomeres were removed from purified virions and the resulting cores were centrifuged to equilibrium in a cesium chloride gradient (Fig. 7). Two populations of cores were found, heavy (H) cores that came from infectious virus and were themselves infectious, and light (L) cores that were not infectious. Analysis of the RNA of the light cores by gel electrophoresis (Fig. 8) showed that the largest segment, number 1, was

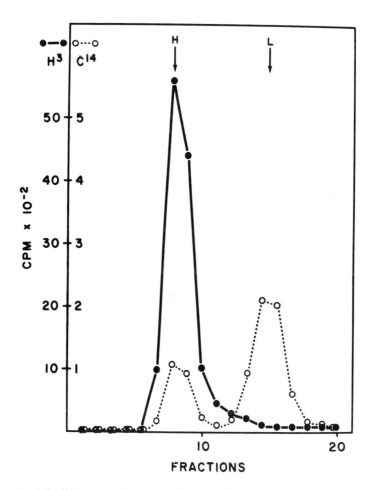

Fig. 7. Isopycnic centrifugation of heavy (H) cores and light (L) defective cores in a cesium chloride gradient. Solid circles, H cores obtained from a viral population that contained no defective virions. Open circles, H and L cores from a population that contained defective virions. Buoyant density of H cores is 1.43 gm/cm^3 and of L cores, 1.415 gm/cm^3.

missing. It was established that the L cores were derived from defective virions that would multiply in the cell only in the presence of infectious virions. During the replicative cycle the infectious virion complements the function missing in the defective virion and enables the defective virion to mature. In passing we might mention that the defective function

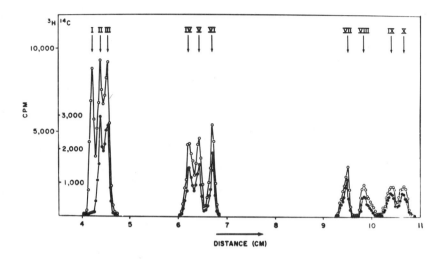

Fig. 8. Analysis of dsRNA from defective cores by electrophoresis on polyacrylamide gel. Closed circles, ^{3}H-labeled dsRNA from L cores. Open circles, ^{14}C-labeled marker dsRNA from a viral population that contained only H cores.

is an early one since it belongs to segment 1. Two important problems arise out of this observation. The first is to find how the defective virions arise in the first place and why all the defective populations we have seen lack the largest genomic segment. The second problem is a more practical one. If we could isolate a pure population of defective virions we could use them in genetic complementation tests with conditional lethal mutants to establish which group of mutants is specifically associated with segment 1 of the genome. By this means one could begin to associate known viral functions with specific areas of the genome and get some real insight into the mechanism of replication of this unusual virus.

The work performed in our laboratory was supported by Public Health Service research grant AI - 02454 from the National Institute of Allergy and Infectious Diseases and by the Medical Research Council of Canada.

REFERENCES

(1) S. Millward and A.F. Graham, in: Comparative Virology, ed. E. Kurstak (Academic Press Inc., New York, 1971) in press.
(2) A.J. Shatkin, Adv. Virus Research, 14 (1969) 63.
(3) C.J. Gauntt and A.F. Graham, in: The Biochemistry of Viruses, ed. H.B. Levy (Marcel Dekker, New York and London, 1969) p. 259.
(4) A.R. Bellamy and L.V. Hole, Virology, 40 (1970) 808.
(5) A.J. Shatkin, J.D. Sipe and P.C. Loh, J. Virology, 2 (1968) 986.
(6) S. Millward and A.F. Graham, Proc. Natl. Acad. Sci. U.S., 65 (1970) 422.
(7) Y. Watanabe and A.F. Graham, J. Virology, 1 (1967) 655.
(8) A.R. Bellamy, L. Shapiro, J.T. August and W.K. Joklik, J. Mol. Biol. 29 (1967) 1.
(9) C. Vasquez and A.K. Kleinschmidt, J. Mol. Biol., 34 (1968) 137.
(10) S. Millward and M. Nonoyama, Cold Spring Harb. Symp. Quant. Biol. 25 (1970) in press.
(11) A.R. Bellamy and W.K. Joklik, J. Mol. Biol., 29 (1967) 19.
(12) P.C. Loh and A.J. Shatkin, J. Virology, 2 (1968) 1353.
(13) R.E. Smith, H.J. Zweerink and W.K. Joklik, Virology, 39 (1969) 791.
(14) H.J. Zweerink and W.K. Joklik, Virology, 41 (1970) 501.
(15) H. Kudo and A.F. Graham, J. Bacteriol., 90 (1965) 936.
(16) A.J. Shatkin and B. Rada, J. Virology, 1 (1967) 24.
(17) L. Prevec and A.F. Graham, Science, 154 (1966) 522.
(18) Y. Watanabe, L. Prevec and A.F. Graham, Proc. Natl. Acad. Sci., U.S., 58 (1967) 1040.
(19) Y. Watanabe, H. Kudo and A.F. Graham, J. Virology, 1 (1967) 36.
(20) J. Borsa and A.F. Graham, Biochem. Biophys. Res. Commun., 33 (1968) 895.
(21) A.J. Shatkin and J.D. Sipe, Proc. Natl. Acad. Sci., U.S., 61 (1968) 1462.
(22) J.J. Skehel and W.K. Joklik, Virology, 39 (1969) 822.
(23) A.K. Banerjee and A.J. Shatkin, J. Virology, 6 (1970) 1.
(24) A.M. Kapuler, Biochemistry, 9 (1970) 4453.
(25) Y. Watanabe, C.J. Gauntt and A.F. Graham, J. Virology, 2 (1968) 869.
(26) B.N. Fields and W.K. Joklik, Virology, 37 (1969) 335.
(27) M. Nonoyama and A.F. Graham, J. Virology, 6 (1970) 226.
(28) M. Nonoyama and A.F. Graham, J. Virology, 6 (1970) 693.
(29) S.C. Silverstein, M. Schonberg, D.H. Levin and G. Acs, Proc. Natl. Acad. Sci., U.S., 67 (1970) 275.

(30) Y. Watanabe, S. Millward and A.F. Graham, J. Mol. Biol., 36 (1968) 107.
(31) M. Kondo, R. Gallerani and C. Weissmann, Nature, 228 (1970) 525.
(32) R. Kamen, Nature, 228 (1970) 527.
(33) N. Ikegami and P. J. Gomatos, Virology, 36 (1968) 447.

DISCUSSION

Dr. H. M. Temin, University of Wisconsin: Are you suggesting, perhaps, that the same mechanism which makes the 5'-tails on the double strands would make all those small oligonucleotides which are in the virus?

Dr. A. F. Graham: If the 5'-tails are put on by some enzyme activity, it would certainly be a candidate for turning out the excess small polynucleotides.

Dr. J. Schultz, Papanicolaou Cancer Research Institute: When a single stranded messenger is combined with a denatured double-stranded segment, and it shows up on your chart as hybridization having taken place, what happens to the second strand of the double stranded segment which doesn't hybridize with the single-stranded messenger? Why shouldn't it show up somewhere on your charts?

Dr. Graham: It is just left out because of the way of doing the experiment. We have highly tritium-labeled messenger RNA and the double-stranded RNA is very lightly labeled with carbon 14. Moreover the messenger is in relatively very low concentration. And so, it's true that one strand of the double-stranded RNA is going to be displaced, but we would never see it, because it is such a small amount.

Dr. M. M. Sigel, University of Miami: Do you think it's possible that the oligonucleotide plays a role in the switch off from the early to the late transcription? Perhaps your first 4 segments can be transcribed in the absence of newly made oligonucleotide, but its synthesis and action (as a primer) is required for the later transcription?

Dr. Graham: I have wondered about this possibility. I just have no answer.

Dr. Temin: How can you be sure that the single-stranded RNA which we've been calling messenger RNA is not also a "replicative intermediate." Can you rigorously exclude that the information transfer into the double strand doesn't go through the single-stranded RNA as opposed to the classical type of scheme you put on in your slide.

Dr. Graham: We thought at one time that it had been excluded because Watanabe did an experiment in which he showed that the single-stranded RNA was not a precursor of the double stranded progeny. We've gone back to look at these experiments and perhaps they are not as clean as we thought. I really think that this should be looked at again.

SOME OBSERVATIONS ON DNA POLYMERASES OF HUMAN NORMAL AND LEUKEMIC CELLS

R. C. GALLO, S. S. YANG, R. C. SMITH, F. HERRERA,
R. C. TING, and S. FUJIOKA
Section on Cellular Control Mechanisms
National Cancer Institute
and
Bionetics Research Laboratories
Bethesda, Maryland

Abstract: (A) An enzyme which synthesizes DNA dependent on RNA, analogous to the Temin-Baltimore enzyme of RNA oncogenic viruses, has been partially purified in peripheral blood of human acute leukemic cells and in a human lymphoma tissue culture cell line. The template was RNase sensitive and DNase resistant, whereas the product was DNase sensitive but RNase and trypsin resistant.

(B) The enzyme utilizes both synthetic and natural RNA templates. No activity has been found with synthetic single-stranded RNA (rA).

(C) With natural RNA templates the enzyme depends on all four deoxynucleoside triphosphates for maximum activity but less so for dGTP. These findings indicate that the enzyme is not a terminal deoxynucleotidyl transferase.

(D) Activity depends on the presence of a divalent cation with Mg^{++} preferred over Mn^{++} with both synthetic and natural RNA templates.

(E) With poly rA.rU and with natural RNA templates DNA polymerase activity was not detected in <u>peripheral blood human</u> lymphocytes, even when mitosis was induced with phytohemagglutinin, nor in a normal human lymphoblast tissue culture cell line. However, an activity with the DNA-RNA hybrid poly dT.rA was found with both. In the case of the fresh cells (peripheral blood lymphocytes) this activity was much lower than that found in leukemic cells.

(F) The RNA-dependent DNA polymerase of human leukemic cells is inhibited by 50 µg/ml of N-demethyl rifampicin but not by rifampicin. In contrast, the DNA-dependent DNA polymerase activities (assayed from normal cells) are not inhibited by the same concentration of either agent.

BACKGROUND

Human leukemogenesis may involve a block in the normal process of leukocyte differentiation and maturation (1-3), perhaps initiated by an oncogenic virus. Molecular hybridization (4) and column co-chromatographic studies (1) have indicated some differences in RNA species between normal and leukemic cells (See Figures 1 and 2), and similar differences have been found between animal tumors and control tissues (5-7). After the reports by Temin (8) and Baltimore (9) of an RNA-dependent DNA polymerase in RNA oncogenic viruses, it was of interest to look for a similar enzyme in human acute leukemic cells, not only because of the possibility that the enzyme might be the same as the virus enzyme and its presence thereby implicate RNA oncogenic viruses in the pathogenesis of human acute leukemia, but also because of the possible role this enzyme reaction might play in biochemical cellular differentiation. Its continued expression in one cell type and repression in another might account for differences in RNA species between the two.

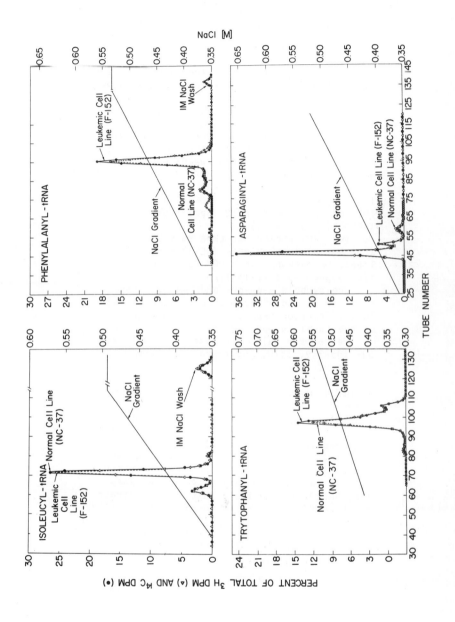

Fig. 1. Legend see next page.

Fig. 1. (a) ^{3}H-labeled Ile-tRNA (3.13 x 10^5 dpm) from normal lymphoblasts (NC-37) and ^{14}C-labeled Ile-tRNA (6 x 10^4 dpm) from leukemic lymphoblasts (F-152). The last peak of the Ile-tRNA was eluted by washing the column with 1 M NaCl.

(b) ^{3}H-labeled Phe-tRNA (2.4 x 10^5 dpm) from normal lymphoblasts (NC-37) and ^{14}C-labeled Phe-tRNA (3.4 x 10^4 dpm) from leukemic lymphoblasts (F-152). The last peak of the Phe-tRNA was eluted by washing the column with 1 M NaCl.

(c) ^{3}H-labeled Trp-tRNA (3 x 10^5 dpm) from normal lymphoblasts (NC-37) and ^{14}C-labeled Trp-tRNA (5 x 10^4 dpm) from leukemic lymphoblasts (F-152).

(d) ^{3}H-labeled Asn-tRNA (3 x 10^5 dpm) from normal lymphoblasts (NC-37) and ^{14}C-labeled Asn-tRNA (5.6 x 10^4 dpm) from leukemic lymphoblasts (F-152).

The symbols --o--o-- and —•—•— represent the ^{3}H and ^{14}C dpm precipitated in each 10 ml. fraction expressed as the percentage of the total ^{3}H and ^{14}C dpm.

Taken from Gallo and Pestka, J. Mol. Biol. 52: 195-219, 1970, and reproduced here with the permission of the publishers.

These aminoacyl-tRNA profiles showed no significant or reproducible differences between normal and leukemic human lymphoblasts. They are illustrated to demonstrate 1) the resoltuion in fractionation of tRNA isoaccepting species and 2) the virtual identity of the prints of individual peaks between the normal and leukemic cells which was the case for the majority of the tRNAs for the 20 amino acids. This is in contrast to the difference in the tyrosyl-tRNA profiles shown in Figure 2.

The small difference (apparent extra peak in the ^{3}H counts) in Phe-tRNA was an artifact of ^{3}H-phenylalanine.

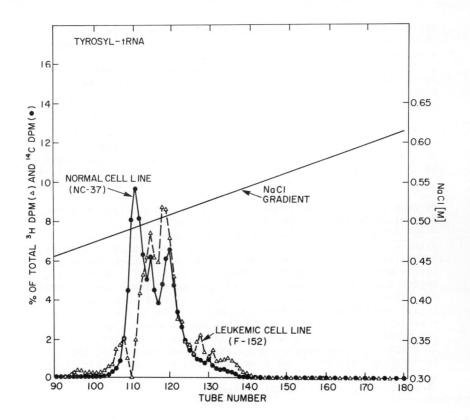

Fig. 2. ^{3}H-labeled Tyr-tRNA (4×10^5 dpm) from leukemic lymphoblasts (F-152) and ^{14}C-labeled Tyr-tRNA (5×10^4 dpm) from normal lymphoblasts (NC-37).

The symbols --o--o-- and —●—●— represent the ^{3}H and ^{14}C dpm precipitated in each 10 ml. fraction expressed as the percentage of the total ^{3}H and ^{14}C dpm.

Taken from Gallo and Pestka, J. Mol. Biol. 52: 195-219, 1970, and reproduced here with the permission of the publishers.

Leukocyte Differentiation

Leukocyte (myeloid and lymphoid) differentiation is a continuously occurring phenomenon, involving the conversion of proliferative blast cells to non-proliferating mature cells. This process takes place in the bone marrow, and under normal circumstances only the most mature cells are found in the peripheral blood. In untreated acute leukemia, leukemic cells enter the peripheral blood and in some respects are often remarkably similar to the "blast" (immature) cells of normal bone marrow.

Figure 3 schematically illustrates a few of the major developments that take place during normal leukocyte maturation. Maturation for myeloid (granulocytic) leukocytes is illustrated by the two top cells, lymphoid differentiation by the two bottom cells. In both cases maturation is from left to right. Morphological changes are obvious and include a marked increase in nuclear pyknosis ("euchromatin" → "heterochromatin"), disappearance of nucleoli, a change in the staining of the cytoplasm (less basophilic), and an increase in lysosomes in the granulocytic cells. Functionally, the cells are converting from metabolically active, proliferative cells to cells which synthesize little RNA, do not synthesize DNA, nor divide (3). Obviously, marked changes in enzyme activities accompany these developments. For example, some enzymes present in relatively high amounts [e.g., thymidylate synthetase, thymidine kinase, and DNA polymerase(s)] in blast cells of the bone marrow are not detected or are markedly reduced in mature cells of the peripheral blood. In contrast, other enzymes (e.g., hydrolytic enzymes of the granulocyte lysosomes) and some phosphorylases may be strikingly increased in mature granulocytes. Some of these findings have been discussed in more detail elsewhere (1-3, 10).

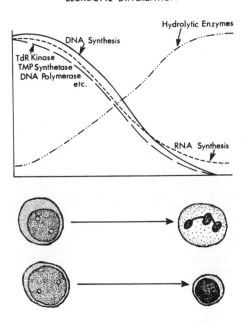

Fig. 3. Schematic illustration of some major biochemical changes in white cell differentiation. Top two cells--myeloblast to granulocyte. Lower two cells--lymphoblast to lymphocyte. For details, see text.

We have examined cells of patients with acute lymphoblastic leukemia, a cell line derived from a patient with Burkitt's lymphoma, and normal controls. In each case a relatively homogeneous population of cells was obtained (over 97% lymphoblasts). These cells contain less lysosomes; therefore, there are much less problems with nucleases than in the granulocytic (myelocytic) leukemias. In addition, relatively comparable normal controls can be obtained, i.e., lymphocytes isolated from peripheral blood of normal donors and stimulated to proliferate by phytohemagglutinin. These cells are a similar cell type and by 72 hours (the time the cells were harvested in these studies) they contain relatively high levels of DNA-dependent DNA polymerase (11) and RNA polymerase activities (12). If a tumor cell contained much higher polymerase activity than control cells (as is the case in comparing any tumor cell to a non-proliferative cell), the possibility of a spurious interpretation of finding a unique polymerase would be more likely.

Without fractionation, peripheral blood human leukocytes ("buffy coat") primarily consist of granulocytes, cells which, as described above, contain little or no DNA polymerase activity and have high nuclease activities (3, 13).

We reported finding an RNA-dependent DNA polymerase in the lymphoblasts of human acute lymphocytic leukemia patients which was not found in normal human lymphocytes (14). An example of the kinetics of a partially purified leukemic cell enzyme is shown in Figure 4. Various methods of purification have been studied, but in general an attempt is made to obtain preparations as free of nucleic acids as possible with minimal or no nuclease activities. The template used for the reaction shown in Figure 4 (o—o) was rat liver RNA. This RNA was prepared by repeated phenol precipitation, DEAE cellulose and gel filtration column chromatography, and extensive treatment with DNase (14). The final preparation was "free" of DNA (less than 0.01%) by diphenylamine determination and consisted of approximately 85% 4S RNA and 15% rRNA.

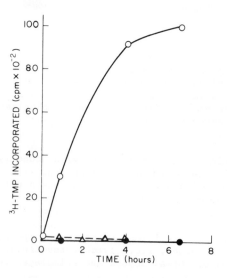

Fig. 4. Legend see next page.

Fig. 4. Kinetics of RNA-dependent DNA polymerase from acute lymphocytic leukemic patient (ALL). Reaction mixtures in a final volume of 1 ml contained: 45 µg of Sephadex G-100 gel filtration purified polymerase; 50 µmoles tris-HCl buffer, pH 8.3; 60 µmoles MgAc; 20 µmoles of dithiothreitol; 60 µmoles of NaCl; 0.8 µmole each of dATP, dCTP, and dGTP; 10 µCi [^{3}H-Methyl]-TTP (New England Nuclear, 7.7 mCi/µmole).

Samples were incubated at 37°C and were withdrawn at intervals and the reactions terminated by the addition of 0.5 ml of cold 0.08 M Na-pyrophosphate with 200 µg of yeast nucleic acid added as a carrier prior to precipitation of macromolecules by adjustment to 12 per cent TCA. The precipitates were then washed extensively with cold 5 per cent TCA on Millipore filters, dried, and dissolved in 2.0 ml of Beckman Bio-solv No. 3. Ten ml of Omnifluor-toluene were added, and the samples were counted in a liquid scintillation counter.

o—o represents the activity of RNA-dependent DNA polymerase from ALL lymphoblasts with 20 µg of rat liver RNA, extensively digested with DNase. ∆—∆ represents the activity of RNA-dependent DNA polymerase from PHA stimulated lymphoblasts of normal donors. •—• represents the activity of RNA-dependent DNA polymerase with RNase (11 µg/ml) digested rat liver RNA as template.

Activity Depends on RNA

Although we have obviously not ruled out the possibility that the RNA is contaminated with DNA, the following points indicate this enzyme is RNA dependent: (a) All activity is lost with RNase pre-treatment of the RNA (see Figure 4); (b) No activity is lost with extensive treatment of the RNA with DNase; (c) Activity is obtained with synthetic double-stranded RNA's; (d) With 10 mg of the RNA, no DNA was detected by the diphenylamine assay. Since sensitivity with this assay is to the µg level, any contamination with DNA would be less than 0.01%. There is really no assurance with any method that a natural RNA is completely free of an oligodeoxyribonucleotide contaminant and, even if a fragment of DNA were present, this could be a mechanism by which this enzyme functions in vivo, i.e., using a small DNA primer and an RNA template. (e) Finally, inhibition of this enzyme activity by derivatives of rifampicin (14, 15) and failure to inhibit DNA-dependent DNA polymerase activities (14, 15) strongly suggests that the two activities are distinct. In this respect the enzyme activity is analogous to that reported by Green with the RNA-dependent DNA polymerase of an RNA oncogenic virus (15).

RNA-Dependent DNA Polymerase Activity in Human Tissue Culture Cells

We initially selected <u>fresh</u> human leukemic and normal cells rather than tissue culture cells because they are the actual cells involved in the disease process. In addition, there is less likelihood of contamination with microorganisms and of developing mutations not related to the disease process. However, there are obvious advantages of a tissue culture system. Therefore, we assayed for RNA-dependent DNA polymerase in a Burkitt lymphoma lymphoblast cell line and in a normal human lymphoblast cell line. Activity was detected in the Burkitt lymphoma cells (Table 1). The approach used for partial purification of this activity (see Table 1) was modified from that used with the fresh leukemic cells. The ratio of activity with RNA as template versus calf thymus DNA as template was followed at each step. In the crude homogenate the ratio is less than 1:10 but after the gel filtration step the ratio is increased to approximately 1:1 (slight variations exist depending on the type of RNA and whether native or denatured DNA is used). However, these results with the Burkitt lymphoma cells should be regarded as preliminary.

TABLE 1

RNA- and DNA-dependent DNA polymerase activities of Burkitt lymphoma cells

Enzyme fraction[a]	Templates	cpm/ml	Net cpm/ml[b]	Ratio of net cpm/ml rat liver RNA to net cpm/ml calf thymus DNA
Ammonium sulfate 0-30%	None	6,690		
	Rat liver RNA	18,040	11,380	0.403
	Calf thymus DNA	34,890	28,200	
Ammonium sulfate 30-50%	None	1,440		
	Rat liver RNA	3,060	1,620	0.048
	Calf thymus DNA	35,090	33,650	
Ammonium sulfate 50-70%	None	29,110		
	Rat liver RNA	32,880	3,770	0.063
	Calf thymus DNA	88,580	59,470	
Sephadex G-100	None	470 (580)		
	Rat liver RNA	2,360	1,890	0.843
	Calf thymus DNA	2,710	2,240	
	Poly rA.dT	2,010 (4,060)	1,540 (3,480)	
	Poly rA.rU	2,530	2,060	
	Poly dAT	4,870	4,400	
	Calf thymus DNA (denatured)	1,960 (7,040)	1,490 (6,460)	

[a] The supernatant obtained after homogenization and centrifugation (105,000 x g for 60 minutes) of Burkitt lymphoma cells was treated with streptomycin sulfate and the supernatant from this fraction was subjected to ammonium sulfate precipitation. The 0-30% ammonium sulfate precipitate was fractionated by Sephadex G-100 column chromatography. The final volume of the

incubation mixture was 0.1 ml, which contained 1 μg of template and 200 μg protein from the ammonium sulfate fraction or 70 μg of protein from the Sephadex G-100 fraction. Incubations were for 1 hour with the ammonium sulfate fractions and 2 hours with the Sephadex G-100 fraction (5 hour values in parentheses).

b cpm/ml "template" minus cpm/ml with no "template".

Nature of the Reaction and Comparison of Some RNAs as "Templates"

Some properties of the enzyme were studied and comparison of several RNAs were made (Table 2). The enzyme has an absolute requirement for Mg^{++}. Omission of a deoxynucleoside triphosphate produces a 90% inhibition of the reaction. Thus, the reaction appears to be a polymerase and not a terminal nucleotidyl transferase.

Although in crude extracts the ratio of activity with DNA to RNA as template is more than 100 to 1, this preparation shows no activity with native or denatured DNA. On the other hand, activity is obtained with poly d(A-T) and with the DNA-RNA hybrid poly dT.rA.

TABLE 2

Comparison of "templates" and nature of the RNA-dependent DNA polymerase from human acute leukemic lymphocytes

Template		Net cpm/ml
Without RNA		0
Without MgCl$_2$ (with rat liver RNA)		0
Without dCTP " " " "		405
Without dATP " " " "		622
Calf thymus DNA (native)		0
Poly dT		0
Poly rA		0
Rat liver RNA	2 µg/ml	8,130
" " "	5 µg/ml	5,806
" " "	40 µg/ml	4,149
23S E. coli rRNA	5 µg/ml	10,600
tRNA isoaccepting species (all at 5 µg/ml)		
tRNA^{f}-met		2,210
tRNA^{arg}		2,280
tRNA^{glu}		11,480
Homologous human lymphocytic leukemia rRNA	5 µg/ml	13,210
Homologous human lymphocytic leukemia tRNA	5 µg/ml	11,160
Poly rA.rU (Miles)	10 µg/ml	7,120
" " " (Biopolymer)	10 µg/ml	9,760

The enzyme was the partially purified preparation described in Figure 4. The reaction conditions were identical to those described in the legend to Figure 4 except that the ^{3}H-TTP was increased to 100 µc/ml. Incubations were for 2 hours.

Mechanism of the Enzyme Reaction and Characterization of the Product

At the present time the enzyme has not been purified to the point where reasonable information can be obtained on the mechanisms of the reaction. Of interest is the template activity with physiological RNA (tRNA and rRNA). The best template activities were obtained with a purified isoaccepting species of tRNA^{glu}, which was generously supplied by Dr. G. D. Novelli.

That the product is DNA is indicated by the incorporation of deoxynucleoside triphosphates and by the specificity of digestion with DNase (Table 3).

TABLE 3

Enzymatic digestions of ^{3}H-TMP-labeled product obtained with lymphocytic leukemic cell extract

Treatments	cpm/ml minus background
Control, none	12,260
RNase, 2 mg/ml	11,640
Trypsin, 1 mg/ml	11,080
*NaOH, 0.5 N	7,670
DNase, 4 mg/ml	560

RNA-dependent DNA polymerase reaction was carried out for 4 hours as described in Figure 4 except that the enzyme preparation did not include the gel filtration step (see text). The sample was then divided into 5 aliquots and treated with either water (control), RNase, DNase, trypsin, or NaOH (0.5 N). The reaction was then continued for an additional 1 hour.

 * The adjustment to 0.5 N NaOH terminated the reaction early and hence yielded lower counts.

We have not, however, yet shown: (a) whether the deoxyribonucleotides are covalently joined to a 3'-OH ending of the RNA or (b) whether they are covalently joined to an oligodeoxynucleotide primer which, in the case of the natural RNAs, accompany the RNA during purification. If the latter is the case, it is likely that RNA functions as a template since no activity has been found with oligodeoxynucleotides and polydeoxynucleotides alone, and all four deoxyribonucleoside triphosphates are required for activity.

We are now in the process of evaluating these questions as well as the size of the DNA synthesized, whether it is single- or double-stranded, and if it specifically hybridizes with the RNAs used in these experiments.

DNA Polymerase Activity in Normal Cells

This RNA-dependent DNA polymerase activity has not been found in non-proliferating adult organs. However, the best cells to use as "normal controls" for proliferating human leukemic lymphoblasts obtained directly from the peripheral blood would seem to be proliferating normal human lymphoblasts obtained directly from human blood. Normal lymphocytes can easily be obtained and purified, however, unless stimulated by a mitogenic agent, they are not proliferating cells. We, therefore, have primarily used human lymphocytes stimulated to enter "cycle" with the mitogenic agent phytohemagglutinin (PHA). Some of the details of our use of this system have been described elsewhere (16, 17). These cells are not, of course, established tissue culture cell lines, but are incubated only for 72 hours with PHA for maximum induction of DNA polymerase (11). With relatively crude extracts we had not found RNA-dependent DNA polymerase in these cells (Figure 4). On the other hand, activities with calf thymus DNA (native or denatured) and with poly d(A-T) were comparable to activities in crude extracts of leukemic cells. <u>It is essential to emphasize that the lack of activity with natural RNAs and with double-stranded synthetic RNAs in these cells could be due to insufficient sensitivity of the assay.</u> If this is the case, then the differences between the leukemic and normal lymphoblasts would be limited to a quantitative difference. In subsequent preliminary experiments we have found some activity with the DNA-RNA hybrid poly dT.rA in these normal lymphoblasts as well as with an established human normal lymphoblast tissue culture cell line (see below). The activity in the PHA stimulated lymphocytes with poly dT.rA was less than 1/5 of the activity found with leukemic cells. If the reaction with poly dT.rA is, in fact, with the same polymerase which utilizes natural RNAs and double-stranded synthetic RNA, then the differences between the normal and neoplastic cells are quantitative.

Instead of being "carried into" the cell by an RNA oncogenic virus, the enzyme may be a cellular enzyme normally repressed as cells differentiate and mature; its continued greater expression in leukemic cells might be involved in the aberrant differentiation characteristic of these cells, as we have emphasized previously (14). A role for this type of enzyme in differentiation and gene amplification has been stressed previously by others (18, 19). If the enzyme of these human cells is identical to the enzyme of the virion, a number of questions immediately arise. Does the virus pick up this enzyme from the cell or did the cell obtain the information originally from the virus, expressing itself only when in a relatively undifferentiated state and after neoplastic transformation analogous to the hypothesis of Heubner (20)? Another possibility is that some information (a subunit of the enzyme?) is viral specific and some specified by the genome of the animal cell.

DNA Polymerase Activities from a "Normal" Human Lymphoblast Established Tissue Culture Cell Line

To attempt to distinguish the RNA-dependent DNA polymerase of leukemic cells from the DNA-dependent DNA polymerase, we are investigating some of the properties of the latter activity from normal and leukemic human lymphoblasts. Investigation of this (these) enzyme(s) is important, since the DNA polymerase of E. coli appears capable of responding to a poly rA.rU template (21). For normal cells we are utilizing both phytohemagglutinin stimulated peripheral blood lymphocytes and a "normal" human lymphoblast cell line as the source of enzymes.

Our preliminary results show at least two DNA polymerase activities in the established cell line (see Figures 5-7). These preparations are relatively crude preparations so that the template specificities cannot be interpreted in a quantitative manner. Calf thymus DNA (native and denatured), natural RNAs, poly rA.rU, poly dT.rA were all tested with both enzymes. As shown below (Figure 6), peak IV from the DEAE cellulose column appears to have preference for native calf thymus DNA compared with denatured DNA. This enzyme, activity, localized to peak III from the DEAE cellulose column (Figure 5) had greater activity with denatured than with native DNA (Figure 7).

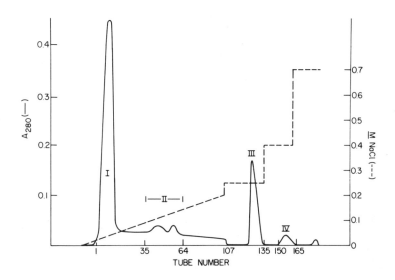

Fig. 5. DEAE cellulose chromatography of homogenate from normal human lymphoblasts. The cells were homogenized and the membrane-nuclear pellet "extracted" as previously described (14). The combined "extract" and supernatant were then centrifuged at 105,000 x g for 2 hours. The supernatant was placed on a 3 x 29 cm DEAE cellulose column equilibrated with buffer (see below) and eluted with 0.2 M NaCl. The eluate was concentrated and dialyzed against 120 volumes of 25 mM tris-HCl, pH 7.5, buffer; 1 mM $MgCl_2$; 1 mM dithiothreitol; and 0.5 mM EDTA. This material (protein concentration 4 mg/ml) was then passed through a 2 x 20.5 cm DEAE column equilibrated with the same buffer and eluted with NaCl. This chromatography is shown in the figure. Subsequent purifications involve phosphocellulose and gel filtration columns.

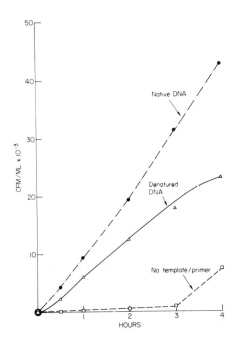

Fig. 6. Kinetics of incorporation of ^{3}H-TMP into DNA with partially purified DNA-dependent DNA polymerase from normal human lymphoblast tissue culture cells. The enzyme is from fraction I of the DEAE cellulose chromatography step (Figure 5). The protein concentration was 550 μg/ml. Native calf thymus DNA and denatured calf thymus DNA templates are compared. No activity was found with natural RNAs or poly rA.rU as "templates."

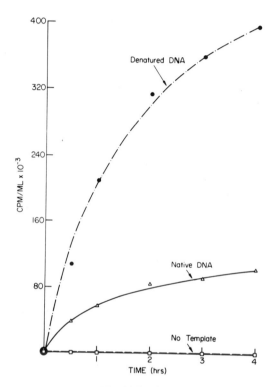

Fig. 7. Kinetics of incorporation of ^{3}H-TMP into DNA with partially purified DNA-dependent DNA polymerase from normal human lymphoblast tissue culture cells. The enzyme is from fraction III of the DEAE cellulose chromatography step (Figure 5). The protein concentration was 230 µg/ml. Native and denatured calf thymus DNA templates are compared. No activity was found with natural RNAs or poly rA.rU as "templates."

The RNA "templates" tested included poly rA.rU, rat liver RNA, tRNA^{glu}, and E. coli 23S rRNA. No activity was found with any of these RNAs. In contrast, some activity was found with the DNA-RNA hybrid poly dT.rA (see Figure 8).

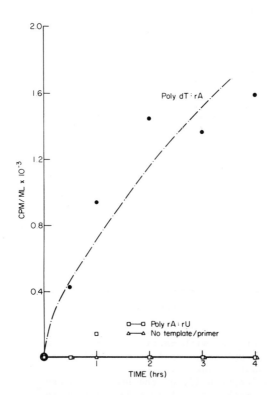

Fig. 8. Kinetics of incorporation of ^{3}H-TMP into DNA dependent on the DNA-RNA hybrid poly dT.rA. The enzyme is a partially purified preparation from a "normal" human lymphoblast established tissue culture cell line (Fraction III) from DEAE cellulose chromatography.

It should be stressed that these cells, although called "normal", are morphologically non-distinguishable from established human lymphoblast cell lines from patients with acute leukemia. There is no way of being certain that these tissue culture cells are not neoplastic.

Differential Inhibition of RNA-Dependent and DNA-Dependent DNA Polymerases

N-demethylrifampicin (40 µg/ml) but not rifampicin (40 µg/ml) inhibits the RNA-dependent DNA polymerase of an RNA oncogenic virus (MSV) (15). We have reported similar results with the human leukemia RNA-dependent DNA polymerase (14) (see also Figure 9). In contrast, no significant inhibition of the DNA-dependent DNA polymerases (from normal lymphoblasts) was found with similar concentrations of N-demethylrifampicin. Extremely high concentrations (400 µg/ml) of rifampicin and N-demethylrifampicin produced slight and equivalent inhibition (about 30%) of the DNA polymerase activity which appears to show preference for denatured DNA (Table 4).

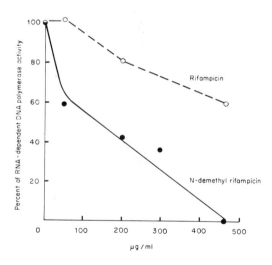

Fig. 9. Legend see next page.

Fig. 9. Effect of rifampicin and N-demethylrifampicin on the RNA-dependent DNA polymerase from lymphocytes of human acute lymphocytic leukemia. Assays were as described in Figure 4 except that only 22 µg of the gel filtration purified polymerase was used per ml of reaction mixture.

TABLE 4

Effect of rifampicin and N-demethylrifampicin on two DNA-dependent DNA polymerases from normal human lymphoblasts*

Enzyme fraction**	Template	Inhibitor	Net cpm/ml	% inhibition
I	Native DNA	0	39,400	--
I	Native DNA	DMSO 44 µl/ml	28,390	0 (control)
I	Native DNA	rifampicin 400 µg/ml	25,700	9
I	Native DNA	N-demethyl-rifampicin 400 µg/ml	30,400	0
III	Denatured DNA	0	366,000	--
III	Denatured DNA	DMSO 44 µl/ml	475,900	0 (control)
III	Denatured DNA	rifampicin 400 µg/ml	317,200	33
III	Denatured DNA	N-demethyl-rifampicin 400 µg/ml	331,400	30

* Conditions were as in the legend to Figure 6 except that inhibitors were used in the concentrations indicated. The inhibitors were dissolved in DMSO. Incubations were for 2 hours.
** Refers to DEAE cellulose column fractions shown in Figures 5, 6, and 7.

Some Interpretations and Questions

The results in this report demonstrate that a DNA polymerase dependent on RNA is present in human acute lymphocytic leukemia cells. The obvious analogy to the enzyme of RNA oncogenic virions does not, of course, prove that the enzyme is identical to the virion enzyme. The observation that natural RNAs may be templates for this polymerase suggests that a role for this enzyme may be in gene amplication and consequently for increasing the amount of specific RNA species. However, before a conclusion can be made regarding this possibility, much more has to be done to characterize the product of the reaction. The evidence reported here indicates that the RNA-dependent reaction carried out by the leukemic DNA polymerase is not an associated function of the much more abundant DNA-dependent DNA polymerase. For example, the RNA but not the DNA-dependent polymerase shows sensitivity to the derivative of rifampicin, N-demethyl-rifampicin.

If the enzyme is identical to the enzyme of the RNA oncogenic viruses, will inhibition of the enzyme activity diminish tumor growth or perhaps more interesting, will it prevent subsequent relapse of leukemic patients in remission, i.e., are at least some relapses in leukemia due to "re-infection" (re-transformation) by an oncogenic virus?

REFERENCES

1. R. C. Gallo and S. Pestka, J. Mol. Biol., 52 (1970) 195.

2. D. H. Riddick and R. C. Gallo, Cancer Res., 20 (1970) 2484.

3. R. C. Gallo, in: Regulation of Hematopoiesis, Vol. 2, ed. A. Gordon (Appleton-Century-Crofts, New York), p. 1238.

4. P. E. Neiman and P. H. Henry, Biochemistry, 8 (1969) 275.

5. B. S. Baliga, E. Borek, I. B. Weinstein and P. R. Srinivasan, Proc. Natl. Acad. Sci. U.S.A., 62 (1969) 899.

6. F. Gonano and V. P. Chiarugi, Exp. Molec. Pathol., 10 (1969) 99.

7. M. W. Taylor, C. A. Burk, G. A. Granger and J. J. Holland, J. Mol. Biol., 33 (1968) 809.

8. H. M. Temin and S. Mizutani, Nature, 226 (1970) 1211.

9. D. Baltimore, Nature, 226 (1970) 1209.

10. R. C. Gallo and J. Whang-Peng, in: Fourth International Miles Symposium on Molecular Biology: Biological Effects of Polynucleotides, (Springer-Verlag, New York), in press.

11. L. A. Loeb, S. S. Agarwal and A. M. Woodside, Proc. Natl. Acad. Sci. U.S.A., 61 (1968) 827.

12. P. Hausen and H. Stein, Europ. J. Biochem., 4 (1968) 401.

13. Y. Rabinowitz, Blood, 27 (1966) 470.

14. R. C. Gallo, S. S. Yang and R. C. Ting, Nature, 228 (1970) 927.

15. M. Green, Nature, in press.

16. R. C. Gallo, J. Whang-Peng and S. Perry, Science, 165 (1969) 400.

17. D. H. Riddick and R. C. Gallo, Blood, in press.

18. H. Temin, Perspectives Biol. Med., Autumn 1970, 11.

19. D. Baltimore, in: Cold Spring Harbor Symposia, in press.

20. R. J. Heubner, G. J. Todaro, P. S. Sarma, J. W. Hartley, A. E. Freeman, R. L. Peters, C. E. Whitmire and H. Meier, in: Proc. 2nd International Symp. on Tumor Viruses, Paris, 1969.

21. S. Lee-Huang and L. Cavalieri, Proc. Natl. Acad. Sci. U.S.A. 50 (1963) 1116.

ACKNOWLEDGMENTS

Rifampicin and N-demethylrifampicin were prepared by the Gruppo Lepetit Co., Milano, Italy. We thank Dr. G. D. Novelli, Oak Ridge National Laboratory, for the isoaccepting species of E. coli tRNA, and Dr. R. D. Wells, University of Wisconsin, for the poly dT.rA. Finally, we thank Mrs. Alva Russell and Miss Carla Davis for their excellent technical assistance.

This work was partially supported by contract number NIH-70-2050 of the Chemotherapy area of the National Cancer Institute and by contract number NIH-71-2025 of the Special Virus Cancer Program of the National Cancer Institute, NIH, USPHS.

DISCUSSION

Dr. Chang, Louisiana State: The question I wanted to point out is that we are working on the nucleotide sequence of glutamic acid tRNA from *E. coli*. The tRNA we're working on is provided by Dr. Novelli and Dr. Kellmers in Oak Ridge National Laboratory, so apparently, we have the same kind of material which you used to test the template activity and I like to take a chance to point out that during our sequential work on this particular tRNA we, of course, are provided with a large enough amount of this RNA so that we can work on the sequence with a method that Dr. Holly used for his yeast alanine tRNA, that is, we used DEAE cellulose chromatography as the method to separate the enzymatic degradated fragment and to sequenced fragments. And we finished the sequence of complete pancreatic fragments and the T_1 ribonuclease fragments. From these results we found that we have difficulty to overlap the pancreatic fragments with the T_1 fragments. Number two, from the T_1 ribonuclease digestion there is a very slow moving material in the DEAE cellulose column which has approximately one fifth of the total absorbancy we put into the column. And this material is not digestible with snake venom phosphodiesterase, not digestible with pancreatic ribonuclease and we don't know what it is. And that has given us tremendous difficulty to work on the sequence of it. Then I decided to put the intact glutamic acid RNA through a benzylated DEAE cellulose column to see if there is any impurity there of the RNA which we obtained from Oak Ridge and found that the material had separated into two peaks - a fast moving material which is approximately one fifth of the quantity we put into the cellulose column has no acceptor activity at all for glutamic acid. I just wonder, the question is which of these components, the impurity or the glutamic tRNA, gives you the template?

Dr. R. C. Gallo: We will try and answer that as soon as we get a little bit of each one of those two components from you.

Dr. G. D. Novelli, Oak Ridge National Laboratory: In defense of material that we are giving away as tRNA, I want to say that it is biologically pure. That is, the amino acid acceptor activity is equivalent to the adenosine terminus. How this 20 per cent of stuff got in there I don't know but we now know how to get it out.

Dr. Gallo: The major concern would be if it were DNA. However, our "template" activity was destroyed by ribonuclease and not by deoxyribonuclease but do you think it could be a

DNA fragment?

Dr. Chang: Of course we will try to characterize more or less somewhat of this fast moving material which has no acceptor activity. And the material that we obtained from Dr. Kellmers says that it is 83 per cent pure.

Dr. Gallo: I'm sorry. Your guess is that it is an RNA fragment?

Dr. Chang: I don't know. I'm sorry.

Dr. S. Spiegelman, Columbia University: I just want to make a comment, I talked this over with Bob privately, but I think it's important to people who are listening to this and trying to evaluate experiments in this field. Anybody who has worked with nucleic acids for some time knows that it is awfully difficult to end up with a preparation which is really completely clean and contains only one, the kind of thing that you're looking for. Now in our hands when we prepare RNA for template activity we have found that DNA-ase treatment of RNA preparations, no matter how extensive, offers no guarantee that you've really got an RNA preparation free of DNA. The obvious reason for this is that first of all DNA-ase doesn't work well towards the end of the game and second, in the presence of large amount of RNA, which should be there, it has even greater difficulty of completing its job.

Dr. Chang: Dr. Novelli, could I say one more thing about the primary structure of this glutamic acid RNA. It may be interesting to you and Dr. Gallo. That is, in this glutamic acid tRNA right before the T4CG tetra sequence, there is a row of eight G's T CG and the completed pancreatic digestion on this tRNA gave a tremendous amount of Cp that means that the eight G's in front of T4CG is based paired with C's. I was just wondering does this row of G-C base pair give any kind of a stimulating effect if it is stimulated by glutamic acid tRNA?

Dr. Gallo: I can't answer the question for certain since there hasn't been an experiment to directly answer that, but my guess would be that it would.

Dr. J. Schultz, Papanicolaou Cancer Research Institute: For clarification of Dr. Spiegelman's remarks, one recognizes the difficulty involved in preparing a pure RNA from natural sources. Nevertheless the findings with synthetic RNAs surely indicate that it is what it is supposed to be and that in spite of the fact that you can't make pure RNA from natural sources, there is still present a RNA dependent DNA polymerase.

Dr. Gallo: Yes, certainly the enzyme(s) also work with synthetic RNA's; for example, we have used poly A:U; poly I:C; and the same DNA-RNA hybrid used by Dr. Spiegelman, dT:rA. Also despite the difficulty raised by Dr. Spiegelman with DNA-ase, the "template" activity was lost with RNA-ase and not DNA-ase.

Dr. Schultz: Have you tried this to look for this enzyme in the bone marrow itself rather than trying to prepare lymphoblasts in culture?

Dr. Gallo: No, we haven't looked at bone marrow. The reason is that the bone marrow has many differentiated cells which contain a lot of nucleases. Furthermore, one cannot obtain the quantity of tissue needed to do any purification. Usually one gets only 0.1 to 0.3 gms of tissue. Therefore, we tended to shy away from marrow. We wanted to deal with lymphoid type tissues at least at the onset. I felt that if we didn't find it in bone marrow, particularly with the present level of sensitivity, it might not be meaningful. However, with the more sensitive DNA-RNA hybrid templates, this may be worthwhile. You refer to our study of lymphoblasts in culture. I would emphasize that most of our studies have not been with established tissue culture cells but leukemic cells obtained directly from the peripheral blood of untreated patients with acute lymphoblastic leukemia. Even the "normals" in most cases were lymphocytes obtained from the blood of healthy volunteers. We've also looked at mature normal organs and have not detected this enzyme activity. With proliferating normal lymphoblasts, the activity was not detected with "natural templates". We have not looked at normal bone marrow or other proliferating tissue like intestinal epithelium.

TRANSCRIPTION OF VIRAL GENES IN CELLS TRANSFORMED BY DNA AND RNA TUMOR VIRUSES

MAURICE GREEN
Institute for Molecular Virology
Saint Louis University School of Medicine

Abstract: Both DNA and RNA tumor viruses may transform cells by integration into cellular chromosomes of viral DNA sequences which are transcribed and translated to proteins which maintain the transformed properties of the cell. Evidence for this mechanism for DNA tumor viruses is shown. The properties of the RNA dependent polymerase of RNA tumor viruses suggest similar mechanisms for RNA tumor viruses.

Cells transformed by groups A, B, and C oncogenic and transforming human adenoviruses possess 20 to 80 copies of the viral genome in the cell nucleus associated with cell chromosomes. Three populations of virus-specific RNA molecules, representing from 5 to 20% of the viral DNA genome (23 million daltons), are transcribed in adenovirus transformed cells--one specific for group A, the second for group B, and the third for group C. Viral RNA sequences in transformed cells are transcribed from "early" viral genes as heterogeneous RNA species, some larger than theoretical viral gene transcripts. Both viral and cellular RNA sequences appear to be present in the same polycistronic RNA molecule, suggesting that transcription is initiated on viral DNA sequences and continued into adjacent cellular sequences or vice versa. In Ad 2 productive infection, three monocistronic viral RNA's are present in polyribosomes and nuclei early after infection and 6-8 are resolved in the cytoplasm late after infection. In addition 4 polycistronic viral RNA precursors of polyribosomal viral mRNA are present in the nucleus late after infection.

RNA tumor viruses possess DNA polymerase activities which utilize endogenous viral RNA as template and synthesize DNA sequences complementary to those of viral RNA. We have studied the RNA and DNA dependent DNA polymerase activities of two murine sarcoma viruses (MSV), avian myeloblastosis virus (strain BAI) (AMV), and feline sarcoma virus (FeSV). The DNA polymerase of AMV and FeSV exhibits two phase kinetics--an initial rapid reaction for four minutes followed by a slower reaction for 20-60 minutes. Equilibrium sedimentation in Cs_2SO_4 density gradients show the rapid formation of RNA-DNA complexes composed of 70S RNA with small DNA strands (molecular weight 150,000), followed by the formation of DNA free of an RNA template; about half the final DNA product is double-stranded. The denatured DNA product of the MSV, FeSV, and AMV DNA polymerase sediments at 5-7S, has a molecular weight of about 150,000 daltons, and bands in CsCl gradients at $\rho = 1.728$ to 1.735.

The RNA dependent DNA polymerase activity of MSV, feline leukemia virus (FeLV), and AMV are inhibited by rifampicin derivatives with modified side chains but not by rifampicin itself. The DNA dependent DNA polymerase activities of these viruses is inhibited by actinomycin D. Mammalian cell and E. coli DNA dependent DNA polymerases are not inhibited by comparable amounts of rifampicin derivatives.

Molecular hybridization between the highly radioactive MSV DNA product and RNA from the nucleus and cytoplasm of MSV transformed cells was performed to detect and quantitate viral RNA in these organelles. A large fraction of the nuclear RNA ($\sim$5%) and cytoplasmic RNA ($\sim$1%) is virus-specific in MSV transformed cells which replicate virus. A much smaller fraction of RNA is virus-specific in MSV transformed cells which are cryptic, i.e., they do not replicate virus. Thus, in contrast to previous reports, large quantities of viral RNA sequences are synthesized in cells replicating or transformed by RNA tumor viruses.

INTRODUCTION

DNA and RNA tumor viruses provide promising experimental systems for studying the synthesis and regulation of macromolecules, cell transformation, and growth control in mammalian cells at the molecular level. I will describe below the results of recent studies in our laboratory on DNA and RNA viruses discussed in the light of our understanding of cell transformation and viral gene transcription.

ONCOGENIC DNA VIRUSES

My colleagues, Kei Fujinaga, J. Thomas Parsons, Hans Caffier, Deane Tsuei, and myself have studied viral gene transcription and the mechanism of oncogenesis by the human adenoviruses. There are three oncogenic and transforming groups of human adenoviruses—highly oncogenic group A, weakly oncogenic group B, and transforming group C (Table 1). Members within each group are closely related, share most of their DNA sequences, and induce common viral RNA sequences in transformed cells. We show below that (a) multiple viral gene copies are present in adenovirus transformed cells, (b) only a small fraction of the viral genome is transcribed in adenovirus transformed cells, (c) adenovirus RNA sequences are transcribed late during productive infection as large nuclear polycistronic precursor polyribosomal viral mRNA, and (d) adenovirus RNA in transformed cells is transcribed as polycistronic mRNA molecules containing both viral and cell RNA sequences.

TABLE 1[a]
ONCOGENIC AND TRANSFORMING HUMAN ADENOVIRUSES

Group	Adenovirus Members	Oncogenicity	Viral DNA % G+C
A	Ad 12, 18, and 31	"Highly oncogenic"[b] in newborn hamsters	48-49
B	Ad 3, 7, 11, 14, 16, and 21	"Weakly oncogenic"[b] in newborn hamsters (all but Ad 11)	49-52
C	Ad 1, 2, 5, and 6	"Nononcogenic" in newborn hamsters but morphologically transforms rat embryo cells in vitro	57-59
	Ad 4, 8, 9, 10, 13, 15, 17, 19, 20, and 22-30	"Nononcogenic"	55-61

[a]From (1).
[b]Highly oncogenic adenoviruses induce tumors in a large proportion of newborn hamsters within two months after injection with purified virus; "weakly oncogenic" in a small proportion after 4-18 months.

CELL TRANSFORMATION BY HUMAN ADENOVIRUSES

During the past five years studies on the mechanism of cell transformation by DNA viruses have shown that transformed cells possess multiple copies of viral DNA genes (Fig. 1). Some viral genes are transcribed in transformed cells as detected by hybridization with viral DNA, a process which isolates viral RNA molecules for further studies.

Multiple viral DNA copies in adenovirus transformed cells

Cells transformed by ten adenoviruses of groups A, B, and C contain functional viral genes since they synthesize virus-specific RNA (3-7). The number of viral gene copies was estimated by hybridization of adenovirus complementary ^{3}H-RNA (cRNA) with DNA from adenovirus transformed cells (Table 2) (Ad cRNA was prepared on an adenovirus DNA template with the E. coli RNA polymerase). Ad cRNA contains mainly early adenovirus gene transcripts (8). From 14 to 97 viral DNA copies are present in Ad 12 (group A), 2 (group C), and 7 (group B) transformed cells (Table 3).

CELL TRANSFORMATION BY DNA VIRUSES

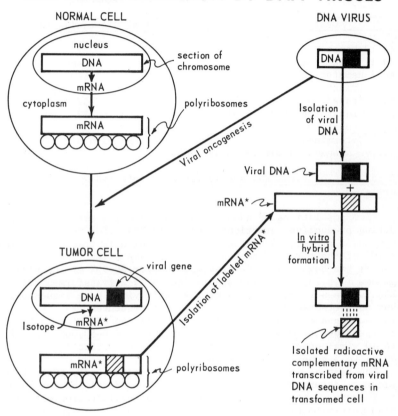

Fig. 1. Scheme depicting cell transformation by DNA viruses. From (2).

TABLE 2*
DETECTION OF VIRAL GENES IN ADENOVIRUS TRANSFORMED CELLS

Transformed cell DNA + Ad[^{3}H]cRNA $\xrightarrow[\text{4 X SSC, 20 hours}]{\text{66 C}}$

DNA[^{3}H]cRNA complex

*From (2)

TABLE 3
NUMBER OF VIRAL DNA COPIES IN CELLS
TRANSFORMED BY HUMAN ADENOVIRUSES*

Cell	Viral DNA equivalents/cell
Ad 12 hamster tumor	22
Ad 12 transformed hamster	53-60
Ad 7 hamster tumor	86-97
Ad 2 transformed rat (8617)	22-30
Ad 2 transformed rat (8625)	37
Ad 2 transformed rat (8629)	29
Ad 2 transformed rat (8638)	14

*Determined by hybridization of cell DNA with highly radioactive adenovirus complementary RNA.

Resolution of viral RNA species in Ad 2 infected and transformed cells

Two populations of adenovirus RNA molecules are synthesized during productive infection: (a) early RNA prior to viral DNA synthesis at six hours after infection which is transcribed from 8-20% of the viral genome, and (b) late RNA after viral DNA synthesis has begun which is transcribed from 80-95% of the viral genome (7,9,10). A third population in Ad 2 transformed cell RNA is transcribed from 50% of the early viral gene sequences, but not detectably from late viral gene sequences (7). Since the Ad 2 genome (23×10^6 daltons of DNA) can code for 23 to 46 polypeptides of average molecular weight 25,000-50,000, from 2 to 10 viral genes are expressed early during productive infection and 1 to 5 genes in transformed cells.

The time of appearance of viral RNA species early after infection was determined by labeling cells with ^{3}H-uridine for 30 min to 2 hours, 2 to 4 hours, 4 to 6 hours after infection with a high multiplicity of Ad 2 (200 PFU/cell) to ensure synchronous infection (11). RNA isolated from the cytoplasmic polyribosomes was resolved on 2.8% polyacrylamide gels, and virus-specific RNA was quantitated by hybridization of RNA extracted from gel slices with viral DNA (12). No virus-specific RNA was detected from 30 min to 2 hours (Fig. 2) but two major viral RNA species (23S and 17S), stable to treatment with dimethylsulfoxide, were found in polyribosomes labeled 2 to 4 and 4 to 6 hours after infection. The 17 and 23S species and increased amounts of a minor 27S viral RNA specie were found in the absence of protein synthesis (Fig. 3).

RNA isolated from Ad 2 transformed rat embryo cells was labeled for 6 hours with ^{3}H-uridine and resolved by electrophoresis on 2.8% polyacrylamide gels (polyribosomal RNA) or by sucrose density gradient centrifugation (nuclear RNA) (Fig. 4). In contrast to distinct viral RNA species obtained early during

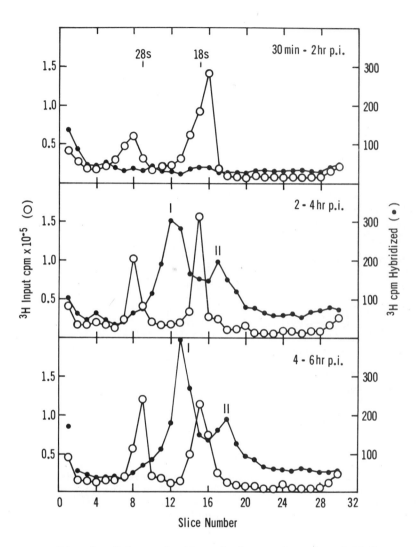

Fig. 2. Polyacrylamide gel electrophoresis of Ad 2 RNA synthesized for 30 min to 2 hours, 2 to 4 hours, and 4 to 6 hours after infection. KB cells infected with Ad 2 were labeled with [^{3}H]uridine for 30 min to 2 hours, 2 to 4 hours and 4 to 6 hours after infection. Polyribosome ^{3}H-RNA was applied to 2.8% polyacrylamide gels, electrophoresed, and the distribution of hybridizable RNA determined. Input ^{3}H cpm per slice (o); ^{3}H cpm hybridized to 2 μg of Ad 2 DNA (•). From (12).

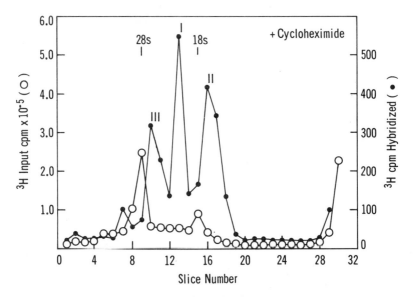

Fig. 3. Polyacrylamide gel electrophoresis of early Ad 2 virus-specific RNA synthesized from 2 to 6 hours after infection in the presence of cycloheximide. Input ^{3}H cpm per slice (o); ^{3}H cpm hybridized to 2 μg of Ad 2 DNA (•).

productive infection, nuclear and polyribosomal RNA viral sequences in transformed cells are present as heterogeneous molecular weight species ranging from 10 to 30S in the cytoplasm and 4 to 45S in the nucleus. These large viral RNA species possess more genetic information than theoretically can be accounted for by the 5 to 10% of the viral genome expressed in Ad 2 transformed cells which at most would represent a 23S viral RNA molecule. These viral RNA species could be transcribed from (a) multiple Ad 2 DNA sequences integrated in tandem or (b) both contiguous host and cell DNA sequences. Evidence supporting the latter possibility is shown in Table 4. Virus-specific RNA from Ad 7 and Ad 12 transformed cells, purified by two cycles of hybrid formation with homologous viral DNA, hybridized with both viral and cell DNA (13). These data suggest that initiation or termination sequences on integrated viral DNA may not be recognized by transformed cell RNA polymerase resulting in the continued transcription from host cell DNA into adjacent viral DNA or from viral DNA into cellular DNA. The size heterogeneity of viral RNA sequences may be a consequence of the integration of viral DNA into different chromosomal sites (14).

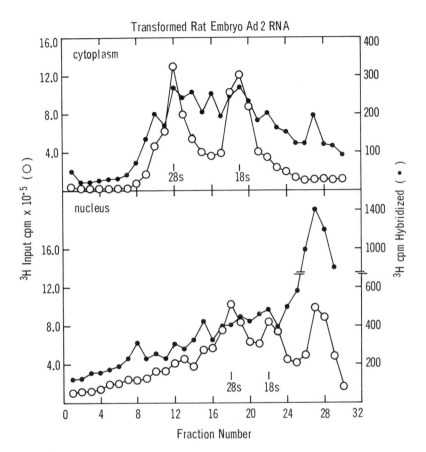

Fig. 4. Polyacrylamide gel electrophoresis of virus-specific RNA from Ad 2 transformed rat embryo cells. Polyribosomal [^{3}H]RNA was applied to duplicate polyacrylamide gels. Nuclear [^{3}H]RNA was layered on a 10-30% sucrose gradient and centrifuged for 5 hours at 38,000 rpm in a SW41 rotor. Input ^{3}H cpm per fraction (o); ^{3}H cpm hybridized to 2 μg of Ad 2 DNA (●).

TABLE 4*
HYBRIDIZATION OF VIRUS-SPECIFIC RNA FROM AD 2 AND AD 7
TRANSFORMED CELLS WITH VIRAL AND CELLULAR DNA

RNA			DNA		Bound radioactivity	
Source	Purification cycle[a]	Input (cpm)	Source	µg/filter	cpm[b]	%
Ad 7 transformed hamster cells (5728)	2	763	Ad 7 hamster cell	5	587	76.5
			(NIL 2E)	50	182	24.0
			E. coli	50	28	3.6
			Blank	0	12	1.5
Ad 2 transformed rat cell (8617)	2	1395	Ad 2	5	523	37.5
			rat cell (9258)	50	254	18.4
			E. coli	50	83	5.9
			Blank	0	37	2.6

*From (13).
[a]Number of purification steps by hybrid formation with homologous viral DNA and elution using the urea procedure for hybridization at 37° without RNase treatment.
[b]Average of duplicate hybridization reactions performed by the urea procedure with RNase treatment.

Resolution of Ad 2 RNA species in the nucleus and cytoplasm late during productive infection

KB cell nuclear RNA labeled with ^{3}H-uridine for 15 and 60 min at 18 hours after infection with Ad 2 was electrophoresed on 2.8% polyacrylamide gels. RNA eluted from gel slices was annealed with Ad 2 DNA (15). Up to 10 viral RNA species larger than 26S were detected (Fig. 5). N_I to N_{IV} were the major species labeled during 15 min (Fig. 5a) while increasing proportions of N_{VI} to N_X and three smaller RNA's were detected after 60 min (Fig. 5b). Treatment with 8 M urea or 95 dimethylsulfoxide did not alter the number or rate of migration of viral RNA species; thus the large nuclear RNA's are not RNA aggregates but intact adenovirus RNA molecules with molecular weights as high as 4×10^6 daltons.

Stable cytoplasmic viral RNA species were isolated by pulse labeling cells at 16 hours after infection for 30 min with ^{3}H-uridine and chasing for 2 hours with cold uridine. Less than 10% of the labeled RNA remained in the nucleus. Six to eight viral RNA species migrating at 12 to 30S were resolved by short electrophoresis (Fig. 6a) and further separated by longer electrophoresis (Fig. 6b). These viral RNA species (L_I to L_{VIII}) most likely are stable viral mRNA molecules corresponding to some of the known virion structural polypeptides (Table 5).

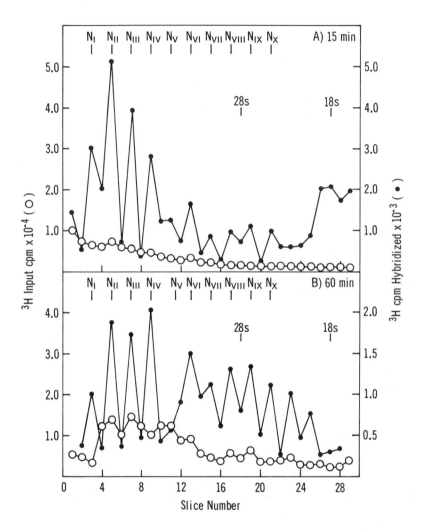

Fig. 5. Resolution of late nuclear Ad 2 RNA species by polyacrylamide gel electrophoresis. KB cells, infected for 18 hours, were labeled with [^{3}H]uridine (4 μc/ml) for 15 min (A) and 60 min (B). Input ^{3}H cpm per slice (o); ^{3}H cpm hybridized to 1 μg Ad 2 DNA (•). From (15).

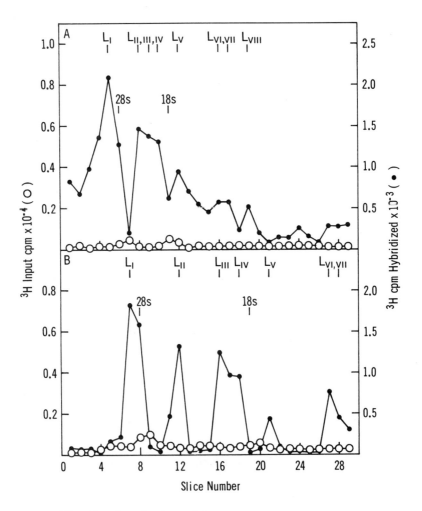

Fig. 6. Resolution of late cytoplasmic pulse-chased Ad 2 RNA species by electrophoresis on polyacrylamide gels. Electrophoresis was carried out for 3 hours (A) and 5 hours (B). Input ^{3}H cpm per slice (o); ^{3}H cpm hybridized to 1 µg Ad 2 DNA (●). From (15).

TABLE 5
CYTOPLASMIC ADENOVIRUS RNA SPECIES SYNTHESIZED
LATE AFTER INFECTION

	Sedimentation value[a]	Estimated molecular weight[b]	Polypeptide equivalent	Adenovirus capsid peptides[c]
L_I	29S	1.82×10^6	200,000	
L_{II}	24S	1.23×10^6	136,000	Hexon -- 130,000
L_{III}	21S	0.93×10^6	103,000	
L_{IV}	19S	0.74×10^6	82,000	Penton -- 70,000
L_V	16S	0.56×10^6	62,000	Fiber -- 62,000
L_{VI}	14S	0.40×10^6	44,000	Core -- 44,000
L_{VII}	12S	0.29×10^6	32,000	Core -- 24,000
L_{VIII}	10S	0.19×10^6	21,000	

[a]Sedimentation values were estimated from the rate of migration of the RNA species in polyacrylamide gels relative to 18S and 28S RNA.
[b]Calculated from molecular weight = $1550 S^{2.1}$.
[c]Molecular weights of the capsid proteins were determined by SDS-polyacrylamide gel electrophoresis (16-19).

To determine whether high molecular weight nuclear RNA species are precursors to cytoplasmic viral RNA's, sequence homology was analyzed by DNA-RNA hybridization-competition experiments. Both cytoplasmic and polyribosomal RNA competed completely with labeled high molecular weight nuclear RNA (greater than 35S) for sites on viral DNA (15). Thus the high molecular weight nuclear viral RNA sequences are present in cytoplasmic and polyribosomal RNA and probably represent precursors to cytoplasmic viral RNA. The experiments presented here provide techniques to identify specific precursors of viral RNA's within the cell nucleus, to follow their fate, and to study the mechanism governing their processing, transport, and translation.

Control of transcription during productive infection and in transformed cells

The above experiments indicate that late gene expression during productive infection is controlled at two levels: (a) transcriptional and (b) post-transcriptional. (a) Transcriptional control involves the specific binding of RNA polymerase to a limited number of initiation regions on viral DNA resulting in the transcription of large precursor viral RNA molecules. For example, assuming that the nuclear RNA N_I, N_{II}, and N_{III} contain different nucleotide sequences, the transcription of these three RNA species would represent 95% of the Ad 2 genome.

(b) Post-transcriptional control involving the cleavage of nuclear precursor viral RNA molecules to smaller RNA molecules prior to translation in cytoplasmic polyribosomes. The size heterogeneity of nuclear viral RNA may reflect the cleavage of precursor RNA to smaller species within the nucleus. Transcription of Ad 2 DNA and post-transcriptional cleavage appear similar to that observed for herpesvirus DNA (20) and SV40 DNA (21). But it is different from the transcription and cleavage of ribosomal RNA (22-25) and nuclear DNA-like RNA (26-30) since the processing of ribosomal and nuclear DNA-like RNA results in the degradation of some initial RNA sequences while all nucleotide sequences are conserved in Ad 2 nuclear 36 to 43S RNA species.

Early viral gene transcription is regulated by the cells DNA dependent RNA polymerase. Three viral RNA species are found which are not cleaved prior to translation. The size of these RNA species correspond to the viral genetic information transcribed early after productive infection.

The regulation of transcription of adenovirus DNA sequences in transformed cells is intriguing. Hybridization-competition analyses show that only early viral genes are transcribed in adenovirus transformed cells. But some late viral gene sequences are transcribed in SV40 transformed cells (31-33). Whether these SV40 sequences are involved in cell transformation is not known. Furthermore host cell viral RNA molecules are transcribed in adenovirus transformed cells, suggesting that RNA is transcribed from contiguous viral and cell DNA sequences in the same molecule. Further understanding on the transcription of viral and cellular genes requires the isolation and characterization of the DNA dependent RNA polymerase enzyme and sigma-like factors, mammalian cell components which are notoriously unstable (34).

TRANSCRIPTION OF VIRAL GENES IN CELLS TRANSFORMED BY RNA TUMOR VIRUSES

Ten RNA tumor viruses as well as the Visna virus of sheep and foamy viruses of monkeys possess RNA dependent DNA polymerase activity (35-43). Several of these viruses also contain DNA dependent DNA polymerase activity (41, 44, 45). I describe below briefly studies by my colleagues, Hinae and Makoto Rokutanda, Kei Fujinaga, Corrado Gurgo, Ranjit Kumar Ray, J. Thomas Parsons, J.W. Beard, Dorothy Beard, and myself on: (a) the properties of the RNA and DNA dependent DNA polymerase activities of MSV, FeSV, and AMV, (b) evidence that the whole viral genome is transcribed in vitro, (c) studies on specific inhibitors of the RNA and DNA dependent DNA polymerase activities of leukemia-sarcoma viruses, and (d) the detection of viral RNA sequences in cells transformed by RNA tumor viruses by a novel molecular hybridization procedure.

The DNA polymerase activities of RNA tumor viruses are schematized in Fig. 7. The initial product is 70S viral RNA (+

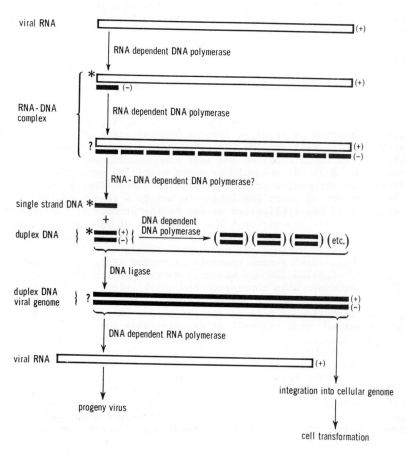

Fig. 7. RNA tumor virus DNA polymerase activities and possible function of DNA product in cell transformation and virus replication. From (46).

strand) complexed with small DNA pieces (- strand) of molecular
weight about 150,000 daltons. Subsequently free single and
double-stranded DNA molecules of the same molecular length are
formed. The whole viral genome is copied to DNA (47, 14). The
viral DNA fragments probably are assembled to a duplex viral DNA
genome in vivo, a possibility consistent with the presence of
ligase activity in the virion (48). The functions of the DNA
product in cell transformation and virus replication are hypothesized in Fig. 7.

RNA dependent DNA polymerase activities of MSV, AMV, and FeSV

The properties of RNA and DNA dependent DNA polymerase activities of purified virions are similar (38, 40, 41, 49). Detergent disruption is required for the demonstration of enzyme
activity. The FeSV and AMV DNA polymerase reactions appear to be
biphasic, a rapid reaction for four min   followed by a slower
reaction. A 7S DNA molecule is found within 30 seconds, and no
larger DNA structures are found at later times (Fig. 8, MSV).
The formation of short DNA fragments may be a result of the
mechanism of DNA replication or may be the product of nuclease
activity within the virion. The rapid rate of DNA synthesis in
vitro is of the same magnitude as DNA synthesis in vivo for HeLa
cells (40 to 50 nucleotides per second (50)).

The whole viral genome probably is copied (Fig. 9). About
85% of P^{32}-labeled 70S MSV RNA will hybridize with the 90 min
MSV DNA product. The copying of the entire viral genome, the
presence of both RNA and DNA dependent polymerase activities,
and the formation of both single and duplex DNA molecules are
all consistent with the role of enzymatically synthesized viral
DNA as a transforming agent. The mechanism of synthesis of single and double-stranded DNA and the number of enzymes or enzyme
activities involved are not known. Furthermore, the presence of
endonuclease and exonuclease activities in Rous sarcoma virus
(48) confounds the interpretation of data on DNA strandedness,
conformation, and size. Moreover, the formation of a DNA
product in cells replicating RNA tumor viruses has not yet been
demonstrated, thus adding the most serious concern on the
role of the RNA dependent DNA polymerase in viruses.

Specific inhibitors of the RNA and DNA dependent DNA polymerase
activities of RNA tumor viruses

Specific inhibitors of the viral enzyme can help analyze its
possible role in neoplasia and might provide drugs for leukemia
and cancer therapy. We have studied the effect on the DNA
polymerase activity of RNA tumor viruses of rifampicin derivatives with modified aminopiperazine side chains (see Fig. 10),
4-N-demethyl rifampicin, 2,6-dimethyl-4-N-benzyl demethyl rifampicin (AF/ABDMP), and 4-N-benzyl demethyl rifampicin (AF/ABP),

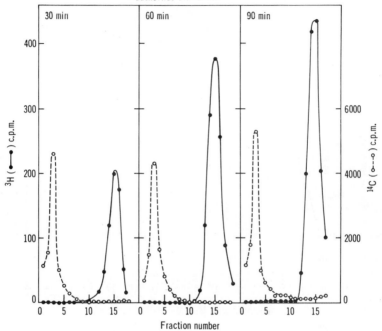

Fig. 8. Zone sedimentation of ^{3}H-DNA from RNA-^{3}H-DNA hybrids in alkaline sucrose density gradients From (38).

three aminopiperazine derivatives, and actinomycin D (Act D). Rifampicin derivatives prepared by condensing 3-formyl rifampicin SV(AF) with the appropriate aminopiperazine derivatives were generously supplied by Drs. L. Silvestri and G. Lancini of Gruppo Lepetit, Milano, Italy.

We show here that rifampicin is without effect, but AF/ABDMP, AF/ABP, and demethyl rifampicin inhibit the DNA polymerase activity of several RNA tumor viruses, including MSV, FeLV, and AMV. At appropriate concentrations of AF/ABDMP, MSV RNA dependent DNA polymerase is selectively inhibited; other DNA dependent DNA polymerase activities including ones from MSV, from E. coli DNA, and from mammalian cells are not appreciably inhibited. In contrast, Act D inhibits the DNA dependent but not RNA dependent DNA polymerase activity.

The rifampicin drugs are of great interest since they inhibit bacterial and mitochondrial RNA polymerases, inhibit the replication of poxvirus (51-57) and adenovirus (55) and block focus

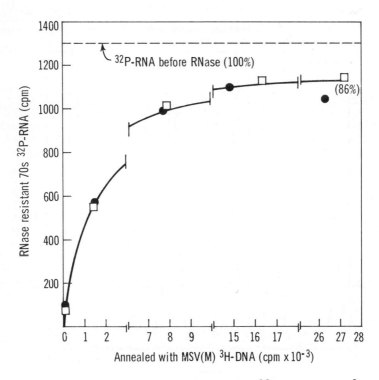

Fig. 9. Saturation of viral 70S ^{32}P-RNA by MSV ^{3}H-DNA product.

formation by Rous sarcoma virus (58). Recent data suggest that rifampicin blocks the cleavage of a polypeptide precursor of an internal protein of vaccinia virus (59). The effect of different levels of rifampicin, N-demethyl rifampicin, AF/ABDMP, and AF/ABP on the DNA polymerase activity of purified MSV(M) was determined (Fig. 11). Rifampicin at 400 µg/ml did not inhibit ^{3}H-TTP incorporation while the three rifampicin derivatives inhibited enzyme activity over 50% at 50 to 100 µg/ml. The most effective inhibitor, AF/ABDMP, at 100 µg/ml reproducibly blocked 95 to 100% of ^{3}H-TTP incorporation into MSV.

Thiry and Lancini (60) reported that the aminopiperazines, AP4, AP5, and AP8, inhibit the growth of vaccinia and herpesviruses, but as shown in Table 6, these compounds are poor inhibitors of MSV DNA polymerase. AP4, the side-chain component of AF/ABDMP, at 500 µg/ml inhibited MSV DNA polymerase 21% while AP5 and AP8 inhibited 50 and 13%. In contrast, demethyl rifampicin at these high concentrations (500 µg/ml) inhibited enzyme activity 99.7%.

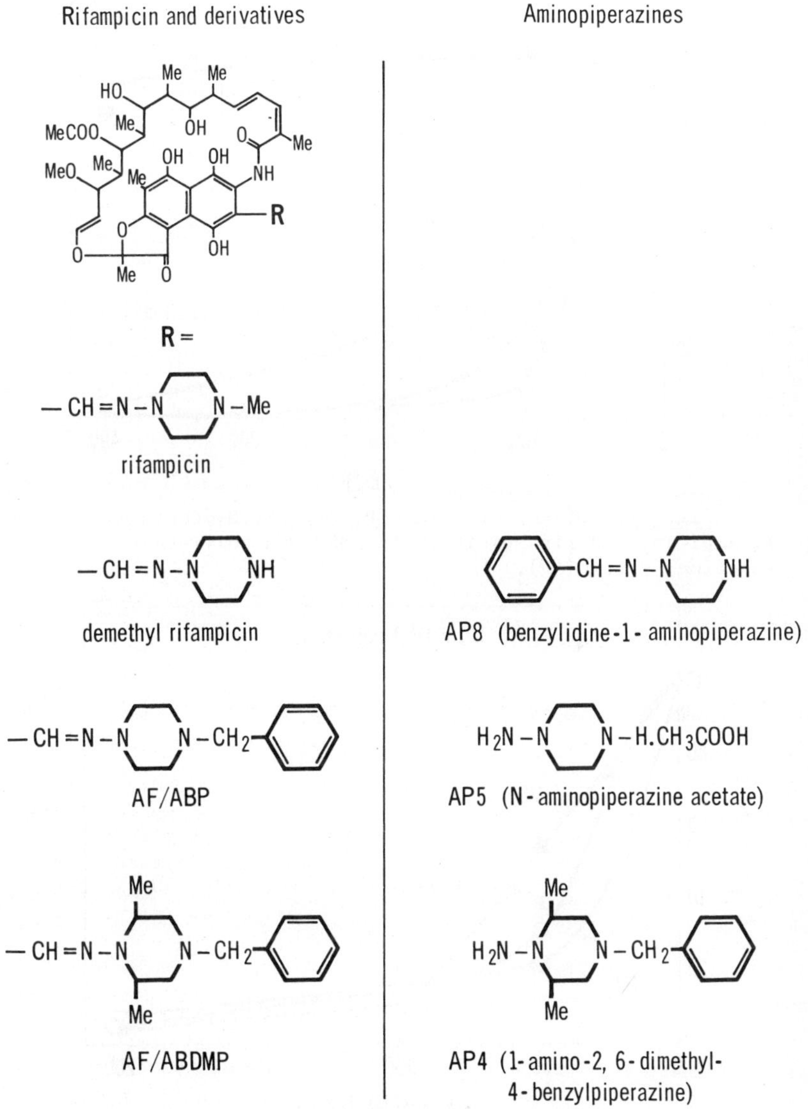

Fig. 10. Structures of rifampicin derivatives and aminopiperazines tested for ability to inhibit the DNA polymerase activities of RNA tumor viruses. From (46).

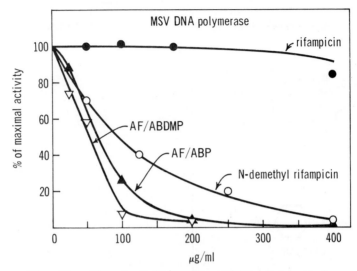

Fig. 11. Effect of AF/ABDMP, AF/ABP, N-demethyl-rifampicin, and rifampicin on the MSV DNA polymerase reaction. From (49).

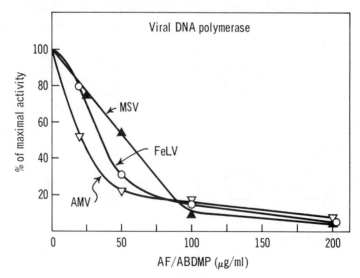

Fig. 12. Inhibition of MSV, FeLV, and AMV DNA polymerase by AF/ABDMP. From (49).

TABLE 6
EFFECT OF N-DEMETHYL RIFAMPICIN AND AMINOPIPERAZINE DERIVATIVES ON MSV AND KB CELL DNA POLYMERASES

	^{3}H-TTP incorporated					
	MSV DNA polymerase			KB cell DNA polymerase		
Treatment	Amount µg	cpm	Per cent inhibition	Amount µg	cpm	Per cent inhibition
None (control)	–	7840	0	–	2531	0
N-demethyl rifampicin	500	27	99.7	1000	–	–
AP4	500	6202	21	1000	1908	25
AP5	500	3792	50	1000	2333	7
AP8	250	6822	13	1000	2277	10

Average of duplicate enzyme assays. From (49).

AF/ABDMP inhibits the DNA polymerase activity not only of MSV but also of FeLV and AMV (Fig. 12). In all these cases, 100 µg/ml AF/ABDMP inhibited enzyme activity more than 80%. Thus the polymerizing system of these three different species of RNA tumor viruses possess a common structural feature recognized by rifampicin derivatives.

Levels of AF/ABDMP that inhibit MSV DNA polymerase over 90% show little if any inhibition of DNA polymerase from an established human cell line (KB) or from E. coli (Fig. 13). (E. coli and mammalian DNA polymerase activities are measured in the presence of a denatured calf thymus DNA template while viral enzyme activity utilizes its endogenous RNA templates.) In addition to their RNA dependent DNA polymerase activity, RNA tumor viruses show some DNA dependent DNA polymerase activity (44,45,41,61) detected by adding duplex DNA as template. When native calf thymus DNA was added (Fig. 13) to the MSV reaction mixture, inhibition was observed with AF/ABDMP, but at somewhat higher levels than were required for the RNA-dependent reaction.

The effect of different levels of AF/ABP on the MSV, KB cell, and E. coli DNA polymerase activities are shown in Fig. 14. MSV DNA polymerase is inhibited 50% at about 75 µg/ml while KB cell and E. coli DNA polymerase are inhibited 50% at about 200-250 µg/ml. Thus AF/ABP is again a more effective inhibitor of endogenous viral polymerase than of bacterial and mammalian cell enzymes.

As shown in Fig. 15, demethyl rifampicin inhibits the MSV, FeLV, and AMV DNA polymerase activities. KB cell DNA polymerase is inhibited less at all concentrations tested. Thus all three rifampicin derivatives are more effective inhibitors of DNA polymerase of RNA tumor viruses than of KB cells.

As shown in Fig. 16, Act D blocks the KB cell and E. coli DNA polymerase 90% (denatured DNA is used as a template); but the maximum inhibition of the MSV polymerase (using endogenous

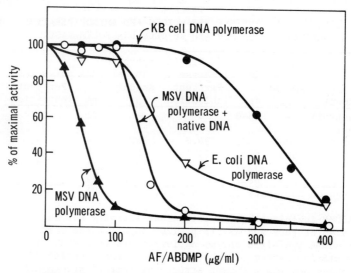

Fig. 13. Effect of AF/ABDMP on the RNA and DNA dependent DNA polymerase activities of MSV, the E. coli DNA polymerase, and the KB cell DNA polymerase. From (49).

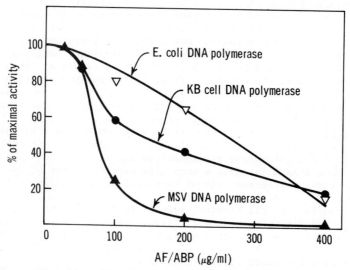

Fig. 14. Effect of AF/ABP on the MSV, E. coli, and KB cell DNA polymerase. From (49).

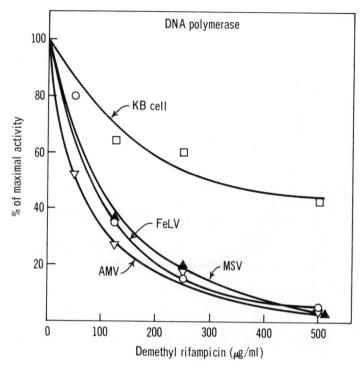

Fig. 15. Effect of demethyl rifampicin on the MSV, FeLV, AMV, and KB cell DNA polymerase. From (49).

template) ranged from 50 to 70% in numerous experiments. Act D is known to inhibit DNA dependent RNA polymerase activity by binding to duplex DNA. The high levels of Act D required to inhibit DNA polymerase activity reflects the relatively poor binding of Act D to the single-stranded DNA template. The partial inhibition of MSV DNA polymerase activity and the fact that Act D binds poorly to RNA-DNA hybrids and to single-stranded RNA (61) suggests that Act D does not inhibit the initial formation of a RNA-DNA hybrid but does block the subsequent DNA dependent DNA polymerase activity. This interpretation is supported by the experiments in Fig. 17. When Act D was added at 0 or 2 min, enzyme activity continued linearly at a rate comparable to untreated controls for about 15 min and then decreased, but when Act D was added at 30 min, enzyme activity ceased abruptly. Further evidence that Act D affects the DNA dependent reaction is given in Table 7. Preliminary treatment of MSV with a nonionic detergent (NP40) and RNase reduces endogenous DNA polymerase activity about 60% (-DNA, column 1), probably due to the partial destruction of the viral RNA template. More important,

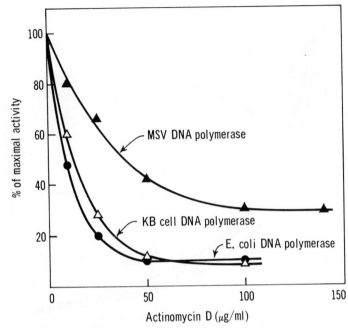

Fig. 16. Effect of Act D on the MSV, KB cell, and E. coli DNA polymerase. From (49).

TABLE 7
EFFECT OF ACTINOMYCIN D AND RNASE ON MSV RNA
AND DNA DEPENDENT DNA POLYMERASE ACTIVITIES

Treatment	^{3}H-TTP incorporated (cpm)*	
	− DNA	+ DNA
None	13,750	90,340
+ Act D (100 μg/ml)	7,150	8,500
+ RNase (100 μg/ml)	6,480	110,000
+ RNase + Act D	2,330	3,430

*Average of duplicate experiments.
From (49).

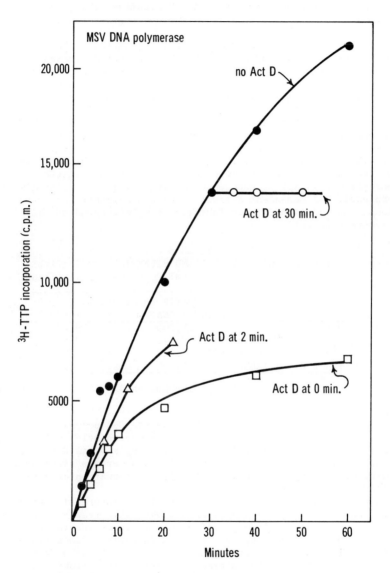

Fig. 17. Effect of time of addition of Act D on MSV DNA polymerase activity. From (49).

the residual enzyme activity is Act D sensitive; i.e., inhibition by RNase and Act D are additive. Further support for this notion is provided by analysis of the DNA dependent DNA polymerase activity (Table 7, second column). As expected, RNase does not inhibit the DNA-dependent reaction, but Act D blocked ^{3}H-TTP incorporation 90%. We conclude that Act D inhibits the DNA dependent DNA polymerase activity of MSV, very possibly by binding to the DNA template generated by the RNA dependent DNA polymerase reaction.

That Act D and rifampicin derivatives affect different MSV DNA polymerase activities is demonstrated further by the data of Table 8. AF/ABDMP inhibits MSV DNA polymerase activity 90% in the absence of added template (column 1) but no inhibition was observed when template DNA was added. In contrast, Act D inhibited only half of endogenous activity while blocking almost completely DNA dependent activity. Higher AF/ABDMP concentrations (200 μg/ml) partially inhibited the DNA dependent DNA polymerase as well.

TABLE 8
EFFECT OF ACTINOMYCIN D AND RIFAMPICIN DERIVATIVES ON MSV RNA AND DNA DEPENDENT POLYMERASE ACTIVITIES

Treatment	^{3}H-TTP incorporated (cpm)*	
	− DNA	+ DNA
None	12,300	89,400
+ Act D (100 μg/ml)	5,670	8,600
+ AF/ABDMP (100 μg/ml)	1,900	113,000
+ Act D + AF/ABDMP	1,300	9,300

*Average of duplicate enzyme assays.
From (49).

Conclusions: We have shown that the two DNA polymerase activities of RNA tumor viruses can be independently inhibited: the formation of an RNA-DNA hybrid utilizing a viral RNA template is inhibited by rifampicin derivatives with certain structural features but not by rifampicin itself; the formation of duplex DNA is inhibited by Act D.

The rifampicin derivatives are especially interesting since they inhibit the unique RNA dependent DNA polymerase activity of RNA tumor viruses. Although rifampicin is inactive, the removal of the 4-N-demethyl group on the aminopiperazine side chain or its substitution by a benzyl group converts it to an effective inhibitor. Based on the information derived from studies with bacterial and mitochondrial RNA polymerase (62-65) one can expect that the common structure of the polymerase-viral RNA complex recognized by the three rifampicin derivatives on the three RNA tumor viruses is a part of the enzyme itself. The macrocyclic ring structure of rifampicin is apparently required for anti-polymerase activity since the aminopiperazine derivatives are inactive. Further modification of the aminopiperazine portion of rifampicin or the macrocyclic structure may provide more effective inhibitors of the RNA tumor virus DNA polymerase.

These results are both tantalizing and reassuring for they show that basic research on the molecular biology of normal and virus-infected bacteria and mammalian cells may at last fulfill the faith that such studies provide a basis for understanding the mechanism of animal virus infection and carcinogenesis. The above results provide the basis for a rational chemotherapy of viral diseases and cancer. Most animal viruses either possess or induce a DNA or RNA polymerase (66, 67). Thus a prime target for chemotherapy is the viral specific polymerase. Chemicals which specifically bind to polymerase enzymes, such as rifampicin derivatives provide the beginning of the specific design of anti-polymerase agents.

Virus-specific RNA in cells transformed by RNA tumor viruses

Attempts to study the replication of RNA tumor viruses have been handicapped by the inability to detect radioactive 70S viral RNA in ^{3}H-uridine labeled infected or transformed cells (66, 68, 69). A report (70) describing the detection of very small amounts of 70S RNA in cycloheximide treated cells is not readily repeated (71). The failure to detect 70S viral RNA has been attributed to a rapid incorporation of intracellular 70S viral RNA into virions at the cell membrane; this could result in an intracellular viral RNA content less than 0.01% of the total cellular RNA. Instead we demonstrate here that transformed cells replicating MSV, while they contain no detectable intracellular 70S viral RNA (69), do possess large amounts of viral RNA sequences both in the nucleus and cytoplasm.

Virus-specific RNA was detected by a novel approach, hybridization of cellular RNA with the ^{3}H-DNA product formed by the RNA-dependent DNA polymerase of MSV; the amount of RNA-DNA hybrids was quantitated by Cs_2SO_4 density gradient centrifugation and by hydroxyapatite chromatography. Using the latter procedure significant quantities of virus-specific RNA were found even in an MSV transformed hamster cell line which synthesizes neither detectable virus nor viral proteins. We conclude that cells transformed by RNA tumor viruses certainly contain large amounts of intracellular viral RNA, but the intracellular viral RNA is in a state different from the 70S RNA of the virion, probably in smaller pieces.

Highly radioactive MSV(H) ^{3}H-DNA, prepared by incorporation of ^{3}H-TTP into detergent-disrupted MSV, was annealed with RNA from MSV(H) (Harvey) virus transformed mouse cells (MEH cell line), or normal mouse cells and the annealed product was sedimented to equilibrium in Cs_2SO_4 density gradients. Upon annealing 4 µg of MSV RNA with MSV ^{3}H-DNA, 70% of ^{3}H-DNA was converted to an RNA-

DNA hybrid as shown by the shift of ^{3}H-DNA from the DNA (adenovirus ^{32}P-DNA marker) to the RNA position (cell RNA marker) of the gradient (Fig. 18a), as expected for hybrid molecules of RNA with a small DNA fragment (40). When 50 μg of RNA from MSV(H) transformed mouse cells was annealed with MSV ^{3}H-DNA, 70% of the DNA again shifted to the hybrid position (Fig. 18b), demonstrating the presence of virus-specific RNA in MSV transformed cells. In controls, normal mouse cell RNA shifted no ^{3}H-DNA to the RNA region (Fig. 18c).

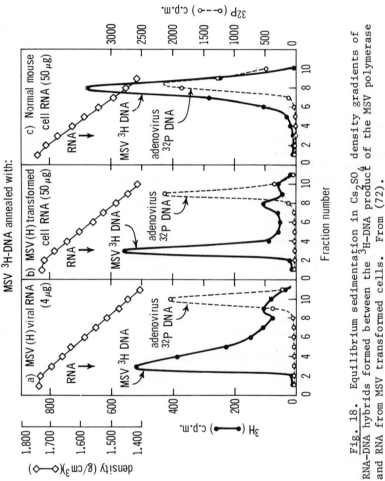

Fig. 18. Equilibrium sedimentation in Cs_2SO_4 density gradients of RNA-DNA hybrids formed between the ^{3}H-DNA product of the MSV polymerase and RNA from MSV transformed cells. From (72).

The virus-specific RNA sequences detected in transformed cell RNA did not originate from virus particles budding from the cell membrane, but were found both in the cytoplasmic fraction and in isolated nuclei. In addition, to exclude the possibility that MSV(H) transformed mouse cells are unusual, e.g., a slow releaser of virus, we analyzed a cell line (78A1) derived from a different animal species (rat) transformed by a different strain of MSV (M, Moloney). As shown in Fig. 19, 70 to 80% of MSV ^{3}H-DNA was converted to an RNA-DNA hybrid by annealing with 50 µg of purified cytoplasmic (S-30) or nuclear RNA from MSV(M) transformed rat cells (Fig. 19a and b). No hybrid was detected with S-30 or nuclear RNA from control Ad 2 transformed rat cells (Fig. 19c and d), or from a variety of other cell lines of mouse, rat, and human origin (not shown).

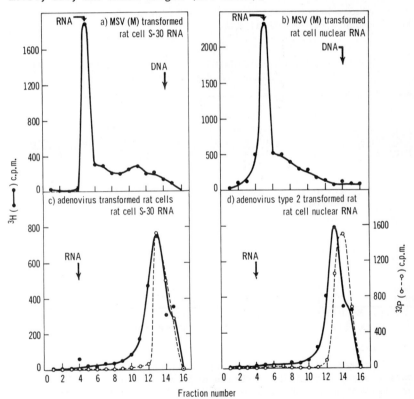

Fig. 19. Equilibrium sedimentation in Cs_2SO_4 density gradients of RNA-DNA hybrids formed between the ^{3}H-DNA product of the MSV polymerase and 50 µg of RNA from the nucleus or cytoplasm (S-30 fraction) of MSV(M) transformed rat cells (a and b) or adenovirus transformed rat cells (c and d). From (72).

To exclude the possibility that the virus-specific RNA in the nuclear fraction is a cytoplasmic contaminant, we purified nuclei further by treatment with the nonionic detergent nonidet P-40 (NP40) and then with deoxycholate (DOC) + Tween 40; the latter removes the outer membrane of the nucleus (73). RNA from both purified nuclei and from the DOC-Tween extract formed hybrids with MSV ^{3}H-DNA.

The amount of hybrid formed by annealing a constant amount of MSV ^{3}H-DNA (2500 cpm) with increasing amounts of RNA from nuclear and S-30 fractions of MSV(M) (Table 9) and MSV(H) (Table 10) transformed cells was determined both by equilibrium centrifugation in Cs_2SO_4 gradients and elution from hydroxyapatite columns. The relationship between the extent of hybrid formation and the virus-specific RNA content of the sample was standardized by simultaneous hybridization reactions between MSV ^{3}H-DNA and known amounts of 70S viral RNA.

TABLE 9
HYBRID FORMATION BETWEEN MSV ^{3}H-DNA AND RNA FROM MSV(M) TRANSFORMED RAT CELLS (78A1)

Source of RNA	Amount µg	Hybrid Formation			
		Hydroxyapatite		Cs_2SO_4 gradient	
		c.p.m. recovered	% RNA-DNA hybrid	c.p.m. recovered	% RNA-DNA hybrid
Nuclei	10	1447	68	3939	62
	5	1496	53	2801	57
	0.94	1740	70	1293	45
	0.20	1716	34	1744	20
	0.19	2766	22		
	0.20	2659	26		
	0.10	2014	13		
Cytoplasm (S-30)	10	1632	48	2942	33
	5	1455	32	3486	24
	1.0	1784	30	1236	17
	0.5	1763	16		
	0.5	1008	16		

Hybridization was performed and the results analyzed on Cs_2SO_4 gradients by hydroxyapatite chromatography. From (72).

Hydroxyapatite chromatography provides a rapid and quantitative assay for RNA-DNA hybrids and is applicable to the analysis of multiple samples. MSV ^{3}H-DNA (alkali denatured and neutralized) after annealing with 50 µg of KB cell RNA (Fig. 20, extreme right) or with no RNA (Fig. 20, extreme left) is eluted quantitatively (96 to 99%) from hydroxyapatite columns by 0.12 M phosphate buffer (PB) (open bars). DNA-RNA hybrids formed by annealing MSV ^{3}H-DNA with viral RNA or RNA from MSV transformed

TABLE 10
HYBRID FORMATION BETWEEN MSV ^{3}H-DNA AND RNA FROM MSV(H) TRANSFORMED MOUSE CELLS (MEH)

| | | Hybrid Formation | | | |
| | | Hydroxyapatite | | Cs_2SO_4 gradient | |
Source of RNA	Amount μg	c.p.m. recovered	% RNA-DNA hybrid	c.p.m. recovered	% RNA-DNA hybrid
Nuclei	50	2054	60	5586	64
	10	1772	52	3597	65
	5	1686	54	3444	54
	1.2	1613	73	1066	40
	0.5	1787	62	882	28
	0.12	1772	32	1589	15
	0.12	1059	24		
	0.05	1665	12		
Cytoplasm (S-30)	50	2374	56	3286	53
	2.2			1083	37
	1.0	1036	14		

From (72).

cells are retained on hydroxyapatite columns in 0.12 M PB, but are eluted by 0.4 M PB (solid bars). These same conditions are also used to distinguish between single- and double-stranded DNA molecules (74, 41). From 15 to 70% RNA-^{3}H-DNA hybrids were eluted with 0.4 M PB when increasing amounts of viral RNA (0.01 to 0.04 μg) or nuclear RNA (0.05 to 0.4 μg) were annealed with 2500 cpm of ^{3}H-DNA (Fig. 20). The amount of hybrid detected was very reproducible but the sensitivity varied by a factor of 2 to 3 with different lots of hydroxyapatite.

Good agreement between Cs_2SO_4 gradient and hydroxyapatite were obtained with the following or higher levels of RNA (Tables 9, 10, 11): 1 μg of nuclear RNA, 5 μg of cytoplasmic RNA (Table 9), and 0.1 μg of viral RNA. With lower RNA concentrations, Cs_2SO_4 gradients gave lower values for RNA-DNA hybrid than did hydroxyapatite. The greater sensitivity of hydroxyapatite suggests that ^{3}H-DNA duplexed to small RNA fragments may be retained on hydroxyapatite in 0.12 M PB but may be missed on Cs_2SO_4 gradients where they would band close to DNA. The maximum amount of hybrid formed was 70%, most likely reflecting the content of MSV DNA(-) strands in the ^{3}H-DNA preparations, i.e., sequences complementary to viral RNA, and the upper limit of the efficiency of RNA-DNA hybridization.

The content of the virus-specific RNA in the nucleus and cytoplasm of MSV(H) and MSV(M) transformed cells was estimated from the hydroxyapatite data. Since 0.1-0.2 μg of nuclear RNA from MSV(M) and MSV(H) transformed cells converted 13 to 34% of MSV ^{3}H-DNA to RNA-DNA hybrids (Tables 9, 10 and Fig. 20) and 0.01 μg of viral RNA yielded similar levels of hybrid (Fig. 20,

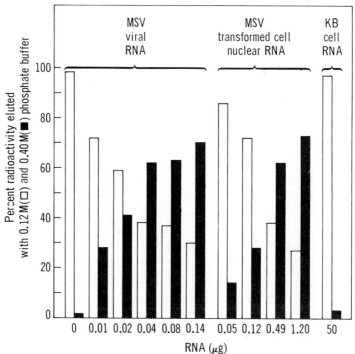

Fig. 20. Hydroxyapatite chromatography of RNA-DNA hybrids formed between the ^{3}H-DNA product of MSV polymerase and MSV viral RNA or 78A1 nuclear RNA. From (72).

Table 11), we estimate that nuclear RNA contains 5-10% virus-specific RNA sequences. Since 1 µg of cytoplasmic RNA (Table 9 and 10) yielded similar hybrid levels, we estimate that cytoplasmic RNA contains 1% virus-specific RNA sequences.

To determine whether cells transformed by RNA tumor viruses which do not replicate virus also contain virus-specific RNA sequences, we analyzed RNA from an MSV transformed hamster tumor cell (HT-1) which contains no detectable infectious virus or viral proteins, but from which MSV can be rescued by co-cultivation with mouse embryo cells in the presence of a murine leukemia virus helper (75). As shown in Table 12, RNA from nuclei, detergent washed nuclei, the detergent wash, and the S-30 fraction, each contain virus-specific RNA sequences. But since 50-

TABLE 11
HYBRID FORMATION BETWEEN MSV ^{3}H-DNA AND VIRAL RNA

Source of viral RNA	Amount µg	Hydroxyapatite c.p.m. recovered	Hydroxyapatite % RNA-DNA hybrid	Cs_2SO_4 gradient c.p.m. recovered	Cs_2SO_4 gradient % RNA-DNA hybrid
MSV(H)	1.05	–	–	–	65
	0.53	–	–	2259	63
	0.21	–	–	2290	59
	0.14	1976	70	–	–
	0.08	1705	63	–	–
	0.06	–	–	2297	65
	0.04	1955	59	–	–
	0.02	1228	43	1273	18
	0.01	1247	30	–	–
MSV(H)	0.04	1071	62, 60	–	–
	0.02	1491	44, 42	–	–
	0.01	1288	28, 26	–	–
	0	1209	2	–	–
MSV(M)	0.16	1920	63	–	–
	0.08	2019	62	–	–
	0.04	1849	53	–	–
	0.02	2060	39	–	–

From (72).

TABLE 12
HYBRID FORMATION BETWEEN MSV(H) ^{3}H-DNA AND RNA FROM A CRYPTIC MSV TRANSFORMED HAMSTER CELL (HT-1)

Source of RNA	µg	c.p.m. recovered	Per cent RNA-DNA hybrid
Nuclei	2	1220	5
	5	1370	16
	10	1110	32
Nuclei (detergent washed)	10	1320	26
Nuclear detergent wash	8	1380	14
	20	1640	31
Cytoplasm (S-30)	2	1370	3
	6	1495	7
	24	1080	15
	40	1080	18

From (72).

100 times as much RNA were required from HT-1 cells as from 78A1 and MEH cells to yield comparable amounts of RNA-DNA hybrids, we conclude that cryptic MSV transformed hamster cells synthesize 1/50th to 1/100th as much virus-specific RNA as do transformed cells that replicate virus. The maximum of 30% hybrid formed (Table 12) may be due to (a) less than saturating amounts of cell RNA, (b) a smaller number of transcribed viral gene sequences, or (c) the possible synthesis of only viral RNA(-) strands.

Conclusions: These results were at first unexpected since until now it has been assumed that extremely small amounts of viral RNA(+) strands are present in cells infected or transformed by RNA tumor viruses. Our findings show that as much as 5-10% of the nuclear RNA and 1% of the cytoplasmic RNA of MSV transformed rat and mouse cells contain virus-specific RNA sequences. But the total amount of virus-specific RNA in the cytoplasm is higher than that in the nucleus, since only 5% (detergent washed nuclei) or 10% (non-detergent washed nuclei) of the total RNA of the cell, as determined by orcinol analysis, is present in the nucleus (14). These viral RNA sequences are mainly if not exclusively (+)RNA strands since at least 70% of the MSV DNA product is the (-) strand.

The failure to detect viral 70S RNA in cells replicating RNA tumor viruses probably means that viral RNA is present in a different form, perhaps as the RNA subunits observed when 70S RNA is denatured (76).

The high content of virus-specific RNA in the cell nucleus suggests that viral RNA is synthesized in the nucleus. The two most likely mechanisms are: (a) viral RNA is transcribed from viral DNA sequences initially copied on parental viral RNA by the RNA dependent DNA polymerase of the virion, or (b) viral RNA is transcribed from complementary RNA (-) sequences initially copied on parental viral RNA by an RNA dependent RNA polymerase. The latter mechanism is familiar for RNA containing bacterial viruses and animal viruses other than RNA tumor viruses. The inhibition of viral RNA synthesis by Act D in cells infected with RNA tumor viruses but not in cells infected by other animal and bacterial RNA viruses (66) favors mechanism (a), the transcription of viral RNA on a DNA template. The report of complementary viral RNA (-) in the nucleus of the 78A1 cell line suggests that mechanism (b) may also be possibly involved in viral RNA (+) synthesis (77).

The procedure described here for the detection of RNA specific for RNA tumor viruses by molecular hybridization coupled with hydroxyapatite chromatography provides a sensitive and specific tool for studying the intracellular synthesis and function of viral RNA and for detecting the possible involvement of virus genetic information in human neoplasia and other diseases.

REFERENCES

1. M. Green, in Recent Results in Cancer Research, Special Supplement, Biology of Amphibian Tumors, M. Mizell (Springer-Verlag, New York, 1969) p. 445.
2. M. Green. Fed. Proc. 29 (1970) 1265.
3. K. Fujinaga and M. Green. Proc. Nat. Acad. Sci. U.S. 55 (1966) 1567.
4. K. Fujinaga and M. Green. Proc. Nat. Acad. Sci. U.S. 57 (1967) 806.
5. K. Fujinaga and M. Green. J. Virol. 1 (1967) 576.
6. K. Fujinaga and M. Green. J. Mol. Biol. 31 (1968) 63.
7. K. Fujinaga and M. Green. Proc. Nat. Acad. Sci. U.S. 64 (1969) 255.
8. M. Green and M. Pina, submitted for publication.
9. D.C. Thomas and M. Green. Virology 39 (1969) 205.
10. K. Fujinaga, S. Mak, and M. Green. Proc. Nat. Acad. Sci. U.S. 60 (1968) 959
11. M. Green and G. Daesch. Virology 31 (1961) 169.
12. J.T. Parsons and M. Green. Submitted for publication.
13. D. Tsuei, K. Fujinaga, and M. Green, unpublished data.
14. M. Green, unpublished data.
15. J.T. Parsons, J. Gardner, and M. Green. Proc. Nat. Acad. Sci. U.S., in press.
16. J.V. Maizel, D.O. White, and M.D. Scharff. Virology 36 (1968) 115.
17. D.O. White, M.D. Scharff, and J.V. Maizel. Virology 38 (1969) 395.
18. M.S. Horwitz, J.V. Maizel, and M.D. Scharff. J. Virol. 6 (1970) 569.
19. H. Caffier, H. Raskas, J.T. Parsons, and M. Green. Nature, in press.
20. E.K. Wagner and B. Roizman. Proc. Nat. Acad. Sci. U.S. 64 (1969) 626.
21. M.A. Martin and J.C. Byrne. J. Virol. 6 (1970) 463.
22. K. Scherrer, H. Latham, and J.E. Darnell. Proc. Nat. Acad. Sci. U.S. 49 (1963) 240.
23. R.A. Weinberg, U. Loening, M. Willems, and S. Penman. Proc. Nat. Acad. Sci. U.S. 58 (1967) 1088.

24. M. Birnstiel, J. Speirs, I. Purdon, K. Jones, and U.E. Loening. Nature 219 (1968) 454.
25. R.A. Weinberg and S. Penman. J. Mol. Biol. 47 (1970) 169.
26. K. Scherrer, L. Marcand, F. Zajdela, I. London, and F. Gros. Proc. Nat. Acad. Sci. U.S. 56 (1966) 1571.
27. G. Attardi, H. Parnas, M. Hwang, and B. Attardi. Proc. Nat. Acad. Sci. U.S. 58 (1966) 1051.
28. J. Warner, R. Soeiro, C. Birnboim, and J. Darnell. J. Mol. Biol. 19 (1966) 349.
29. S. Penman, J. Smith, and E. Holtzman. Science 154 (1966) 786.
30. J. Stevenin, P. Mandel, and M. Jacob. Proc. Nat. Acad. Sci. U.S. 62 (1969) 490.
31. Y. Aloni, E. Winocour, and L. Sachs. J. Mol. Biol. 31 (1968) 415.
32. K. Oda and R. Dulbecco. Proc. Nat. Acad. Sci. U.S. 60 (1968) 525.
33. G. Sauer and J.R. Kidwai. Proc. Nat. Acad. Sci. U.S. 61 (1968) 1256.
34. M. Green, J.T. Parsons, M. Pina, K. Fujinaga, H. Caffier, and I. Landgraf-Leurs. Cold Spring Harbor Sym. Quant. Biol., in press.
35. H. Temin and S. Mizutani. Nature 226 (1970) 1211.
36. D. Baltimore. Nature 226 (1970) 1209.
37. S. Spiegelman, A. Burny, M.R. Das, J. Keydar, J. Schlom, M. Travnicek, and K. Watson. Nature 227 (1970) 563.
38. M. Green, M. Rokutanda, K. Fujinaga, R.K. Ray, H. Rokutanda, and C. Gurgo. Proc. Nat. Acad. Sci. U.S. 67 (1970) 385.
39. M. Hatanaka, R.J. Huebner, and R.M. Gilden. Proc. Nat. Acad. Sci. U.S. 67 (1970) 143.
40. M. Rokutanda, H. Rokutanda, M. Green, K. Fujinaga, R.K. Ray and C. Gurgo. Nature 227 (1970) 1026.
41. K. Fujinaga, J.T. Parsons, J.W. Beard, D. Beard, and M. Green. Proc. Nat. Acad. Sci. U.S. 67 (1970) 1432.
42. A.C. Garapin, J.P. McDonnell, W. Levinson, N. Quintrell, L. Fanshier, and J.M. Bishop. J. Virol. 6 (1970) 589.
43. E.M. Scolnick, S.A. Aaronson, and G.J. Todaro. Proc. Nat. Acad. Sci. U.S. 67 (1970) 1034.

44. S. Spiegelman, A. Burny, M.R. Das, J. Keydar, J. Schlom, M. Travnicek, and K. Watson. Nature 227 (1970) 1029.
45. S. Mizutani, D. Boettiger, and H. Temin. Nature 227 (1970) 424.
46. M. Green, M. Rokutanda, K. Fujinaga, H. Rokutanda, C. Gurgo, R.K. Ray, and J.T. Parsons. Gruppo Lepetit Symposium (1970) in press.
47. P. Duesberg, personal communication.
48. H. Temin, personal communication.
49. C. Gurgo, R.K. Ray, L. Thiry, and M. Green. Nature 229 (1971) 111.
50. J.A. Huberman and A.D. Riggs. J. Mol. Biol. 32 (1968) 327.
51. Z. Ben-Ishai, E. Heller, N. Goldblum, and Y. Becker. Nature 224 (1969) 29.
52. E. Heller, M. Argaman, H. Levy, and N. Goldblum. Nature 222 (1969) 273.
53. B.R. McAuslan. Biochem. Biophys. Res. Comm. 37 (1969) 289.
54. B. Moss, E.N. Rosenblum, E. Katz, and P.M. Grimley. Nature 224 (1969) 1280.
55. J.H. Subak-Sharpe, M.C. Timbury, and J.F. Williams. Nature 222 (1969) 341.
56. Z. Zakay-Rones and Y. Becker. Nature 226 (1970) 1162.
57. A. Nagayama, B.G.T. Pogo, and S. Dales. Virology 40 (1970) 1039.
58. H. Diggelmann and C. Weissmann. Nature 224 (1969) 1277.
59. E. Katz and B. Moss. Proc. Nat. Acad. Sci. U.S. 66 (1970) 677.
60. L. Thiry and G. Lancini, unpublished data.
61. R. Haselkorn. Science 143 (1964) 682.
62. E. diMauro, L. Snyder, P. Marino, A. Lamberti, G.P. Tocchini-Valentini. Nature 222 (1969) 533.
63. A. Sippel and G. Hartmann. Biochim. Biophys. Acta 157 (1968) 218.
64. E.P. Geiduschek and J. Sklar. Nature 221 (1969) 833.
65. R. Haselkorn, M. Vogel, and R.D. Brown. Nature 211 (1969) 836.
66. M. Green. Ann. Rev. Biochem. 39 (1970) 701.
67. M. Green. Ann. Rev. Microbiol. 20 (1966) 189.

68. M. Baluda and D. Nayak. Sym. Biol. Large RNA Viruses, 1969, Cambridge, England, in press.
69. M. Rokutanda, H. Rokutanda, and M. Green, unpublished data.
70. N. Biswal, M.B. Grizzard, R.M. McCombs, and M. Benyesh-Melnick. J. Virol. 2 (1968) 1346.
71. J.P. Bader. Virology 40 (1970) 494.
72. M. Green, M. Rokutanda, and H. Rokutanda, submitted for publication.
73. S. Penman. J. Mol. Biol. 17 (1966) 117.
74. R.J. Britten and D.E. Kohne. Science 161 (1968) 529.
75. R.J. Huebner, J.W. Hartley, W.P. Rowe, W.T. Lane, and W.I. Capps. Proc. Nat. Acad. Sci. U.S. 56 (1966) 1164.
76. P.H. Duesberg and P.K. Vogt. Proc. Nat. Acad. Sci. U.S. 67 (1970) 1673.
77. N. Biswal and M. Benyesh-Melnick. Proc. Nat. Acad. Sci. U.S. 64 (1969) 1372.

These investigations were supported by United States Public Health Service grant AI-01725, contract PH43-64-928 from the National Institute of Allergy and Infectious Diseases, Vaccine Development Branch, National Institutes of Health, and contract PH43-67-692 from the National Cancer Institute, Viral Carcinogenesis Branch, Etiology Area, National Institutes of Health, and Research Career Award Grant 5-K6-AI-4739 from the National Institutes of Health.

DISCUSSION

Dr. H. M. Temin, University of Wisconsin: I've seen the effects of these rifampicin derivatives and the fairly high levels involved reminded me of experiments we had done with gliotoxin and I just mention this for people who are interested in inhibitors. Gliotoxin, I guess I don't even remember the chemistry of it, inhibits at fairly low levels from other RNA viruses. So we tried it with the Rous sarcoma virus polymerases and found that levels similar to those that Dr. Green just showed for rifampicin derivatives, the active one with 20, 50, and 100 micrograms inhibiting, that gliotoxin will inhibit the polymerase step of the RNA-dependent to DNA and not as levels up to 200 micrograms per mil inhibits the DNA dependent set. So if people are interested in inhibitors this is another class that can be looked at but, again as with these rifampicin derivatives, the levels are quite high, so one has to be careful about using them.

Dr. S. Kit, Baylor College of Medicine: In the enzymatic synthesis for detection of viral RNA in transformed cells, are you also copying the sequence of the 4S RNA that are found in virions?

Dr. M. Green: If you look at the product, all of it sediments with 70S RNA.

Dr. Kit: No, what I'm trying to ask you is whether you are synthesizing a DNA replicate of the amino acid acceptor portion of the 4S RNA found in the virions. Now this is relevant to your detection of RNA in transformed cells; because if the 4S RNA is a cellular RNA and you are making a DNA replica of it, then you may also be hybridizing this "4S" DNA with the nuclear RNA from transformed cells.

Dr. Green: No. Well, this is not possible because if you use 50 micrograms of any of a number of control cell RNAs there is no detectable hybrid formation. While if you use 0.2 µg of nuclear RNA from a transformed cell, you obtain 70% hybridization. Thus you are dealing with very large differences, of the order of 500 to 1000.

Dr. Kit: I take it that after you have hybridized the RNA with the enzymatically made DNA, that you have not melted the hybrid and attempted to re-anneal the RNA portion with either 70S or 4S RNAs from the virion.

Dr. Green: The trouble with 4S RNA from the virion is it's not only 4S RNA, but viral 70S breakdown products. In these products the hybrid formed is mainly with 70S RNA, from which we obtain 70% hybridization. Fifty micrograms of nuclear or cytoplasmic RNA from a variety of normal cells will give no detectable hybrid formation.

Dr. Kit: How about transformed cells?

Dr. Green: This is not a problem, unless there is unique viron species of 4S RNA. If that's so, it's even more interesting.

Dr. D. Baltimore, M.I.T.: Have you looked at the size of RNA in transformed cells which hybridizes with the DNA?

Dr. Green: We're looking at it now.

Dr. G. D. Novelli, Oak Ridge National Laboratory: From your hybridization studies and comparing a cell line that's productive and one that's transformed and not productive, can you say anything about why?

Dr. Green: With the RNA or DNA tumor viruses?

Dr. Novelli: Either one of them.

Dr. Green: Transformation by DNA tumor viruses may involve the DNA dependent RNA polymerase since transformed cells cannot transcribe late genes, e.g., hamster cells and rat cells do not have the proper polymerase, or factors necessary to transcribe late genes. With RNA tumor viruses, cryptic cells are not common, but at least now we know that RNA is transcribed but whether the whole viral genome is transcribed or not is not yet known. This is just the beginning of our studies on the mechanism of RNA tumor virus transformation.

Dr. H. S. Ginsberg, University of Pennsylvania School of Medicine: Of the three possibilities that you listed for the requirement for DNA, one of them is certainly the least interesting: that is, the DNA is merely redundant material. This possibility could be easily ruled in or ruled out by doing hybridization and melting experiments. Have you done any of these?

Dr. Green: Yes, we have done that and we find that about one-third to one-half of the DNA-RNA sequences that you anneal will melt at low temperatures. But still this amounts to about 5,000-10,000 gene sequences in every mammalian cell which will hybridize with viral RNA.

STRUCTURE AND SYNTHESIS OF ADENOVIRUS CAPSID PROTEINS*

HAROLD S. GINSBERG
Department of Microbiology
University of Pennsylvania School of Medicine

Abstract: The hexon and fiber capsid proteins of type 5 adenovirus have been purified and analyzed by chemical and physical means. The hexon is a multimeric protein consisting of 3 asymmetric units; each asymmetric unit contains 2 polypeptide chains of approximately 70,000 and 21,000 daltons, respectively. The fiber is a monomeric protein whose polypeptide chain has a molecular weight of 61,000 daltons. The polypeptides of the capsid proteins are synthesized in the cytoplasm, and immediately after release from polyribosomes the polypeptides are transported into the nucleus. The nascent polypeptide chains have sedimentation coefficients of about 3 S, the same as the fiber and the large polypeptide chains of the hexons. After transport to the nucleus, the nascent proteins are rapidly assembled into mature hexons, penton bases, penton fibers, and mature pentons.

Once upon a time, when our studies of adenoviruses began, this agent seemed relatively ideal for the investigation of a DNA-containing virus: it was smaller than vaccinia or herpes simplex virus; it could be propagated in relatively large quantities; and it seemed to have a simple structure, containing only DNA and protein. Electron microscopic examination showed it to be an isometric virus (1), and it was said that the capsids of such viruses are composed of multiples of a single monomeric protein (2).

---

* I take pleasure in acknowledging the major contributions of my colleagues in these studies. This paper was possible because of the efforts of Preston H. Dorsett, Rudi Scherz, Leland F. Velicer, Gary Cornick, and Paul F. Sigler. I also wish to express my appreciation to Professor Henri Isliker, Institute of Biochemistry, University of Lausanne. These investigations were initiated during my sabbatical leave in his laboratory: his sound advise, warm encouragement, excellent facilities, and generous support were graciously extended to me.

Immunological studies, however, suggested that the virion contains at least 3 antigenic structures (3,4), and high resolution electron micrographs confirmed the notion that the viral capsid consists of 3 morphological subunits: the hexons, the penton bases, and the penton fibers (5, 6). Biosynthesis of the capsid proteins requires prior replication of viral DNA, and they are temporally synthesized as "late" proteins (7). The virion also contains 2-4 internal proteins associated with the viral DNA (8,9,10).

It is the purpose of this paper to describe the chemical and physical structure of two of the capsid proteins (the hexons and fibers), the synthesis of the viral polypeptide chains, and the morphogenesis of the capsid structural units. Type 5 adenovirus was employed.

EXPERIMENTAL

Structure of the hexon protein. Milligram quantities from infected cells of highly purified hexon can be readily obtained by multiple fluorocarbon extractions, DNase and RNase treatments, streptomycin and ammonium sulfate precipitations, and sequential chromatography on hydroxylapatite and DEAE-cellulose columns (11). The product of this purification can be crystallized (12,13,14), and its purity demonstrated by immunological and biochemical techniques (Fig. 1).

The hexon is the major structural component of the virion and comprises about 65% of the particle's total protein. It is immunologically complex serving as the immunogen which induces type-specific neutralizing antibodies as well as antibodies which in the complement-fixation or precipitin assay react with hexons of all other adenoviruses.

Chemical and physical analyses of the hexon protein revealed it to be unique in several features. Reduction and alkylation did not disrupt its multimeric structure and even 8 M urea and 1% sodium dodecyl sulfate (SDS) at pH 7.2 did not denature it completely. Addition of 5 M guanidine·HCl (recrystallized) yielded proteins with sedimentation coefficients of 3 S and 6 S in sucrose gradients (native hexon has a sedimentation coefficient of 12 S in sucrose gradients). When iodoacetamide, 2-mercaptoethanol, or dithiothreitol was added simultaneously with 5 M guanidine· HCl a single protein peak with a sedimentation coefficient of 3 S was obtained. However, when crystallized hexon was reduced and alkylated in 6 M guanidine·HCl, and chromatographed on Sepharose 4B columns in 5 M guanidine·HCl, polypeptides of approximately 70,000 and 21,000 daltons were obtained. Assuming uniform isotopic labeling, there was

Characteristics of Type 5 Adenovirus Hexon Protein

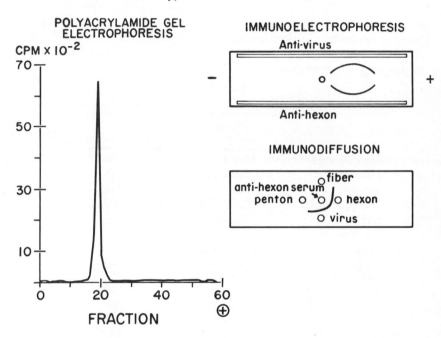

Fig. 1. Examination of the purity of type 5 adenovirus hexon protein. 1) Polyacrylamide gel electrophoresis: untreated purified hexon was electrophoresed on 5% acrylamide gels containing 0.1% SDS and 0.5 M urea at pH 9.0 (22). 2) Immunoelectrophoresis: antiserum in the upper trough was prepared by immunization of purified type 5 adenovirus; antiserum in the lower trough was prepared by immunization of rabbits with purified type 5 adenovirus hexon protein. 3) Immunodiffusion: the center well contained antiserum to purified type 5 hexon; the surrounding well contained purified antigens as noted. Identical results were obtained when antiserum prepared with twice crystallized hexon was used.

approximately 3 times more of the larger polypeptide chain than the smaller one. Or stated in another way, the 2 species of polypeptide chains were present in equal numbers.

Amino acid analysis (Table I) reveals the explanations for these somewhat unusual characteristics: 1) the very high content of hydrophobic amino acids probably are responsible for the extensive denaturing conditions required for disruption; and 2) the presence of free sulfhydryl groups results in disulfide interchange when only 5 M guanidine·HCl was added to the protein — simultaneous alkylation prevented the formation of disulfide bonds; reduction and alkylation assured that the proteins were random coils for determination of molecular weights.

Table I
Amino Acid Analysis of Hexon Protein from Type 5 Adenovirus

	Moles Per Cent	Residues per Hexon
Aspartic acid	14.0 ± 0.2	326
Threonine*	7.0 ± 0.05	158
Serine*	6.5 ± 0.02	149
Glutamic acid	9.5 ± 0.2	230
Proline	6.3 ± 0.3	134
Glycine	6.9 ± 0.4	172
Alanine	6.8 ± 0.1	156
Valine	4.9 ± 0.3	125
Methionine	2.7 ± 0.1	65
Isoleucine	4.0 ± 0.2	102
Leucine	8.0 ± 0.1	183
Tyrosine	5.4 ± 0.2	132
Phenylalanine	5.0 ± 0.4	120
Lysine	4.8 ± 0.1	111
Tryptophan**	1.3 ± 0.4	57
Histidine	1.7 ± 0.3	43
Arginine	4.8 ± 0.1	110
Cysteic acid ***	0.55 ± 0.2	12

*   Values of 24 hr and 72 hr hydrolysates extrapolated to zero time.
** Determined by UV absorption.
*** Oxidized sample.

Fifty-four to 56 peptides were identified in peptide maps prepared from trypsin digests of purified hexon. These data, according to amino acid analyses (Table 1), indicate that the larger protein could not be a simple aggregate of the small polypeptide chain, as suggested from preliminary data (15). N-terminal amino acid analyses were done to obtain additional confirmatory evidence for the presence of 2 species of polypeptides, but unfortunately the N-terminal residues were blocked.

To obtain more detailed data of the hexon structure at atomic resolution the tetrahedral crystals were studied by X-ray diffraction with G. Cornick and P. Sigler (Department of Biophysics, University of Chicago). The characteristics of the type 5 adenovirus hexon crystals are summarized in Table 2 (13). From these crystallographic data one can conclude that the hexon protein is composed of 3 asymmetric units, each with a protein mass of 83,000 ± 8000 daltons. This evidence is consistent with the biochemical findings and implies that each asymmetric unit consists of a large and a small polypeptide chain, and that the hexon is assembled from 3 polypeptide chains of approximately 70,000 daltons and 3 polypeptides of about 21,000 daltons. Based on biochemical evidence the intact hexon would have a molecular weight of 273,000 daltons, a value within the range set by crystallographic analysis.

Table 2

Crystallographic Data for Type 5 Adenovirus Hexon Crystals

Space group	$P 2_1 3$
Edge of the unit cell	149.9 ± 0.1 Å
Crystal Density	1.173 ± 0.004 g/cc
Density of protein-free supernatant solution	1.070 ± 0.002 g/cc
Mass of protein in the asymmetric unit*	$83,000 \pm 8000$ daltons
Volume fraction of protein*	0.37 ± 0.03
Crystal volume per unit of protein mass*	3.37 ± 0.29 Å^3/dalton

* Assuming a partial specific volume of 0.740 ± 0.004 cc/g (From G. Cornick, P. F. Sigler, and H. S. Ginsberg [13]).

It must be noted, however, that there are marked discrepancies in reported molecular weights of adenovirus hexons, ranging from 200,000 to 400,000 daltons (6,12,14,16-19), in X-ray crystallographic data (12,13,14), and in the purported characteristics of the hexon's polypeptide chains (15,20,21). These differences will be analyzed in the Discussion.

Structure of the fiber protein. Purified fiber protein, like hexon, can be obtained in milligram quantities although additional physical-chemical procedures are required to obtain a highly purified product. In addition to the steps used for purification of hexon protein, rate zonal sedimentation in a linear sucrose gradient was used to separate free fiber from residual intact penton, and finally, the fiber was electrophoresed in an ampholine-sucrose gradient pH 5.8 (isoelectric point of the purified fiber was 6.2) (11). The protein obtained was immunologically identified as fiber, it was a single antigen by immunoelectrophoresis and immuno-gel diffusion, it electrophoresed as a single band in acid and alkaline acrylamide gels, it sedimented as a single protein in the Model E analytical ultracentrifuge, and it was free of labeled host material (1×10^6 cpm) which had been added to the original infected cell homogenate. The purified fiber had a sedimentation coefficient ($S^o_{w,20}$) of 5.82, a diffusion coefficient (D^o_{20}) of 8.8×10^{-7}, and a molecular weight of 60,500 daltons.

Procedures necessary to denature the fiber were not quite so demanding as those essential for degradation of the hexon to its polypeptide chains. Thus, 5 M guanidine·HCl at pH 8.6, 10 M urea at pH 2.7, or 8 M urea with 1% SDS at pH 7.6 converted the 6 S fibers to molecules which had a sedimentation coefficient of 2.8 - 3.0 S. Unlike hexon protein, unfolding the native protein did not effect a disulfide interchange, and alkylating agents were not necessary to obtain single polypeptide chains. The amino acid content of the fiber protein (Table 3) probably explains the relative ease with which it was denatured as compared to the hexon: hydrophobic bonds probably maintain the stable form of both proteins, and therefore, the fewer number of hydrophobic amino acid residues in the fiber make it more readily unfolded.

The fiber polypeptide chains traveled as a single homogeneous species when electrophoresed in polyacrylamide gel containing 5 M urea and 0.1% SDS at pH 9.0 (22), and when chromatographed on Sepharose 4B columns equilibrated with 5 M guanidine·HCl (23). The molecular weight was approximately 61,000 as determined by both methods.

Each fiber, therefore, consists of only a single polypeptide chain. The unique morphological structure of the fiber must be established by intramolecular interactions; the preferred folding of the protein must also confer some of the fiber's type-specific (4) and sub-group specific immunological properties (24) as well as its biological characteristics (25,26).

Table 3

Amino Acid Composition of the Fiber Protein[+]

Amino Acid	Moles Per Cent	Residues per Polypeptide Chain
Aspartic acid	12.4	57
Threonine*	11.9	55
Serine*	8.4	38
Glutamic acid	6.6	30
Proline	6.5	30
Glycine	9.2	52
Alanine	8.2	38
Valine	5.8	27
Cystine	---**	--**
Methionine	1.6	7
Isoleucine	4.4	20
Leucine	13.7	63
Tyrosine	2.3	11
Phenylalanine	4.0	18
Tryptophan[++]	1.1	5
Lysine	6.0	28
Histidine	1.3	6
Arginine	1.4	6

+ Mean of determinations on 2 samples.
* Value extrapolated to zero hydrolysis.
** Concentration too low to measure.
++ Determined by method of Beaven and Holiday.

Synthesis and Morphogenesis of Capsid Proteins. Just as the disruption of a protein and examination of its constituent parts reveal characteristics of the "supramacromolecule", so too, a study of the biosynthesis of the polypeptide chains and their assembly into the native multimeric structure can furnish evidence on the subunits of the functional proteins. Therefore, an investigation of the synthesis and morphogenesis of the adenovirus capsid proteins accompanied the structural studies described above.

Viral capsid proteins, identified with immunocytologic techniques can only be detected in the nucleus (27,28) and virions assemble in the nucleus (29,30,31). Nevertheless, viral proteins, like cellular proteins, are synthesized on polyribosomes in the cytoplasm (32). Fortunately, host protein synthesis ceases within 20 hrs after type 5 adenovirus infection (33), so that production of viral proteins is relatively simple to study. A one minute pulse of ^{14}C-labeled amino acids revealed that the nascent viral polypeptide chains were synthesized on a relatively uniform population of polyribosomes with maximum production on polyribosomes having a sedimentation coefficient of 200 S. In contrast, in uninfected cells the maximum synthesis of proteins took place on polyribosomes of about 280 S. The nascent polypeptide chains, released artificially from infected cell polyribosomes after incubation with 1% sodium decyl sulfate (SDeS) and RNase, had a sedimentation coefficient of 3.2 - 3.4 S. The viral polypeptide chains in vivo were synthesized on and released from polyribosomes within 2 min after which they were immediately transported into the nucleus and concomitantly developed immunological maturity (15).

The nascent polypeptide chains released from polyribosomes corresponded in size to the fiber polypeptide chain and to one class of the hexon polypeptide chains (the characteristics of the polypeptide constituents of purified penton base have not yet been determined). It seemed likely, therefore, that the newly synthesized polypeptide chains were precursors of the viral structural proteins. To test this possibility the morphogenesis of the capsid proteins was studied during a "chase" after a one minute pulse with labeled amino acids. Serial sucrose gradient analyses were made with infected cell extracts during the chase period to observe the formation of the hexon, penton base, penton fiber, and intact penton. Labeled hexon (sedimentation coefficient of 12 S) or labeled fiber (sedimentation coefficient of 6 S) was used as a marker to identify the mature structures. The results of a representative experiment are summarized in Fig. 2. After a 1 min pulse only 17% of the nascent protein was soluble in the gradient; the remainder of the labeled protein was in the pellet in aggregates larger than 15 S. The majority of soluble counts after the 1 min pulse sedimented in the upper portion of the gradient with a maximum in the region of 3 S protein; there were also small peaks of 12 S and 9 S material. Following a 1 min and 3 min chase the peak of 12 S protein, which cosedimented with the hexon marker, and the 9 S peak became progressively larger and a distinct 6 S peak appeared. With increasing time during the chase, the 12 S and 6 S peaks became progressively larger, the 3 S and smaller material diminished, and the 9 S peak began to disappear, apparently blending into the 12 S peak. After a one hour chase only 2 protein peaks could be identified in the sucrose gradients: the 6 S peak, which cosedimented with purified fiber and was immunologically identified as fiber; and the 12 S peak, which became progressively skewed on its trailing limb as the 9 S peak appeared to fuse with it, cosedimented with purified hexon and was shown immunologically to contain hexon protein. The large 12 S peak was not homogeneous, however: hexon antiserum precipitated 72 - 84% of the protein and fiber antiserum, which reacts with penton, precipitated 14 - 25% of the protein; and the trailing half of the 12 S peak chromatographed on DEAE-cellulose was separated into hexon and penton proteins. The penton had a sedimentation coefficient of 10.6 S, it was trypsin sensitive, and it was precipited by fiber antiserum. Sodium decyl sulfate (0.25%) rapidly disrupted the penton into intact fiber with a sedimentation coefficient of 6 S and 3.4 S polypeptide chains (15).

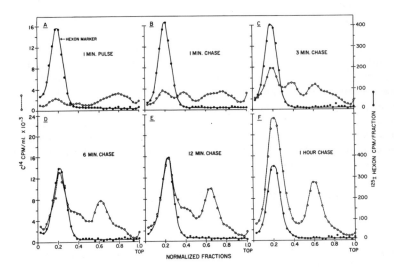

Fig. 2. Morphogenesis of type 5 adenovirus capsid proteins 20 hrs after infection (200 PFU/cell). ^{14}C-labeled amino acids were added for 1 min; a sample was taken immediately after the pulse; the remainder of the cells were washed and cultured in isotope-free medium containing an excess of amino acids. Homogenates prepared at indicated times were treated with RNase (225 µg/ml), ^{125}I-hexon marker was added, and the mixtures were layered on 5 to 20% sucrose gradients. The gradients were centrifuged in a Spinco SW 39 rotor at 20 C for 18 hr at 30,000 rev/min. (From L. F. Velicer and H. S. Ginsberg [15]).

The 9 S protein was trypsin-sensitive, it was composed of 3.2 - 3.4 S polypeptide chains, and it reacted with antibodies to purified type 5 adenovirus, but not to hexon antibodies. These data imply that the protein having a sedimentation coefficient of 9 S is the penton base which after its assembly is joined with a fiber to form the mature penton (15).

DISCUSSION

Sufficient data have now emerged to permit the development of structural models of the adenovirus capsid proteins. Perspectively, the relationship between structure and function of the virion's structural components should be uncovered.

The major viral protein, the hexon, has been the most extensively studied and the data are the most contentious. Even the reported molecular weights vary as much as two-fold, from 200,000 to 400,000 daltons (16,17,18,19). Although consensus does not establish truth, the majority of determinations range from 250,000 to 320,000 daltons. A variety of procedures have been employed for purification of the hexon and determination of its size; undoubtedly the major differences noted arise from these divergencies. According to analytical ultracentrifugation (Archibald procedure) we obtained a molecular weight of 310,000 (11); X-ray diffraction yielded a value of 249,000 ± 24,000 daltons (13).

It is recognized that the hexon is a multimeric protein, but there is no agreement upon the number and characteristics of the polypeptide chains. According to our data, obtained by X-ray diffraction and chemical techniques, the hexon is assembled from 6 polypeptide chains which are arranged as pairs in 3 asymmetric units; each asymmetric unit contains 2 polypeptide chains of approximately 70,000 and 21,000 daltons. According to the arginine and lysine content of the hexon the number of peptides identified in a tryptic digest (54-56) is consistent with a protein mass of only about 70,000 daltons if all regions of the 2 species of polypeptides were unique. In contrast to these data, Pettersson and Philipson report that the hexon consists of 6 identical polypeptide chains of 60,000 daltons each (21). These investigators used a purification procedure in which preparative polyacrylamide electrophoresis was a major step; the hexon obtained had immunological characteristics which were different from the hexon purified in our laboratory (18). But their methods to disrupt the hexon and to measure the size of the polypeptide chains were identical to those which we employed (21). In addition, to the purification procedure used by Pettersson and Philipson, one other procedure may explain the differences observed: to identify consistently the large and small polypeptide chains it was necessary to employ crystallized hexon, a procedure which was not followed by Pettersson and Philipson (21). When we employed hexon obtained from DEAE-cellulose columns, the step prior to crystallization, reaction with 6 M guanidine·HCl yielded only a single species of polypeptide chains of approximately 70,000 daltons. The reasons for these differences must be clarified.

Variations also exist in data obtained by X-ray diffraction. Franklin et al. have examined type 2 adenovirus hexon crystals (14). The type 2 and type 5 crystals are isomorphous; the type 2 hexon, like the type 5 crystal, contains 3 asymmetric units; and the edge of the unit cell of both hexon crystals is 149.9 Å. But the mass of the asymmetric unit is 83,000 ± 8,000 daltons according to our data (13), and approximately 120,000 daltons from the data of Franklin et al. (14). The reasons for these extreme differences are ambiguous. It is striking, however, that in each laboratory the chemical and X-ray diffraction data are mutually consistent.

Additional divergent data are available. Maizel et al. reported that the hexon consists of 3 polypeptide chains of 120,000 daltons each (20). The hexon they employed was simply purified by DEAE-cellulose chromatography. The partially purified hexon was either: 1) denatured in SDS, urea, and 2-mercaptoethanol, and measured by electrophoresis on polyacrylamide gels containing SDS (20); or 2) incubated in 6 M guanidine·HCl and analyzed in the analytical ultracentrifuge (35). It appears that the hexon was not completely denatured, but the reasons for the failure to obtain total disruption are not obvious.

The structure of the fiber is simpler and its characteristics more easily determined. The fiber consists of a single polypeptide chain of 61,000 daltons. These data are identical to those of Maizel et al. who investigated type 2 adenovirus (20). It was surprising to find that the fiber which consistently has a terminal knob when observed in electron micrographs (5,6), and which has type-specific (4,24,25) as well as sub-group immunological reactivity (24), should consist of only a single polypeptide chain. The development of intrachain bonds, probably chiefly hydrophobic bonds, must be responsible for the native structure. The disruption of these bonds by guanidine·HCl or urea, collapses the molecule and converts it from a fiber into a random coil.

The penton base, the fiber's partner in the virion, has not yet been explored. The evidence presented (Fig. 2), however, suggests that it is comprised of polypeptide chains of essentially the same size as those present in the fiber and the hexon (i.e. 60-70,000 daltons). The penton base, unlike the hexon or fiber, is relatively unstable (15) and readily digested by trypsin. Characterization of its chemical structure should reveal the basis for these differences.

Like so many biological agents which off-hand appear simple, you must know adenoviruses to recognize their complexities. As with many seemingly complex biological systems, however, knowledge of their structures may reveal the basis of their functions - thus, they become simpler. We hope to approach this stage with adenoviruses. That is, when we clearly know the physical and chemical structures of the hexon, the fiber, and the penton base, we should understand in molecular terms the unique immunogenicity of the hexon which induces synthesis of type-specific neutralizing and family cross-reactive antibodies, or the biological activities of the penton base which can function as a DNA endonuclease (36) or as a substance that induces cytopathic alterations of cells (37,38).

REFERENCES

1. R. W. Horne, S. Brenner, A. P. Waterson, and P. Wildy, J. Mol. Biol., 1 (1959) 84.

2. F. H. C. Crick and J. D. Watson, Nature, 177 (1956) 473.

3. H. G. Klemperer, and H. G. Pereira, Virology, 9 (1959) 536.

4. W. C. Wilcox and H. S. Ginsberg, Proc. Natl. Acad. Sci. U.S. 47 (1961) 512.

5. W. C. Wilcox, H. S. Ginsberg, and T. F. Anderson, J. Exptl. Med., 118 (1963) 307.

6. R. C. Valentine and H. G. Pereira, J. Mol. Biol., 13 (1965) 13.

7. J. F. Flanagan and H. S. Ginsberg, J. Exptl. Med., 116 (1962) 141.

8. W. G. Laver, H. G. Pereira, W. C. Russell, and R. C. Valentine. J. Mol. Biol., 37 (1968) 379.

9. L. Prage, U. Pettersson, and L. Philipson, Virology, 36 (1968) 508.

10. W. G. Laver, Virology, 41 (1970) 488.

11. H. S. Ginsberg, R. Scherz, and P. H. Dorsett (to be published).

12. W. M. Macintyre, H. G. Pereira, and W. C. Russell, Nature, 222 (1969) 1165.

13. G. Cornick, P. B. Sigler, and H. S. Ginsberg. J. Mol. Biol., (in press).

14. R. M. Franklin, U. Pettersson, K. Akervall, B. Strandberg, and L. Philipson, J. Mol. Biol., (in press).

15. L. F. Velicer and H. S. Ginsberg, J. Virol. 5 (1970) 338.

16. E. H. Wasmuth and A. A. Tytell, Life Sciences, 6 (1967) 1063.

17. K. Köhler, Z. Naturforschg, 206 (1965) 747.

18. U. Pettersson, L. Philipson, and S. Höglund, Virology, 33 (1967) 575.

19. K. F. Shortridge and F. Biddle, Orch. f. die gesamte Virusforschg, 29 (1970) 1.

20. J. V. Maizel, Jr., D. O. White, and M. D. Scharff, Virology, 36 (1968) 126.

21. U. Pettersson and L. Philipson. (Personal communication).

22. E. D. Kiehn and J. J. Holland, J. Virol., 5 (1970) 358.

23. W. W. Fish, K. G. Mann, and C. Tanford, J. Biol. Chem., 244 (1969) 4989.

24. H. G. Pereira and M. V. T. de Figueinedo, Virology, 18 (1962) 1.

25. L. Rosen, Virology, 5 (1958) 574.

26. A. J. Levine and H. S. Ginsberg, J. Virol., 1 (1967) 747.

27. G. S. Boyer, F. W. Denny, Jr. and H. S. Ginsberg, J. Exptl. Med., 110 (1959) 827.

28. H. G. Pereira, A. C. Allison, and B. Balfour, Virology, 7 (1959) 300.

29. L. Kjellen, G. Sagermolm, A. Svedmyr, and K. Thorsson, Nature 175 (1955) 505.

30. C. Harford, A. Hamlin, E. Parker, and T. van Ravensway, J. Exptl. Med., 104 (1956) 443.

31. C. Morgan, C. Howe, H. M. Rose, and D. H. Moore, J. Biophys. Biochem. Cytol. 2 (1956) 351.

32. L. F. Velicer and H. S. Ginsberg, Proc. Natl. Acad. Sci. U.S., 61 (1968) 1264.

33. L. J. Bello and H. S. Ginsberg, J. Virol., 1 (1967) 843.

34. M. S. Horowitz, J. V. Maizel, and M. D. Scharff, J. Virol., 6 (1970) 569.

35. B. T. Burlingham, U. Pettersson, W. Doerfler, and L. Philipson, Bact. Proc., (1970) 172.

36. S. F. Everett and H. S. Ginsberg, Virology, 6 (1958) 720.

37. H. G. Pereira, Virology, 6 (1958) 601.

These investigations were supported in part by Public Health Service grants from the National Institute of Allergy and Infectious Diseases. They were also under the sponsorship of the Commission on Acute Respiratory Diseases of the Armed Forces Epidemiological Board, and were supported by the U.S. Army Medical Research & Development Command, Department of the Army.

DISCUSSION

Dr. H. Raskas, St. Louis University: First, I would like to make one comment. In collaboration with Maurice Green and other people in our department we have established a system to study adenovirus protein synthesis *in vitro*. G. Shanmugam has found that even though the hexon polypeptides are synthesized and released from the ribosomes *in vitro*, they are not assembled into capsomeres. Thus your comment that the *in vivo* nascent and newly synthesized peptides already have hexon immunological activity may be explained by reasons other than the newly synthesized proteins already being in the final capsomere configuration. That being the comment, I'd like to ask two questions. Are there other proteins aside from the hexon that are not fully dissociated on SDS acrylamide gels? Second, could you describe what information you have concerning the end group of the hexon?

Dr. H. S. Ginsberg: I have heard through the grape vine that there are other proteins that will not disrupt by the Mazell technique, but I do not know the details as to the exact proteins implicated; there does appear to be a variety of enzymes and other proteins that apparently are not disrupted. In answer to your second question it was found in our laboratory as well as in Phillipson and Peterson's laboratory that the N-terminal amino acids of the hexon are blocked.

Dr. K. B. Jacobson, Oak Ridge: We've been working with an enzyme which consists of subunits and using Mazell's procedure for dissociating it. His procedure calls for urea, SDS, and mercaptoethanol to be used during dissociation. We find more complete dissociation if we omit the urea. There's a case of another enzyme, being worked on by Tauster, which exhibits this same phenomenon of being more completely dissociated in the absence of urea than with it.

Dr. Ginsberg: The hexon protein in the absence of urea but with SDS and mercaptoethanol we essentially do not get disruption or obverse aggregates. I'd like to make a comment relative to Raskas' earlier statement. With Dr. James Wilhelm we also have been studying protein synthesis *in vitro*. Our studies used polyribosomes from infected cells. We have demonstrated fidelity of translation, and *in vitro* synthesis of immunologically active proteins

Dr. D. Baltimore, M.I.T.: We've also had difficulty with urea in the same way and I think that the problem is that urea which has been in solution for a little while generates compounds which may cross link proteins.

Dr. Ginsberg: We have only used fresh recrystallized urea. This is not only necessary for urea but also it is true for quanidine HCl. If you do not either use quanidine recrystallized yourself or use the Mann ultra pure crystals, you get into similar grave problems.

Dr. Raskas: How active is the antibody to hexon capsomere on the hexon polypeptide?

Dr. Ginsberg: We cannot say that the single polypeptide chain is immunologically active because one has to remove artificially the nascent peptides from the polyribosomes. At that time it may begin to aggregate in the test tube. So, we have no evidence that single polypeptide chains are immunologically active -- only that there are being made on the polyribosome proteins which are or rapidly become immunologically active.

Dr. Raskas: So what you are seeing could possibly be peptides on the ribosomes and not capsomere?

Dr. Ginsberg: Oh yes. We are not saying that they are capsomere.

Dr. M. M. Sigel, University of Miami: In this connection, I was wondering if you had pulled off enough of the different peaks of the 70S and the 20S to tell whether one of them is the group specific and the other one is the type specific antigen?

Dr. Ginsberg: We are now doing these experiments and the evidence is still too fragmentary to discuss.

THE ROLE OF HERPESVIRUS GLYCOPROTEINS IN THE MODIFICATION OF MEMBRANES OF INFECTED CELLS

BERNARD ROIZMAN and PATRICIA G. SPEAR
Department of Microbiology
The University of Chicago

ABSTRACT

Both internal and external membranes of animal cells become modified after infection with herpesviruses. The modified internal membranes furnish the envelope for nucleocapsids and perhaps assist in the egress of virions from infected cells. The role of modified external membranes in virus multiplication is unknown. However, the modification of the plasma membrane is reflected in its new immunologic specificity and in changes in the social behavior of infected cells. Studies of purified fractions of membranes from infected cells have shown the following: (1) After infection, the synthesis and glycosylation of host membrane proteins cease and are replaced by the synthesis and glycosylation of new proteins. (2) The new glycoproteins are specified by the virus. (3) Two lines of evidence indicate that the virus-specific membrane glycoproteins determine the social behavior of infected cells.

INTRODUCTION

Herpesviruses are a widespread group of relatively large DNA viruses recently implicated in a number of tumors of man and animals. The outstanding characteristics of this group are that viral DNA is made in the nucleus of the cell, viral proteins are made in the cytoplasm, and that viral assembly begins in the nucleus and terminates as the virion buds through a membrane of the cell. The cell producing virions is invariably killed. This paper concerns the involvement of cellular membranes in virus multiplication and consists of two parts. The first part reviews the role of cellular membranes in the multiplication of herpesviruses. The second part deals with current status of the studies on the glycoproteins specified by the virus and their role in altering the function of these membranes.

THE ROLE OF CELLULAR MEMBRANES IN HERPESVIRUS MULTIPLICATION

Current studies indicate that cellular membranes become structurally and functionally altered between 3 and 7 hours post infection. It is convenient to subdivide the membranes of the cells into two groups, i.e., internal membranes involved in virus multiplication and the external membranes which mediate changes in the interaction of cells among themselves.

Internal Membranes

Electron microscopic studies indicate that internal membranes perform two functions in viral multiplication; i.e., they envelope the nucleocapsid and assist in the egress of the enveloped nucleocapsids from infected cells.

Electron microscopic studies rely on the presence of partially enveloped nucleocapsids lined in apposition to membranes as evidence for the site and mechanism of envelopment. This line of evidence (1, 2, 3) indicates that early in infection most herpesviruses are enveloped by the inner lamella of the nuclear membrane whereas late in infection envelopment takes place at all membranes of the cell. Three apparent exceptions to this rule have been reported to date. First, the genital strain of herpes simplex has been reported to be enveloped by both the nuclear and cytoplasmic membranes not only late but also early in infection (3). Second Stackpole (4) published electron microscopic data which he interpreted as indicating that the herpesviruses associated with Lucké adenocarcinoma migrate from the nucleus into the cytoplasm by being successively enveloped and unenveloped by the inner and outer lamellae of the nuclear membrane respectively. According to this scheme, the nucleocapsid is then enveloped by the internal cytoplasmic membranes of the frog kidney cell. The third apparent exception is that of Marek's disease herpesvirus whose envelope is derived from the inner lamella of the nuclear membrane and becomes further modified in a cytoplasmic inclusion (5).

The kinetics and molecular basis of envelopment are not known. Available information may be summarized as follows: (i) There is suggestive evidence that in order to become enveloped, the nucleocapsid must acquire on its surface a structural component designated as the inner envelope and containing lipids. The evidence for the existence of this component is based on three observations (6). First, an electron translucent shell with an ordered structure can be seen surrounding nucleocapsids in apposition to the nuclear membrane. Second, infectious virus particles from nuclei of infected cells sediment in sucrose density gradients more rapidly than naked nucleocapsids but more slowly than enveloped nucleocapsids from extracellular fluid and the cytoplasm of infected cells. The increase in the hydrodynamic size of infectious nuclear particles is too small to be accounted for by a complete envelope similar to that present on cytoplasmic and extracellular particles. Lastly, whereas both the nuclear and cytoplasmic infectious particles are rapidly inactivated by lipases and lipid solvents, the nuclear particles are far less stable in cesium chloride solutions than the cytoplasmic particles. (ii) Several observations suggest that envelopment takes place in two steps. The first step is a generalized modification of all of the membranes involved in envelopment. The evidence for this step is based largely on the observation by Nii et al. (2, 7) that virus specific antigens line all internal membranes of the cells. The second step is a secondary, topologically limited modification of the membrane directly involved in envelopment; it occurs at the time the nucleocapsid comes in apposition to the membranes. The nuclear membrane at this stage manifests two changes i.e., first, the inner and outer lamellae dissociate and segments of the inner lamella become thicker and exhibit a greater affinity for heavy metal stains (8). In general three lines of evidence suggest that the second step does in fact take place. First, the modification takes place only at the site of attachment of particles to membranes; distant sites do not become altered. The modified region resembles in structure and affinity for heavy metal stain the appearance of the envelope of the extracellular and cytoplasmic virus. Second, similar or identical modifications take place at the site of nucleocapsid envelopment at all of the membranes of the cell (Figure 1). Lastly, in DK cells abortively infected with herpes simplex virus the nucleocapsids do not come in apposition to the nuclear membranes, the nuclear membrane does not become altered and envelopment does not ensue (9). The molecular nature of the alterations is not known. It is conceivable that the modification reflects the change in membrane structure resulting from the expulsion of host membrane proteins.

The enveloped particles accumulate between the inner and outer lamellae of the nuclear membrane and in the tubules and vesicles of the endoplasmic reticulum (Figure 2). There is suggestive evidence that the internal cytoplasmic membranes function as a special compartment which shields the virus from degradation in the cytoplasm--the site of initial uncoating immediately after infection--and transports the virus from the site of envelopment to the extracellular fluid (7, 8, 10). The evidence is based on two observations. First, enveloped virions are usually found inside a tubule or vesicle (Figure 2) and only very rarely are they free in the cytoplasm (3, 8, 11, 12). Second, enveloped particles begin to accumulate in the extracellular fluid as early as 5-6 hours post infection i.e., long before the death and dissolution of the cells. Opinion varies as to whether egress is by way of tubules of the endoplasmic reticulum which communicate with the extracellular fluid (8), or by way of vacuoles which break off from the endoplasmic reticulum and transport the virus to the plasma membrane (11).

External Membranes

There is no specific evidence indicating that external membranes play a defined, unique role in virus multiplication other than to

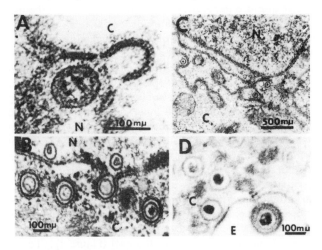

FIGURE 1. HEp-2 membranes and herpes simplex virus envelopment: Electron microscopic studies. A. Nucleocapsid (F strain) in apposition to the inner lamella of the nuclear membrane; the photomicrograph shows details of a structure surrounding the nucleocapsid. B. Enveloped, unenveloped and partially enveloped nucleocapsids (strain F). The nucleocapsids in the cytoplasm are enveloped and moreover, they are surrounded by an additional membrane. The unenveloped nucleocapsids are in the nucleus. One nucleocapsid is partially enveloped by a thickened membrane continuous with the inner lamella of the nuclear membrane. C. Unenveloped nucleocapsids in the cytoplasm of cells infected with G strain. Note that the nucleocapsids are in apposition to endoplasmic reticulum. D. Unenveloped nucleocapsid in apposition to thickened cytoplasmic membrane of cells infected with G strain (3, 6, 8).

Abbreviations: c-cytoplasm, n-nucleus, e-extracellular fluid.

maintain the integrity of the cell. Nevertheless there is considerable evidence that herpesvirus infection causes extensive structural, functional and immunological modifications of the plasma membrane and that at least some if not all of these alterations are due to virus-specific products incorporated into the membranes. The evidence dealing specifically with the external membranes may be summarized as follows:

(i) At least two laboratories have reported that infected cells leak macromolecules (13, 14). The leakage of macromolecules is intensified after the onset of viral DNA and structural protein synthesis. It seems very likely that the increased permeability of the membranes is responsible for the ultimate cessation of all macromolecular synthesis in the infected cells and that the phenomenon is due to an alteration in the

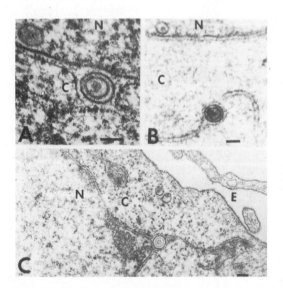

FIGURE 2. HEp-2 membranes and the egress of herpes simplex virus (MP strain) from HEp-2 cells. Electron microscopic studies. A. Enveloped nucleocapsid between two lamellae of the nuclear membrane at the point of origin of the endoplasmic reticulum in cells 8 hours post infection. B. Enveloped nucleocapsid in a cytoplasmic tubule 12 hours post infection. C. Enveloped nucleocapsid between the lamellae of the nuclear membrane at the origin of a tubule which appears to communicate with the cytoplasmic membrane 8 hours post infection.
    Abbreviations: c-cytoplasm, n-nucleus, e-extracellular fluid.
    Scale Marker: 100mµ (8).

structure and function of the plasma membrane.

(ii) A characteristic of herpesvirus infection of cells in culture is that the interaction of cells among themselves becomes altered. The nature of the change in the interaction--henceforth designated as the social behavior of cells--varies depending on the virus strain. Some strains cause the cells to form clumps of various dimensions and

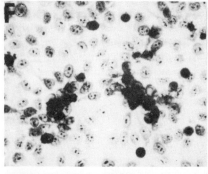
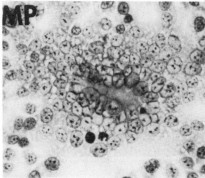
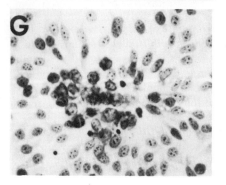

FIGURE 3. The social behavior of HEp-2 cells infected with 2 major subtypes (F, prototype of subtype 1 and G, prototype of subtype 2) and a mutant (MP) of subtype 1. HEp-2 cells were infected at a multiplicity of 1 p.f.u./1,000 cells and stained with Giemsa. Twenty four hours after infection F, MP and G cause the formation of tight small clumps of rounded cells, polykaryocytes and large, loose clumps of cells, respectively. Photomicrographs were taken and printed at the same magnification. The size of clumps is representative; the polykaryocyte is the smallest found in the culture.

adhesiveness. Other strains cause cells to fuse (Figure 3). The changes in the social behavior of cells require, at least in our hands, macromolecular synthesis corresponding in time to the onset of synthesis of structural components of the virus (15, 16). In general the type of social behavior induced by a particular variant is reproducible and readily differentiated from that produced by another variant. The alteration of the social behavior of cells is therefore genetically determined by the virus.

(iii) The change in the social behavior of cells is necessarily mediated by a change in the structure of the plasma membrane. To paraphrase, any change in the function of the cytoplasmic membrane should necessarily be reflected in a change in its structure and hence in its immunologic specificity. The evidence that infected cells acquire a new immunologic specificity was obtained with the aid of a test designed to measure immune injury of somatic cells by antibody and complement. The test is based on the observation that infected cells fail to form plaques when seeded on monolayers of uninfected cells if they had been previously treated with antibody and complement (17). The sensitivity of the test stems from the fact that very few cells are needed because nearly every infected cell produces a plaque. The plaque assay is simple and reproducible. The experiment designed to test whether cells exhibiting an altered social behavior also manifest an altered immunologic specificity were done with 2, 24 and 48 hour-infected cells and rabbit sera prepared against infected cells (15, 18). The results showed that complement and unabsorbed anti-infected cell serum preclude the formation of plaques by cells infected for 2, 24 and 48 hours. However, complement and anti-infected cell serum absorbed with uninfected cells were active only against 24 and 48 hour-infected cells; two hour-infected cells were unaffected. Clearly the 24 and 48 hour-infected cells contained on their surface one or more antigens not present on the surface of uninfected cells. The basic conclusion presented here, namely that the membranes of infected cells become altered with respect to structure and immunologic specificity, was corroborated by Watkins (19). It is now clear that not only cells infected by herpes simplex virus but also cells infected with the virus associated with Burkitt lymphoma (20) exhibit new determinant antigens.

(iv) Several lines of evidence indicate that the antigen on the surface of cells exhibiting an altered social behavior is a structural component of the envelope of the herpesvirion. First, absorption of anti-infected cell serum with partially purified virus depletes the cytolytic activity of the serum (18). Second, as indicated earlier in the text, DK cells infected with the macroplaque strain of herpes simplex produce nucleocapsids which are not enveloped. Anti-sera made in rabbits against infected DK cells have very low cytolytic and virus neutralizing titers (9). In cell lysis-competition tests it has been established that infected DK cells compete very poorly with infected HEp-2 cells for cytolytic antibody. In these series of experiments the absence of envelopment correlated well with the absence of surface antigens and with the inability of infected DK cell lysates to induce both cytolytic and neutralizing antibody in rabbits (9, 21). The third line of evidence relating the surface antigen to a structural component of the virus is based on correlation between neutralizing and cytolytic titers of various sera. We have produced anti-sera against a variety of viral preparations treated with reagents designed to degrade the virion. The potencies of the anti-sera in neutralization and in cytolytic tests vary considerably. However, there was a very good correlation between the neutralizing and cytolytic titers of various anti-sera (21).

(v) On the basis of the evidence that (a) the social behavior of infected cells varies depending on the virus strain and (b) the cells acquire on their surface a new determinant antigen similar or identical to that on the surface of the virion, it could be predicted that strains differing with respect to their effects on the social behavior of cells should also differ with respect to properties and characteristics related to their envelopes. This is precisely what was found. Extensive studies in our laboratory (17, 22, 23, 24, 25) have shown that herpesviruses differing with respect to their effects on the social behavior of cells also differ with respect to immunologic specificity, elution profiles from Brushite columns, stability at 40°C, and other properties related to their surface. These studies led to the conclusion that in the course of virus multiplication one or more structural components of the envelope of the herpesvirion bind to the plasma membrane and that these components are responsible for the structural and functional modification of the membranes.

THE CHEMICAL NATURE OF THE MEMBRANE-BOUND MACROMOLECULES SPECIFIED BY HERPESVIRUSES

The preceding section enumerated the evidence indicating that both internal and external membranes of cells become modified after infection by the addition of new antigenic determinants specified by the virus. The initial objectives of the experiments described in this section were to determine the nature of the virus-specific components present in the envelope of the herpes virion and in the membranes of infected cells. The rationale behind these objectives was two-fold. First, we were interested in determining whether the envelope of the virus contained unique structural components specified by the virus. Second, since the envelope of some strains of virus is derived from an internal membrane of the cell, the question arose whether membrane specific components bound only to the membrane precursor of the envelope or to all membranes of the cell. These experimental objectives required purified membranes free of virus particles and, conversely, enveloped virus particles free of membranes. The procedures for the isolation and purification of the membranes and of the virus were designed to meet these requirements. It is convenient to present the results of these experiments in three sections dealing respectively with viral envelopes, membranes, and with the chemical nature of the changes in the membranes underlying the social behavior of infected cells.

Purification and Analysis of Enveloped Virions

Currently there are not experimental procedures capable of yielding purified envelopes free of other structural components of the virion. These experiments were designed to obtain purified enveloped nucleocapsids and naked nucleocapsids; by subtraction, the proteins present in the enveloped nucleocapsids but not in the naked nucleocapsids would necessarily correspond to the structural components of the envelope. We have purified enveloped and unenveloped nucleocapsids by one of two procedures. In the early experiments the infected cells were Dounce homogenized in saline buffered with 0.05M tris at pH 7.2 and then centrifuged at low speed to remove large debris. The cell sap was then centrifuged on a 10-50% w/w sucrose density gradient for approximately 45-60 minutes at 20,000rpm in an SW25.3 rotor. On centrifugation virus particles formed three bands designated from the top as No. 1, 2 and 3 and containing largely but not exclusively empty nucleocapsids, full nucleocapsids and enveloped full nucleocapsids respectively (Figure 4). In general the naked nucleocapsids (band 2) were by electron microscopic and other criteria quite pure. Band 1 was heavily contaminated by soluble proteins and was not purified further. Band 3 was contaminated by membranes. A higher degree of purification of enveloped particles in band 3 was obtained in the third step. The material in band 3 was rendered 45% w/w with respect to sucrose, placed in the bottom of the tube, overlayed by a 20-35% w/w sucrose density gradient, then centrifuged to equilibrium. Enveloped particles floated to form a band above the 35-45% sucrose interface. It should be pointed out that although this band contained fewer impurities than band 3, free membranes and unenveloped or partially enveloped particles were still present. In more recent experiments we have purified enveloped nucleocapsids by centrifugation of cell lysates in Tris-buffered dextran solutions. This procedure yields much purer enveloped nucleocapsids; the procedure however is not suitable for purifying naked nucleocapsids.

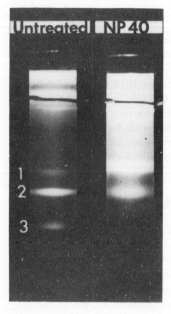

FIGURE 4. Banding of herpes simplex virus particles extracted from the cytoplasm of 18 hour-infected HEp-2 cells. Cytoplasmic extract was prepared by homogenization of infected cells in a Dounce homogenizer in .15M NaCl, .05M tris buffer, pH 7.2. The nuclei were removed by centrifugation at 2,000rpm for 5 minutes. The non-ionic detergent Nonidet P-40 (Shell Oil Co.) was added to approximately one-half of the cytoplasmic extract in sufficient amounts to make 0.5%. The treated and untreated extracts were centrifuged on a linear 10-50% sucrose density gradient at 20,000rpm for 40 minutes in the SW25.3 rotor. Band 1 contained predominantly unenveloped empty nucleocapsids. Band 2 contained predominantly unenveloped full nucleocapsids. Band 3 formed only in the tube containing the untreated preparation and consisted of enveloped nucleocapsids contaminated by membranes.

The electrophoretic analyses in acrylamide gels were done on viral proteins labeled with ^{3}H-amino acids (leu, isoleu, val) and ^{14}C-glucosamine. The electropherograms of proteins contained in naked nucleocapsids (Figure 5) of strain MP (band 2 of sucrose density gradients shown in Figure 4)revealed 6 bands. Similar bands were obtained from virions treated with Nonidet P-40 which strips the envelope leaving the nucleocapsid intact. The proteins contained in the 6 bands were not glycosylated extensively if at all; this conclusion is based on the fact that the amounts of glucosamine which migrated with these bands was very much less than that present in the proteins extracted from purified enveloped virus shown in Figure 6. The difference between electrophoretic profiles of the intact virions and naked nucleocapsids which must be attributed to the envelope consist of 4 major and one minor band of highly glycosylated proteins. Preliminary evidence furthermore suggests that one of the bands may contain at least two glycoproteins.

Purification and Analysis of Proteins in the Membranes of Infected Cells

In these experiments we employed the procedure described by Bosmann et al. (26) for extraction and purification of cellular membranes. The products obtained by this procedure satisfied the rigorous requirement that the purified cellular membranes be free of both naked and enveloped nucleocapsids. The most important step in this procedure is the equilibrium flotation of the membranes through a discontinuous sucrose density gradient from a sample layer containing 45% w/w sucrose. This step separates membranes according to buoyant density rather than cellular topology and yields 4 bands. Of these bands No. 2 consistently contained smooth membranes free of virus, ribosomes, soluble proteins and the other constituents of the cell. Analysis of membranes extracted from infected and uninfected cells revealed the following: (a) The electrophoretic profiles of glycoproteins made after infection and extracted from membranes of infected cells revealed two major and several minor

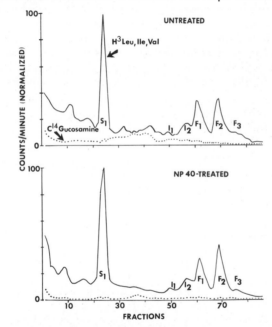

FIGURE 5. Electropherograms of proteins from nucleocapsids in band 2 (Nonidet treated and untreated) as shown in Figure 4. The cells were labeled with ^{3}H-amino acids (leu, isoleu, val) and ^{14}C-glucosamine. The nucleocapsids were solubilized in 0.1% sodium dodecyl sulfate, 0.5M urea, 0.01% β-mercaptoethanol and electrophoresed. The data were normalized as follows: The amino acid counts were normalized with respect to the amino acid content of protein S_1. The glucosamine was normalized with respect to the glucosamine content in fully enveloped particles. The differences in the electrophoretic mobilities of the nucleocapsid proteins of F, MP and G strains are minor (Roizman and Spear, unpublished studies). The direction of electrophoresis is from left to right.

bands. The electrophoretic profiles of membrane glycoproteins from infected cells were consistently different from those of uninfected cells which contained too many bands to be resolved in the acrylamide gels used (Figure 7). This finding is consistent with the observations that the virus inhibits not only the synthesis of host proteins but also further glycosylation of the proteins following infection. (b) The binding of the new glycoproteins to membranes is sufficiently strong to withstand considerable hydrodynamic stress (Figure 8). Thus exposure of the membranes to anti-infected cell antibody increases their density and precludes the membrane from floating to the top of the sucrose density gradient (28). (c) To date incomplete analyses of the sugars present in membrane glycoproteins have revealed the presence of galactosamine, mannose, fucose and galactose in addition to glucosamine. Of these sugars fucose is quantitatively incorporated into the membranes as fucose. Most of the glucosamine is incorporated into the glycoproteins as glucosamine and galactosamine (Keller, unpublished observations). (d) Experiments

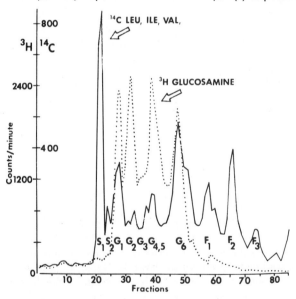

FIGURE 6. Electropherogram of enveloped nucleocapsids (MP strain) prepared from HEp-2 cells labeled between 4 and 18 hours post infection with ^{14}C-amino acids and ^{3}H-glucosamine. The enveloped nucleocapsids were prepared by isopycnic flotation of band 3 similar to that shown in Figure 4. The solubilization and electrophoresis was done as described in legends to Figure 5. The glycoproteins are labeled G_1 through G_6.

designed to determine the site of glycosylation of these proteins have shown that the glucosamine is not incorporated into nascent peptides on polyribosomes, that glucosamine is incorporated into acid-insoluble material in the presence of puromycin and finally that membrane proteins are largely glycosylated in the membranes. As shown in the pulse-chase experiment illustrated in Figure 9, concurrently with glycosylation

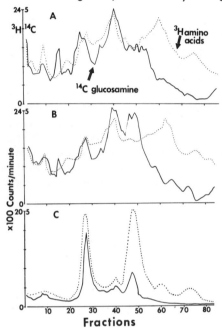

FIGURE 7. Electropherograms of labeled proteins and glycoproteins in purified membranes from infected and uninfected HEp-2 cells labeled with ^{3}H-glucosamine and ^{14}C-amino acids (leu, isoleu, val). A-electropherogram of membrane proteins synthesized in uninfected cells. B-electropherogram of membrane proteins extracted from cells labeled for 12 hours before infection, then incubated for 3 hours prior to infection and 24 hours after infection in medium containing unlabeled precursor. C-electropherogram of membrane proteins extracted from cells labeled after infection. The details concerning the solubilization and acrylamide gel electrophoresis of proteins, scintillation counting of acrylamide gel slices, etc., were described elsewhere (29). The smooth membranes (band 2 obtained by flotation of cytoplasmic extracts through a sucrose density gradient) were prepared according to the procedures of Bosmann et al. (26) as described by Spear et al. (30). The migration of proteins is from left to right.

(amino acid label is conserved during the chase while glucosamine label increases) membrane proteins increase in apparent molecular weight and homogeneity. This is evidenced by the observations that after the chase some of the proteins migrate more slowly in acrylamide gels and in sharper peaks with complete coincidence of amino acid and glucosamine label (27).

Membrane Glycoproteins and the Social Behavior of Infected Cells

As enumerated in the preceding section, there is considerable evidence indicating a correlation between the surface properties of infectious herpes simplex virions and the social behavior of infected cells. It was also shown that herpes virions have on their surface an envelope containing several glycoproteins. It could be predicted therefore that, if the glycoproteins are the macromolecular species involved in determination of the antigenic specificity of the virus and of the surface properties of the virion, viruses differing with respect to their effects on the social behavior of cells would specify (a) different membrane glycoproteins and (b) different envelope glycoproteins. In general two lines of evidence indicate that this prediction is correct. First and directly to

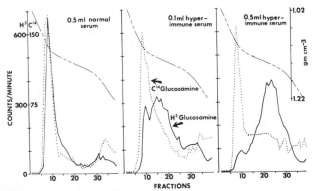

FIGURE 8. Flotation of mixtures of infected and uninfected cell smooth membranes in sucrose density gradients after incubation for 4 hours with normal rabbit serum or with varying amounts of rabbit hyperimmune serum prepared against herpes simplex virus. The incubation mixtures were made 45% (w/w) with respect to sucrose, overlayed with linear gradients of 10-35% (w/w) sucrose and then 3ml saline and centrifuged for 20 hours at 25,000rpm in a Spinco SW27 rotor. The top of the tube is at the left. ────── infected cell membranes labeled with ^{3}H-glucosamine; ········ uninfected cell membranes labeled with ^{14}C-glucosamine. From Roizman and Spear (28).

the point, herpes simplex virus strains differing with respect to their effects on the social behavior of cells also differ with respect to the membrane and envelope glycoproteins they specify (31). As shown in Figure 10 strains MP and G specify not only different viral envelope glycoproteins but also different smooth membrane glycoproteins. Parenthetically a very interesting observation which emerged from this study is that the binding of glycoproteins to membranes is ordered and not random. This is deduced from the fact that in cells infected with the MP and G strains, the envelope glycoproteins differ from those associated with

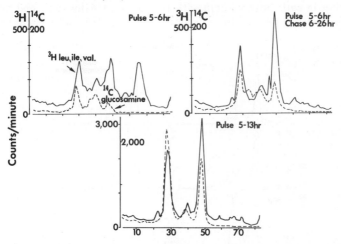

FIGURE 9. Electropherograms of radioactive proteins present in smooth membranes purified (band 2 obtained by flotation of a cytoplasmic extract through a discontinuous sucrose gradient as in Figure 7) immediately after a 1 hour pulse from 5 to 6 hours post infection or after further incubation in non-radioactive medium. Conditions for solubilization of proteins and electrophoresis were as described in Reference 27. Direction of migration was from left to right and the numbers of the abscissa represent gel slices. The cells were labeled with ^{3}H-amino acids (leu, isoleu, val) and ^{14}C-glucosamine. Amino acid label was conserved during the chase while glucosamine label increased. Amino acid label from the non-glycosylated band present after the pulse apparently distributed among the glycosylated bands during the chase. That further glycosylation of the membrane bound proteins occurs during the chase is indicated by the observations that (1) glucosamine peaks trail the broad amino acid peaks after the pulse, (2) after the chase the amino acid peaks are sharper indicating increased homogeneity, and migrate more slowly in coincidence with the glucosamine peaks. Data from Spear and Roizman (27).

smooth membranes of the cells. This observation is particularly significant for the MP strain which becomes enveloped at the nuclear membrane. The data suggest that the glycoproteins binding to the nuclear membrane--the site of envelopment--are different from those binding to smooth cytoplasmic membranes.

The second observation deals specifically with the significance of the apparent correlation between membrane and envelope glycoproteins, and the social behavior of infected cells. Several years ago we reported (15) that in cells infected both with mP, a virus similar to F, and MP strains, both viruses multiplied equally well, but the doubly-infected

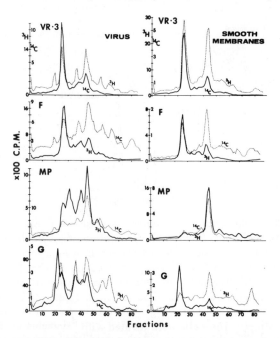

FIGURE 10. Electropherograms of membrane and viral proteins prepared from infected HEp-2 cells labeled simultaneously with glucosamine and a mixture of amino acids (leu, isoleu, val). The smooth membranes and partially purified virus were prepared from the same batch of infected cells. Solid line-profile of labeled glucosamine incorporated into glycoproteins. Dashed line-profile of labeled amino acids incorporated into proteins. Radioactive markers are as shown. The cells were labeled between 4 and 22 hours after infection with the strain of virus as indicated. Data from Keller et al. (31).

cells exhibited the social behavior of the cells infected with the mP strain alone. Recent analyses of the envelope and membrane glycoproteins in singly and doubly-infected cells revealed that, in doubly infected cells both membrane and envelope glycoproteins were those of the dominant F strain (Figure 11). The data indicated that the F strain specified both the social behavior of cells and the composition of virus-specific glycoproteins binding to cellular membranes. It is of interest to note that F strain specifies one major membrane glycoprotein with an approximate molecular weight of 100,000 daltons; this glycoprotein is either absent or present in minute amounts in the membranes extracted from cells infected by the MP strain. The reason for the absence of the high molecular weight glycoprotein in cells infected with the MP strain is not known. One possibility is that the MP strain has lost the genetic information for the synthesis of this protein. This possibility emerges from a recent finding in our laboratory (Kieff, Bachenheimer and

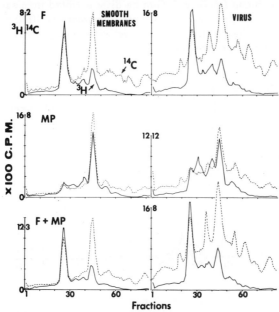

FIGURE 11. Electropherograms of membrane and viral proteins prepared from cells singly and doubly-infected with virus strains differing with respect to their effects on the social behavior of cells. The doubly-infected cells were incubated with F and MP strains at identical multiplicities of infection. All infected cells were labeled between 4 and 22 hours post infection in medium containing ^{14}C-amino acid mixtures (leu, iso-leu, val) and ^{3}H-glucosamine (Keller, Roizman and Spear cited in Ref. 10).

Roizman, unpublished data) that the molecular weight of MP virus DNA is some 5% smaller than that of the F strain.

One of the questions that remains to be answered relates to the fact that the membrane studies described in this section were done on smooth membranes of the cell which are characterized by density rather than cellular topology. Since social behavior of cells is specifically mediated by the plasma membrane, we have initiated studies designed to demonstrate whether viruses differing with respect to the social behavior of cells also differ with respect to the plasma membrane glycoproteins they specify. In attempt to answer this question we have fractionated cells infected with the F strain by procedures designed to yield purified plasma membranes (32) and purified internal cytoplasmic membranes [combination of procedures published by Boone et al. (32) and Bosmann et al. (26)]. The electrophoretic profiles of viral proteins present in these two membrane fractions cannot be differentiated from each other or from those proteins present in the smooth membranes purified by flotation (Figure 12). Studies of the enzymatic activities and physical properties of these membranes are now in progress.

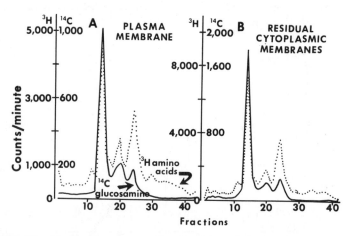

FIGURE 12. Electropherograms of radioactive glycoproteins present in membranes purified from infected cells by 2 different procedures. The cells were incubated with ^{3}H-amino acids (leu, isoleu, val) and ^{14}C-glucosamine from 4 to 18 hours post infection. A. Presumptive plasma membranes purified by the procedure of Boone et al. (32). B. Membranes obtained by sucrose density gradient flotation (26) from a cytoplasmic supernatant after sedimentation of the plasma membrane ghosts obtained by the Boone procedure (32).

It should be pointed out that, to date, the various techniques and procedures used to purify membranes from infected cells failed to yield preparations lacking the viral glycoproteins. These observations suggest that viral glycoproteins bind to all membranes of infected cells and may account for the various modifications of these membranes described in this paper.

ACKNOWLEDGEMENTS

This investigation was aided from time to time by grants from the United States Public Health Service (CA 08494), the American Cancer Society (E314F), the Whitehall Foundation and the Leukemia Research Foundation. One of us (P. G. S.) is a United States Public Health Service postdoctoral trainee (TO I HD 00297-02).

BIBLIOGRAPHY

1. B. Roizman, in: Current Topics in Microbiology and Immunology, Vol. 49 (Springer-Verlag, Heidelberg, 1969) p. 1.

2. S. Nii, C. Morgan and H. M. Rose, J. Virology, 2 (1968) 517.

3. J. Schwartz and B. Roizman, J. Virology, 4 (1969) 879.

4. C. W. Stackpole, J. Virology, 4 (1969) 75.

5. K. Nazerian and R. L. Witter, J. Virology, 5 (1970) 388.

6. B. Roizman, S. B. Spring and J. Schwartz, Fed. Proc., 28 (1969) 1890.

7. S. Nii, C. Morgan, H. M. Rose and K. C. Hsu, J. Virology, 2 (1968) 1172.

8. J. Schwartz and B. Roizman, Virology 38 (1969) 42.

9. S. B. Spring, B. Roizman and S. Schwartz, J. Virology, 2 (1968) 384.

10. B. Roizman, in: Proc. of the 3rd Int. Symp. on Applied and Medical Virology (Warren Green Publishers, St. Louis, Mo., 1970) p. 37.

11. C. Morgan, H. M. Rose, M. Holden and E. P. Jones, J. Exptl. Med., 110 (1959) 643.

12. B. Felluga, Ann. Sclavo, 5 (1963) 412.

13. E. K. Wagner and B. Roizman, J. Virology 4 (1969) 36.

14. T. Kamiya, T. Ben-Porat and A. S. Kaplan, Virology, 26 (1965) 577.

15. B. Roizman, in: Cold Spring Harbor Symp. on Quant. Biol., 27 (1962) 327.

16. B. Roizman, in: Perspectives in Virology IV, ed. M. Pollard, (Hoeber Medical Division, Harper Row Publishers, New York, 1966) p. 283.

17. B. Roizman and P. R. Roane, Jr., J. Immunol., 87 (1961) 714.

18. P. R. Roane, Jr. and B. Roizman, Virology, 22 (1964) 1.

19. J. F. Watkins, Nature, 202 (1964) 1364.

20. G. Pearson, F. Dewey, G. Klein, G. Henle and W. Henle, J. Nat. Cancer Inst., 45 (1970) 989.

21. B. Roizman and S. B. Spring, in: Proc. Conf. on Cross Reacting Antigens and Neoantigens, ed. J. J. Trentin, (Williams & Wilkins Co., Baltimore, 1967) p. 85.

22. B. Roizman and P. R. Roane, Jr., Virology, 19 (1963) 198.

23. B. Roizman and L. Aurelian, J. Mol. Biol., 11 (1965) 528.

24. P. M. Ejercito, E. D. Kieff and B. Roizman, J. Gen. Virology, 2 (1968) 357.

25. M. Terni and B. Roizman, J. Infect. Dis. 121 (1970) 212.

26. H. B. Bosmann, A. Hagopian and E. H. Eylar, Arch. Biochem. Biophys., 128 (1968) 51.

27. P. G. Spear and B. Roizman, Proc. Nat. Acad. Sci., 66 (1970) 730.

28. B. Roizman and P. G. Spear, Science, in press (1970).

29. P. G. Spear and B. Roizman, Virology, 36 (1968) 545.

30. P. G. Spear, J. M. Keller and B. Roizman, J. Virology, 5 (1970) 123.

31. J. M. Keller, P. G. Spear and B. Roizman, Proc. Nat. Acad. Sci. 65 (1970) 865.

32. C. W. Boone, L. E. Ford, H. E. Bond, D. C. Stuart and D. Lorenz, J. Cell Biol., 41 (1969) 378.

DISCUSSION

Dr. M. M. Sigel, University of Miami: This paper is now open for discussion.

Dr. H. M. Temin, University of Wisconsin: You mentioned that you found glycosylation in the presence of puromycin.

Dr. Roizman: Yes....

Dr. Temin: What was the experiment? Do the results say anything about what enzymes were carrying out the glycosylation, and whether the enzymes are virus-specified or host-specified?

Dr. Roizman: The experiment involving puromycin was done as follows: to a culture of infected cells puromycin was added at a concentration sufficient to stop the incorporation of amino acids immediately upon addition of the drug. However, under those conditions glucosamine incorporation continued. This experiment indicated that protein synthesis can be dissociated from glycosylation of the proteins and suggested that glycosylation does not take place on polyribosomes. We verified this conclusion by analyzing for the presence of labeled glucosamine in nascent peptides of free and bound polyribosomes. This experiment also indicated that the proteins are not glycosylated on polyribosomes. In the same series of experiments (Spear and Roizman, PNAS 66: 730-737, 1970) we pulse-labeled with amino acids and glucosamine then chased; in this experiment we found that the amounts of amino acid labeled proteins in the membranes immediately after the pulse and after the chase were exactly the same indicating that the chase was effective for the amino acids. On the other hand, the glucosamine counts increased during the chase indicating that the chase of the glucosamine pool was not effective. The most important part of the experiment was that the bulk of the proteins made and bound to the membrane during the pulse actually became glycosylated during the chase. Furthermore, the electropherograms showed that as the proteins become glycosylated, they migrate in polyacrylamide gels more slowly than in non-glycosylated proteins. The data clearly indicated that the glycosylation of the proteins takes place in the membranes. Now as to your second question; we have done a limited number of experiments attempting to determine whether glycosylation is carried out by enzymes specified by the virus or by the cell. The idea behind these experiments was that if glycosylation is solely determined by the amino acid sequence of the protein and if the virus specifies its own transferases, it would be expected that the glycosylation profile of virus-specific proteins made in different cell lines would be identical. On the other hand, if virus-specific proteins are glycosylated at least in part by transferases specified by the cell, the glycosylation profile of viral specific glycoproteins made in different cells might be different--not necessarily, but might be. We were fortunate to find that the minor glycoproteins specified by the virus are also glycosylated more extensively in Vero cells than in HEp-2 cells. We cannot draw too many conclusions from this data. What the data suggests is that the proteins specified by the virus can be glycosylated by cellular enzymes. It doesn't necessarily mean that the glycosylation be cellular enzymes is required for the synthesis of infectious virus. Parenthetically I should like to point out that herpesviruses are rather large and that the DNA of the virus is sufficient to coat for at least 140 proteins 50,000 daltons in molecular weight. Certainly there is enough genetic information to specify a number of transferases.

Dr. Sigel: Bernie, did I understand you correctly that whenever virus failed to impart to the membrane some of its antigens, you would obtain a yield of unenveloped virus; is that correct?

Dr. Roizman: This is correct.

Dr. Sigel: Would you care to comment about the assembly of virus in relation to this particular phenomenon?

Dr. Roizman: The studies to which you allude dealt with abortive infection of dog kidney cells with the macroplaque strain of herpes simplex virus. In abortively infected dog kidney cells nucleocapsid are assembled in the nucleus but do not become enveloped. Analysis of the abortively infected cells showed failure of synthesis of virus-specific surface antigens. Moreover, in the same series of experiments lysates of these cells did not induce either neutralizing antibody or cytolytic antibody in rabbits. These data suggested to us that the components of the envelope of the virus which are made in the cytoplasm probably bind to all the membranes of the cell and modify not only the nuclear membrane which gives rise to the envelope but also the plasma membrane. In consequence, the plasma membrane acquires the same immunologic specificity as the envelope of the virion. If the glycoproteins are not synthesized, there is no envelopment and no change in the surface antigenicity of the cells. With regard to your second question that you asked; our information concerning assembly and envelopment is somewhat limited. From what we know, and these data are based on information obtained in a number of laboratories, it appears that the proteins of the virus are made in the cytoplasm, that DNA is made in the nucleus, that the proteins are transported from the cytoplasm into the nucleus, and that the nucleocapsid is assembled in the nucleus. For some viruses the envelopment takes place at the nuclear membrane. Some herpesviruses appear to be enveloped not only at the nuclear membrane but also at the cytoplasmic membrane. Just how they get to the cytoplasmic membranes is not known. We do not know very much about the process of envelopment. One of the problems that very much interests us is why herpesvirus nucleocapsids become enveloped whereas adenovirus nucleocapsids do not. I do not think that the synthesis of virus-specific glycoproteins is itself the answer to this problem. It is conceivable that the answer lies in part also in the structure of the surface of the herpes nucleocapsid as compared to that of the adenovirus nucleocapsid. In any event, we do not know what determines the specificity of the interaction between the surface of the nucleocapsid and the membrane of the nucleus.

Dr. G. Cohen, Albert Einstein College of Medicine: I would like to make some comments about examining infected cell membranes for the presence of virus proteins and then drawing conclusions with respect to the meanfulness of this association *in vivo*. A paper by Holland and Kiehn [1970. Science 167: 2027] reports that not only are influenza virus proteins recovered with the plasma membranes of infected cells, but also with nuclear membrane fractions, mitochondrial membrane fractions, endoplasmic reticulum, etc. In our laboratory, we have been working with an RNA-containing rhabdovirus, vesicular stomatitis virus (VSV), which also matures by budding from the host cell plasma membrane. We have been isolating the plasma membranes during infection and looking for virus proteins. As part of these studies, we isolated uninfected plasma membrane ghosts from HeLa cells and combined these membranes at $0^{o}C$ with a mixture of radioactive VSV proteins contained in an infected-cell lysate made virus-free and also "membrane-free" [as defined in Cohen, Atkinson, and Summers, 1971. Nature, in press]. Under these conditions we found that the envelope protein of the virus tightly associates with the uninfected HeLa plasma membranes. Clearly these types of *in vitro* recombination experiments have to be carried out to assess further the significance of virus proteins recovered with the membrane fractions of infected cells.

Dr. Roizman: In effect Dr. Cohen asked the question whether the presence of herpes-specific glycoproteins in the membranes might be fortuitous and due to some non-specific binding of the glycoproteins specified by the virus to the membranes. The experiment described by her could not be done with herpesvirus glycoproteins for the simple reason that herpesvirus glycoproteins cannot be found in cell lysates free of membranes or of virus particles. However, I do not think that the binding of virus-specific proteins to the membranes is a non-specific phenomenon similar to that observed in vesicular stomatitis virus-infected cells. The evidence in support of this conclusion is based on several observations which may be summarized as follows: First, electropherograms of the lysates of infected cell but not in uninfected cells. Yet analyses of the membranes show that only a small fraction of these proteins actually bind to membranes. In fact the structural viral proteins contained in the nucleocapsid do not bind to the membranes of the infected cells. These observations indicate that there is no generalized binding of proteins to membranes and that the binding of proteins to membranes is selective. The second point that I would like to make shows that not only is there selectivity but also specificity of binding of virus-specific proteins to membranes. This conclusion is based on analysis of the glycoproteins binding to membranes in cells infected with different herpes variants. Specifically in cells infected with the F prototype of subtype 1 virus and glycoproteins in the virions which comes from the nuclear membrane, in the plasma membrane, and in the smooth membranes are exactly the same as far as we can tell. However, in cells infected with the G prototype of subtype 2 strain and in cells infected with the macroplaque variant the glycoproteins in the virion are different from those present in the smooth membranes of the infected cell. These data indicate that the binding of the proteins to the membranes is highly specific. The third point that I would like to make concerns the remote possibility that the glycoproteins bind to one membrane and in the course of manipulative procedures they jump from one membrane to another. We tested this point in our experiments with the antibody. In fact, this was one of the reasons for doing this experiment. In these experiments as you recall we mixed infected and uninfected cell membranes and then added antibody. The mixture was incubated for several hours then floated to equilibrium in sucrose density gradients. In these experiments we expected to get fairly good separation of infected and uninfected cell membranes on the basis of immunologic specificity. The way the experiments were designed, it was possible to determine the maximum amount of labeled glycoproteins from infected cells which might have become dissociated from the infected cell membranes and become associated with the uninfected cell membranes. These experiments showed that at most only 10% of the host membranes became contaminated with infected cell glycoproteins. On the other hand the experiments clearly showed that the binding of virus-specific glycoproteins to membranes is sufficiently strong to withstand hydrodynamic stress applied to the membranes during the isopycnic centrifugation. In summary, the selectivity of the proteins binding to membranes, the specificity of binding of virus-specific glycoproteins to various membranes of the cell, and the tenacity with which these proteins continued to adhere to the membranes even when subjected to hydrodynamic stress after exposure to antibody, all of these observations indicated that the finding of herpesvirus-specific glycoproteins bound to cellular membranes is meaningful.

Dr. Cohen: I would also like to add that this *in vitro* binding behavior may be significant to the virus assembly process *in vivo* despite the possibility that when the cell is disrupted, some of the virus proteins then associate with the cell membranes.

Dr. Roizman: We cannot perform experiments similar to your *in vitro* experiments for the simple reason that we cannot find glycosylated proteins as specified by herpesvirus in the cytoplasm of the infected cells. As I have described earlier these proteins become bound and are glycosylated in membranes and cannot in fact be found free in the cytoplasm. I should also like to point out however, that the *in vitro* binding experiments are not terribly convincing for two reasons. First, if proper precautions are not taken the glycoproteins may become sufficiently denatured to tend to bind any membranes. In fact, the experiments are meaningful only if you can show that the glycoproteins recovered in the cytoplasmic extract have not been denatured in any fashion. The second point concerning the experiments you described is that it is a natural function of the surface proteins of the vesicular stomatitis virion to play a role in the absorption of the virion to uninfected cells. It is conceivable that even if the glycoproteins are not denatured they could still have a very high affinity for the membranes of uninfected cells simply because it is their function to mediate in the absorption of the virion to the surface of the cell it infects.

SV40 DNA REPLICATION IN NORMAL AND TRANSFORMED CELLS

SAUL KIT and D.R. DUBBS

Division of Biochemical Virology
Baylor College of Medicine
Houston, Texas 77025

INTRODUCTION

Simian virus 40 (SV40) replicates extensively in primary and established lines of African green monkey kidney (AGMK) cells, replicates to a lesser extent in human cells, and undergoes an abortive infection in mouse or hamster cells. Infection is initiated in mouse and hamster cells, as shown by the fact that SV40-specific RNA and T antigen are made. However, there is little, if any, replication of superhelical viral DNA or capsid proteins. Some of the abortively infected mouse and hamster cells undergo transformation, a process which probably involves the "integration" of the viral DNA with that of the cell.

Although human and monkey cells are susceptible to productive infection by SV40, they can also occasionally be transformed. So long as the transformed state is maintained, superhelical SV40 DNA is not detectable in the transformed cells, as would be expected if the viral DNA were "integrated." However, infectious virus can be recovered from certain transformed cell cultures. The recovery process is associated with the activation of superhelical SV40 DNA replication.

The mechanism of replication of SV40 DNA and of the DNA of closely related polyoma virus will be discussed in this article. Three conditions essential for replication have been identified: (i) concurrent protein synthesis throughout the infectious cycle; (ii) expression of a viral gene specifically concerned with viral DNA replication; and (iii) availability of essential cellular replication factors.

REPLICATION OF PAPOVAVIRUS DNA IN PRODUCTIVELY INFECTED CELLS

The DNAs extracted from papovaviruses, SV40 and polyoma, are superhelical double-stranded molecules, approximately 1.5 to 1.7 microns in length, with molecular weights of about 2.8 to 3.2 x 10^6 daltons (1-3). Through polyribonucleotide binding studies, it has been shown that SV40 DNA contains several dAT-rich sequences, but few dCG-rich sequences (4). Polyoma DNA contains both dAT-rich and dCG-rich sequences (5). Electron microscopic studies of partially denatured polyoma DNA confirm that there are 3 to 5 dAT-rich regions in polyoma DNA (1,6). Kubinski and Rose (4) have suggested that dAT-rich regions may facilitate recombination between host and viral DNAs during transformation. It remains to be seen whether this interesting suggestion can be substantiated.

By studying the incorporation of 5-bromodeoxyuridine into the DNA of polyoma-infected cells, Hirt (7) has shown that polyoma DNA replicates by a semi-conservative mechanism. Hirt (7) observed that superhelical polyoma DNA molecules were formed in which 5-bromodeoxyuridine replaced thymidine in one strand (hybrid density) or both strands (heavy density). Since polyoma DNA is a circular helix with separately continuous strands, semi-conservative replication requires the introduction of at least one break in one of the strands.

Hirt (8) has also demonstrated that replicating polyoma DNA molecules form a theta (θ)-shaped structure with two branch points and three branches, much like that observed with replicating lambda, φX174, and mitochondrial DNA molecules. Electron micrographs of replicating SV40 molecules also demonstrate that they are θ-shaped structures (3).

^{3}H-thymidine incorporation studies and assays of infectious DNA in SV40-infected monkey kidney cells indicate that SV40 DNA replication first begins 12 to 16 hr after infection and rises to a maximum rate of synthesis at about 30 hr. DNA synthesis continues at a rapid rate until 50 hr after infection. The production of infectious virus begins 18 to 24 hr after infection and continues to increase until about 72 hr (3,9,10).

The sedimentation behaviour of replicating polyoma and SV40 DNA has been studied by Bourgaux et al. (11) and Levine et al. (3). Replicating viral DNA was isolated from virus-infected cultures that were labelled with ^{3}H-thymidine for 2 to 7.5 min. The replicating DNA had the following properties: (i) When centrifuged in a cesium chloride-ethidium bromide (CsCl-EtBr) density gradient it had a light density indicative of nicked-circular or linear DNA (12). (ii) After velocity sedimentation in sucrose gradients, it sedimented at about 25S. This sedimentation rate is faster than that of nicked-circular (Form II) SV40 or polyoma DNA which sediment at about 16 to 18S. (iii) Unlike mature viral closed-circular DNA, the replicative intermediates were completely denatured by alkali. (iv) The chromatographic properties of replicating DNA on benzoylated-naphthoylated-DEAE-cellulose suggest that they contain single-stranded regions. (v) In pulse-chase experiments, the radioactive label was first incorporated into 25S material, but after a chase period, most of the label was found co-sedimenting with superhelical viral DNA (heavy density in CsCl-EtBr; sedimentation coefficient of about 21S in sucrose gradients).

The size of the strands in the replicative intermediate have been studied by velocity sedimentation in alkaline sucrose gradients. Although the individual single-stranded molecules varied in size, none appeared to be longer than the strands present in mature viral DNA (11). This finding contraindicates a rolling-circle mechanism for papovavirus DNA replication, for the rolling-circle model predicts that strands longer than unit length are generated during replication.

Recently, Bourgaux (13) has found that the <u>Neurospora</u> endonuclease, that has a specificity for single-stranded DNA, converts the θ-shaped polyoma replicative DNA to linear forms and to "φ"-shaped forms. This suggests that both branch points contain single-stranded regions and that polyoma DNA replicates through a bi-directional mechanism. Evidence for a bi-directional DNA replication mechanism has previously been presented for mammalian nuclear DNA (14) and for lambda DNA (15).

EFFECT OF INHIBITING PROTEIN SYNTHESIS ON SV40 DNA REPLICATION

After inhibition of protein synthesis, for example with chloramphenicol or cycloheximide, the rate of DNA synthesis is markedly reduced in uninfected and virus-infected bacterial and animal cells. In <u>Escherichia coli</u> treated with chloramphenicol,

DNA synthesis proceeds until the completion of the current cycle of duplication and then stops. In <u>Proteus mirabilis</u> and in mouse fibroblast mitochondria, DNA dimers and oligomers accumulate after protein synthesis is inhibited by drug treatment or amino acid starvation. It has been proposed that the generation of multiple circular forms may result from an imbalance in the formation or concentration of the enzymes responsible for DNA duplication (16,17).

Inhibition of protein synthesis by cycloheximide inhibits the rate of SV40 DNA replication when the drug is added early or late in infection (9). Cycloheximide also inhibits the rate of cellular DNA synthesis in either uninfected or SV40-infected CV-1 (monkey kidney) cultures.

Inhibition of the rate of SV40 DNA replication early in infection could signify that a cycloheximide-sensitive protein (e.g., T antigen or DNA polymerase) must be made to initiate DNA replication. To rule out this possibility, experiments were carried out in which 1-β-D-arabinofuranosyl-cytosine (ara-C) was added to cultures immediately after SV40 infection. Ara-C blocks DNA synthesis but not the synthesis of T antigen and other "early" proteins. At 24 hr, DNA synthesis was initiated by adding deoxycytidine to the ara-C block. Nevertheless, addition of cycloheximide at any time between 24 and 46 hr after infection reduced the rate of SV40 and of cellular DNA synthesis by about 80%. These experiments indicate that concurrent protein synthesis is required for optimal rates of SV40 DNA duplication throughout the infectious cycle.

The possibility was considered that cycloheximide treatment caused an inactivation of essential enzymes or a degradation of SV40 DNA. Enzyme assays showed, however, that the activities of thymidine kinase, DNA polymerase, deoxycytidylate deaminase, and thymidylate kinase remained high for at least 9 hr in the presence of cycloheximide. The rate of DNA synthesis was inhibited in less than 1 hr. Also, SV40 DNA pre-labelled with ^{3}H-thymidine before the addition of cycloheximide was relatively stable during the time required for cycloheximide to inhibit DNA replication.

It was of interest to learn whether cycloheximide treatment inhibited the conversion of nicked molecular forms to the superhelical form of SV40 DNA, a step postulated by Levine et al.(3) to be rate-limiting in SV40 DNA replication. Inhibition of the conversion of nicked forms to superhelical DNA would be expected, for example, if the concentration of polynucleotide ligase were inadequate. It was also of interest to learn whether oligomeric circular forms of SV40 DNA would increase in cycloheximide inhibited cultures.

To answer these questions, pulse-chase experiments were carried out. Confluent monolayer cultures of SV40-infected CV-1 cells were synchronized by ara-C treatment from 2 to 24 hr. Then fresh media containing deoxycytidine was added to reverse the ara-C block and to initiate DNA replication. At 36 hr, cycloheximide was added to experimental cultures and at 36.75 hr, 5-fluorodeoxyuridine and uridine were added to

Table 1. Effect of cycloheximide (CH) on the pulse-labelling and chase of SV40 DNA at 37 hr after infection of ara-C-pretreated CV-1 cell cultures.[a]

Group	Hr PI CH added	^{3}H-dT for 7.5 min at: (hr PI)	Chase (hr)	cpm/culture x 10^{-3}		% ^{3}H after CsCl-EtBr cent. of Hirt extract	
				Hirt pellet	Hirt extract	Heavy	Light
1a	none	37	none	73.4	47.5	8.8	85.5
1b	36	37	none	10.2	6.5	21.9	74.9
2a	none	37	1	145.5	48.4	85.7	14.0
2b	36	37	1	16.1	8.0	91.0	6.6

[a] CV-1 cells (11.6 x 10^6 cells/culture) were infected with SV40 and then incubated at 36.5 C from 2 until 24 hr post-infection (PI) with ara-C. At 24 hr, new media containing deoxycytidine (dC) were added. Experimental cultures (1b and 2b) were treated with CH at 36 hr PI. At 36.75 hr PI, all cultures were treated with 5-fluorodeoxyuridine (25 µg/ml) and uridine (12.2 µg/ml) to deplete endogenous thymidine (dT) pools. At 37 hr PI, all cultures were pulse-labelled with ^{3}H-dT (25 µc/ml) for 7.5 min and groups 1a and 1b were harvested. The media of groups 2a and 2b were removed and the cultures were washed with prewarmed media containing dT (100 µg/ml) and dC (10 µg/ml). The wash of group 2b also contained CH (25 µg/ml). New media containing dT and dC (group 2a) or dT, dC, and CH (group 2b) were then added and the cultures were further incubated for 1 hr at 36.5 C.

deplete the endogenous thymidine pools. At 37 hr, control and cycloheximide-treated cultures were pulse-labelled for 7.5 min with ^{3}H-thymidine. The cultures of group 1 (Table 1) were harvested at this time. The media were removed from the cultures of group 2 (Table 1) and these cultures were washed with prewarmed media containing thymidine and deoxycytidine. The wash media of the cycloheximide-pretreated cultures (group 2b) also contained cycloheximide. New media with thymidine and deoxycytidine, or thymidine, deoxycytidine, and cycloheximide were added to the control and cycloheximide-pretreated cultures, respectively, and the cultures were further incubated for 1 hr.

Table 1 shows that cycloheximide treatment inhibited the incorporation of ^{3}H-thymidine into the DNA of sodium dodecylsulfate-Molar sodium chloride (Hirt) extracts, which contain SV40 DNA, and of the Hirt pellet fraction (high molecular weight nuclear DNA) by 83 to 89%. About 2 to 3 times as much radioactivity was found in the Hirt pellet fraction as in the Hirt extract. During the 60 min chase period, the presence of excess nonradioactive thymidine did not completely prevent the incorporation of ^{3}H-thymidine into DNA of the Hirt pellet fraction. However, the total radioactivity of the Hirt extracts did not increase significantly during the chase period.

Analysis of the Hirt extracts by CsCl-EtBr equilibrium centrifugation demonstrated that 75 to 86% of the total radioactivity was in the light density fraction after the 7.5 min pulse (Table 1 and Fig. 1). However, after the 60 min chase, only 7 to 14% of the total radioactivity was found in the light fraction, and 86 to 91% of the label was found in the heavy fraction.

Sucrose sedimentation analyses of the heavy (groups 2a, 2b) and the light (group 1a) CsCl-EtBr fractions of Table 1 are shown in Fig. 2. The remaining samples did not have sufficient radioactivity for accurate analyses by sucrose gradient centrifugation. Fig. 2b shows that the light CsCl-EtBr fraction from the 7.5 min ^{3}H-thymidine pulse consisted mostly of DNA sedimenting more rapidly than Form II SV40 DNA. Thus, the Hirt extracts of group 1a, Table 1, contained mostly nicked forms of DNA sedimenting faster than the nicked-circular form of SV40 DNA. Since the radioactivity was chased into Form I DNA, it is apparent that much of this radioactivity was in the form of replicative intermediates. The heavy CsCl-EtBr fractions obtained after the chase period, with or without cycloheximide, sedimented in the position of Form I SV40 DNA (21S). These experiments

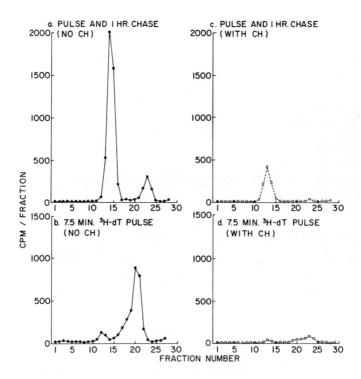

Fig. 1. Equilibrium centrifugation in CsCl-EtBr gradients of sodium dodecylsulfate-Molar sodium chloride extracts from SV40-infected CV-1 cells pulse labelled with ^{3}H-thymidine in the presence or absence of cycloheximide (CH). (See Table 1.)

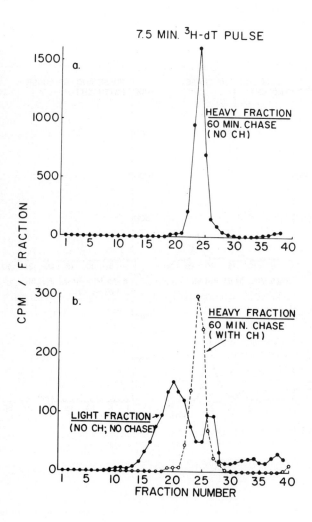

Fig. 2. Velocity sedimentation in nuetral sucrose gradients (5 to 30% w/v) of heavy and light CsCl-EtBr fractions from Table 1 and Fig. 1.

demonstrate that radioactivity can be chased from replicative intermediates to superhelical SV40 DNA in the presence or in the absence of cycloheximide.

In further experiments, it was shown that most of the SV40 ^{3}H-DNA formed in cycloheximide-inhibited cultures during longer labelling periods (2 to 4 hr) was superhelical. Form I SV40 ^{3}H-DNA was generated in cycloheximide-inhibited cultures shortly after SV40 DNA synthesis was started by reversing the ara-C block (24 to 28 hr) and also at later times (28 to 40 hr). These results suggest that reduced polynucleotide ligase activities do not account for the inhibition in the rate of SV40 DNA replication in cycloheximide-treated cultures.

In control and in cycloheximide-treated cultures, essentially all of the superhelical ^{3}H-DNA sedimented at 21S. Thus, in contrast to the results obtained with Col E_1 DNA and mitochondrial DNA, oligomeric forms of SV40 DNA did not accumulate when protein synthesis was inhibited.

The previous experiments suggest that concurrent protein synthesis is required for optimal rates of SV40 and cellular DNA synthesis. The function of the cycloheximide-sensitive protein is unknown. It appears that this protein is not an enzyme of DNA metabolism. Four possibilities may be suggested. The cycloheximide-sensitive protein may be: (i) a protein analogous to the bacteriophage gene 32 product which binds to single-stranded DNA and is needed in stoichiometric amounts for T4 DNA replication (18); (ii) an initiator of new rounds of SV40 DNA replication; (iii) a component of newly generated SV40 DNA replication sites; or (iv) a maturation protein which might accelerate the removal of completed SV40 DNA superhelices from active replication sites.

PAPOVAVIRUS GENE FUNCTION ESSENTIAL FOR VIRAL DNA SYNTHESIS

Studies of temperature-sensitive mutants demonstrate that at least one viral gene function is required for papovavirus DNA replication. Fried (19) has isolated a polyoma virus mutant, TS-a, which grows normally in susceptible mouse cells at the permissive temperature (31.5 C), but not at the nonpermissive temperature (38.5 C). Transformation of hamster cells by mutant TS-a is also inhibited at 38.5 C. Once transformed, however, the cells maintain their transformed phenotype at either temperature.

Under nonpermissive conditions, mutant TS-a is able to initiate a temporary change in hamster cells leading to growth in suspension, that is, abortive transformation. This further indicates that the gene which is affected in the TS-a mutant is not concerned with the expression of the transformed phenotype, but rather with the events leading to the stable perpetuation of the viral genome.

The temperature sensitivity of TS-a is due not to an increased heat lability of the structural components (capsid protein and DNA) of the completed virus particles, but to an increased heat sensitivity of some intracellular process in viral development. This temperature-sensitive process takes place after uncoating of the infecting virus particle, occurring at the time of, and being required for, the production of virus-sized DNA. The induction of host cell DNA synthesis is not markedly inhibited in TS-a infected mouse kidney cells at 38.5 C.

Mouse (3T3) cells transformed by polyoma mutant TS-a have been isolated by Vogt (20). The transformed mouse cells maintain their transformed phenotype if propagated at the nonpermissive temperature. If shifted to the temperature permissive for virus growth, a variable proportion of the mouse cells in the culture produce virus. Polyoma DNA synthesis is detectable in the activated cell population 24 hr after the shift to low temperature and steadily increases for several days (21). Following activation at 31 C, cultures continue to make viral DNA for at least 24 hr after they are shifted back to high temperature. This suggests that induction of viral multiplication involves the asynchronous occurrence of a unique event, after which DNA synthesis can continue even under nonpermissive conditions.

The properties of polyoma mutant TS-a suggest that the viral function required for transformation is also required for infectious viral DNA synthesis. It cannot be excluded, however, that TS-a is a double mutant, temperature-sensitive in one function required for viral DNA synthesis, and temperature-sensitive in a different function required for transformation. Consistent with the latter possibility is the fact that Eckhart (22) has isolated another temperature-sensitive polyoma virus mutant which is defective in the synthesis of infectious viral DNA, but not in transforming ability.

INFECTION OF CELLS BY SV40 DNA

Monkey kidney cells can be productively infected by either SV40 particles or SV40 DNA. Infection by SV40 DNA is relatively inefficient. The plaque-forming activity of SV40 DNA is only about 0.1 to 1.0% that of SV40 particles and under the best conditions so far obtained, no more than 5% of monkey kidney cells produce virus. In contrast, 90 to 100% of monkey kidney cells can easily be infected   productively by SV40 particles.

DEAE-dextran enhances the infectivity of SV40 DNA, probably by increasing the permeability of cells to the DNA and by reducing inactivation of the DNA by nucleases. At a concentration of 1 µg/ml, ethidium bromide also triples the plaquing efficiency of SV40 DNA, but not of SV40 particles. It is possible that ethidium bromide likewise protects infectivity by binding to the SV40 DNA. At ethidium bromide concentrations in excess of 25 µg/ml, however, SV40 DNA infectivity is reduced.

Exponentially growing cultures are more susceptible to SV40 DNA infection than are stationary phase cultures. Several modifications of the conditions for infecting cells with SV40 DNA have been studied, but these modifications did not significantly affect the efficiency of the process. The conditions studied include: (i) use of fetal calf serum in place of calf serum, (ii) change in adsorption temperature from 25 C to 37 C, or (iii) increased DNA adsorption time from 30 min to 4 hr.

The low infectivity of SV40 DNA molecules cannot be ascribed to poor adsorption of SV40 DNA. In one hr, approximately 38, 39, and 25% of superhelical SV40 ^{3}H-DNA is adsorbed by CV-1 (monkey), W98 VaD (human), and primary mouse kidney cells, respectively (23). However, the superhelical Form I SV40 ^{3}H-DNA is rapidly converted to nicked molecular forms (Fig. 3). Thus, by 2 hr after adsorption in the presence of DEAE-dextran, 62, 42, and 29% of the input SV40 ^{3}H-DNA is converted to nicked molecular forms.

Minimal nicking of superhelical SV40 DNA does not in itself result in loss of infectivity. To demonstrate this, the specific infectivities (PFU/cpm) of superhelical and nicked molecular forms of SV40 DNA were studied. Superhelical Form I and nicked forms of SV40 ^{3}H-DNA were extracted from virus-infected cells and pulse-labelled with ^{3}H-thymidine at 37 to 39 hr after infection (Fig. 4). The DNA was purifed by CsCl-EtBr equilibrium

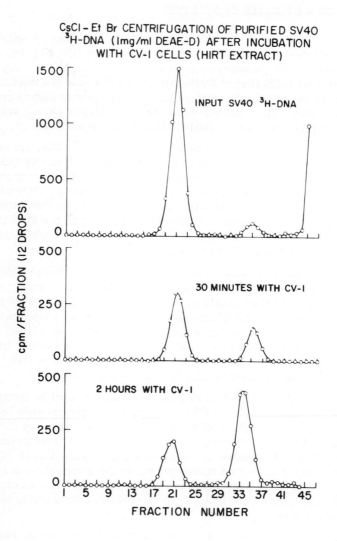

Fig. 3. Equilibrium centrifugation in CsCl-EtBr gradients of Hirt extracts from CV-1 cells incubated with superhelical SV40 ^{3}H-DNA in the presence of DEAE-dextran at 1 mg/ml. The radioactivity in fractions 17 to 21 and in fractions 30 to 37 consist respectively, of "heavy" density Form I SV40 ^{3}H-DNA and "light" density nicked molecular forms of SV40 DNA.

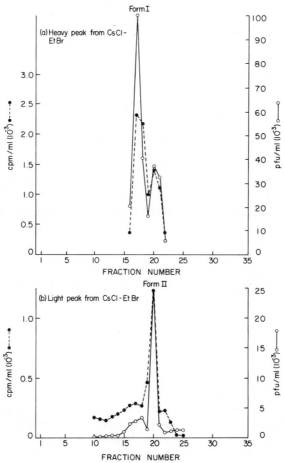

Fig. 4. Velocity sedimentation in neutral sucrose gradients (5 to 30% w/v) of SV40 ^{3}H-DNA isolated from cells infected with SV40 particles and pulse-labelled with ^{3}H-thymidine at 37 to 39 hr after infection. Sucrose sedimentation of the "heavy" peak from CsCl-EtBr gradients is shown in Fig. 4a, that of the "light" peak in Fig. 4b. Radioactivity is shown by -------; infectivity is shown by ———.

centrifugation and velocity sedimentation in sucrose gradients, and assayed for infectivity. The nicked (Form II) SV40 DNA obtained from infected cells was about half as infective as superhelical (Form I) SV40 DNA (Fig. 4). Also, after adsorption of SV40 ^{3}H-DNA to CV-1 cells and subsequent extraction of the labelled DNA, the nicked (Form II) DNA was about as infective as the input superhelical SV40 ^{3}H-DNA, despite the fact that as much as 69% of the extracted DNA represented nicked molecular forms (23). These results suggest that some single-strand breaks can be repaired by polynucleotide ligase action. Furthermore, at least one single-strand break in the SV40 DNA chain is obligatory for the production of replicative intermediates.

When SV40 ^{3}H-DNA was incubated with CV-1 cells for 6, 10.5, 14, and 24 hr, respectively, about 90, 91, 93, and 97% of the DNA was converted to nicked molecular forms. At these times, the specific infectivities fell from 14.1 PFU/cpm to 2.1, 1.5, 0.7, and 0.3 PFU/cpm, respectively. Considering that the conversion of Form I SV40 DNA to open-circular Form II can result from the scission of a single phosphodiester bond, it can be calculated that for 90% conversion of Form I to nicked forms, about 21% of the DNA molecules have 4 or more nicks; for 99% conversion, 68% of the molecules have 4 or more nicks. Thus, much of the DNA is multiply nicked, and there is an increasing probability of nicking both chains opposite to each other (chopping) and of producing linear DNA. There would be diminished probability of forming a plaque by multiply nicked DNA, since additional nicking probably occurs during plaque assay. Much of the adsorbed SV40 DNA is degraded and only that DNA which survives nuclease chopping forms infectious virus and produces plaques (23).

As stated previously, SV40 DNA is less efficient than SV40 particle infection of African green monkey kidney cells. However, SV40 DNA may be used advantageously to initiate infection of cells resistant to virion penetration and uncoating. Several examples illustrate this point. Diderholm et al. (24) described transformation by SV40 DNA in mass cultures of bovine embryo cells, which were very resistant to transformation by high-titered intact virus. Black and Rowe (25) reported T antigen induction by SV40 DNA in hamster BHK21 cells, which were also insensitive to virus infection. Aaronson and Todaro (26) used SV40 DNA to induce T antigen synthesis in marsupial rat kangaroo cells, in which there is a complete block to infection by whole virus. Aaronson and Todaro (26) observed that SV40 DNA was about 1000-fold more efficient per infectious unit than virions in transforming human cell lines. Moreover, Aaronson (27) found marked differences in the susceptibility of human fibroblasts to transformation by SV40 particles. Highly susceptible strains were derived from patients with Fanconi's anemia and Down's syndrome. The differences in transformation frequency among cell strains and with whole virus were eliminated by the use of SV40 DNA, suggesting that the relative resistance of the human cell strains to transformation by whole virus was due to a block at an early step in infection.

SV40 ESSENTIAL REPLICATION FACTORS (SERF)

SV40 particles and SV40 DNA can induce SV40 T antigen formation in mouse cells. Nevertheless, there is little, if any, replication of viral DNA or infectious virus (28). These observations suggest that cellular factors essential for SV40 DNA replication are available in monkey cells, but not in mouse cells. The experiments to be presented provide evidence for this conclusion.

Somatic Hybrids of Monkey and Mouse Cells

Experiments with somatic hybrids of monkey-mouse cells suggest that a monkey cell determinant is required for SV40 DNA replication. The monkey-mouse hybrids were isolated from survivors of mixed cultures of CV-1 monkey kidney and mKS-BU100 mouse kidney cells in selective medium containing hypoxanthine, aminopterin, thymidine, and glycine (HATG) (29). The parental mKS-BU100 cells are a thymidine kinase-deficient cell line transformed by SV40 and fail to survive when cultured in HATG. Co-cultivation of the CV-1 and mKS-BU100 cells activates synthesis of infectious SV40 and results in destruction of those parental CV-1 cells which do not undergo fusion. Thus, extensive SV40 cytopathic effects develop around the twelfth day after seeding.

Surviving colonies of monkey-mouse cells were selected 15 to 18 days after seeding and then cloned in HATG medium. The surviving clones of monkey-mouse hybrid cells contained a substantial complement of mouse chromosomes and few monkey chromosomes. The monkey-mouse hybrid cells contained the SV40 specific T and transplantation antigens and infectious SV40 could be rescued from the hybrid cells by fusing them with additional CV-1 cells. However, the hybrid cells were resistant to superinfection with SV40 particles or SV40 DNA. In this respect, they resembled the parental mKS-BU100 cells.

The monkey-mouse hybrid cells contained thymidine kinase activity, probably derived from a monkey determinant (29). However, the few monkey chromosomes in the hybrid cell did not supply the monkey cell factor(s) which permits SV40 DNA replication. This factor was available during the early phase of co-cultivation of CV-1 and mKS-BU100 cells, as shown by the extensive production of infectious SV40. The factor was also supplied by fusing the surviving monkey-mouse hybrids with additional CV-1 cells, as shown by the rescue of infectious virus from the hybrid cells.

Availability of SERF in Normal Monkey and Human Cells

Besides primary and established lines of African green monkey kidney cells, normal human cell lines contain effective concentrations of SERF. This is shown by the fact that monkey and human cell lines are susceptible to SV40 particle and SV40 DNA infections. Furthermore, SV40 can be rescued from transformed mouse and transformed hamster cells by co-cultivating them with normal monkey or human cells, provided that the transformed cells contain nondefective SV40 genomes (30-32).

The sequence of events during rescue of infectious SV40 from transformed hamster (TSV-5) cells is shown in Fig. 5. Treatment of mixtures of CV-1 and TSV-5 cells with ultraviolet-irradiated Sendai virus (UV-Sendai) induces heterokaryon formation. SERF, available in the CV-1 cells and deficient in the transformed hamster cells, is then found in the common protoplasm. The SV40 genome is released from integration. By 20 hr after fusion, infectious SV40 DNA can be detected in the heterokaryons; by about 29 hr, infectious virus is detectable (32,33). SV40 particles are first detected in the TSV-5 nuclei (40 hr post fusion) (34). Infectious virus and/or infectious DNA is released from the TSV-5 nuclei and initiates a secondary infection of the CV-1 nuclei in the heterokaryons. At 68 to 72 hr after fusion, infectious virus is associated with both TSV-5 and CV-1 nuclei.

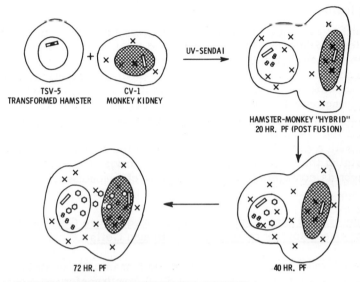

Fig. 5. The sequence of events during SV40 rescue after fusion of transformed hamster (TSV-5) and normal monkey kidney (CV-1) cells.

Virus rescue apparently does not depend upon the transfer of SV40 DNA to the CV-1 nucleus, since the transformed nucleus is the primary site of virus production.

In the sequence of events depicted in Fig. 5, it is postulated that SERF functions after the resident SV40 genome has been released from integration and not in the excision step, per se. An interesting experiment by Swetly et al. (35) supports this conclusion. Nonpermissive mouse cells were infected with SV40 DNA and 24 hr later, the mouse cells were fused with susceptible AGMK cells. Heterokaryon cultures derived from fusion of AGMK cells with the nonpermissive mouse cells produced infectious SV40 when tested for virus production 7 days after fusion.

Further evidence that human cell factors function in the replication of SV40 DNA has been obtained through study of man-mouse somatic hybrids (35). Thus, three man-mouse hybrids containing 8 to 15 human chromosomes produced infectious virus after exposure to SV40 DNA. In contrast, 3 man-mouse hybrids containing only 1 to 5 human chromosomes and 2 man-mouse hybrids containing 5 to 10 human chromosomes failed to produce virus after SV40 DNA infection. Most likely, a human chromosome with a gene for SERF was present in the man-mouse hybrids containing the larger number of human chromosomes, but not in the hybrids containing the fewest human chromosomes.

Rescue of SV40 after Fusion of Transformed Mouse Cells with Transformed Monkey or Transformed Human Cells

Transformed monkey and transformed human cells contain sufficient SERF to permit rescue of SV40 from transformed mouse cells (30,36). Table 2 demonstrates the rescue of SV40 after fusion of transformed mouse cells with normal monkey, SV40-transformed monkey cells, and cells transformed by the adenovirus type 7-SV40 "hybrid." In the experiments of Table 2, only the SV40 genome resident in the transformed mouse cells was recovered during rescue. Thus, fusion of mKS-BU100 or mKS-U13 cells with normal or transformed monkey lines yielded only the large-clear plaque strain of SV40. This is the virus strain which was initially used to transform these cells. Similarly, 3T3(4-88)J-3 and 3T3(U4), which were transformed by the small-clear and fuzzy type plaque strains of SV40, respectively, yielded only small-clear and fuzzy type SV40 after fusion with the normal and transformed monkey cells. Also, when transformed mouse lines containing marker strains of SV40 were fused with transformed human skin lines, W98 VaD and W98 VaH, only the SV40 genome resident in the transformed mouse lines was rescued (35). The transformed monkey lines shown in Table 2 and transformed human lines, W98 VaD and W98 VaH, have not yielded infectious SV40 after fusion with CV-1 monkey cells. These lines contain either a defective SV40 genome or an SV40 genome which is seldom released from integration.

Table 2. Rescue of SV40 after fusion of transformed mouse cells with normal or transformed African green monkey kidney (AGMK) cells.

Monkey cell line[b]	Virus yield (PFU/culture) 7 days after fusion with SV40-transformed mouse line:[a]			
	mKS-U13	mKS-BU100	3T3(4-88)J-3[c]	3T3(U4)[d]
CV-1	8.2×10^4 2.0×10^5	5.8×10^3	4.8×10^3	8.0×10^4
Vero	3.3×10^3	1.1×10^3		
AGMK	1.9×10^3			
BSC/SV40 cl 1-1	3.1×10^2		3.2×10^2	3.9×10^3
BSC/SV40 (NCM4)	9.8×10^3	3.1×10^2		
CV-1 LL-E46	3.4×10^4	5.5×10^1		

[a] Virus yields were determined by assay on CV-1 monolayers.
[b] CV-1, Vero, and AGMK are normal monkey cells; BSC/SV40 cl 1-1 and BSC/SV40(NCM4) were transformed by SV40; CV-1 LL-E46 was transformed by adenovirus 7-SV40 "hybrid."
[c] The virus recovered from the cells formed only small-clear plaques, typical of the SV40 used to transform these 3T3 cells. In contrast, the virus recovered from mKS-U13 or mKS-BU100 cells forms large-clear plaques.
[d] The virus recovered from these cells formed only fuzzy-type plaques, typical of the SV40 used to transform these 3T3 cells.

Fig. 6 depicts the rescue of SV40 after fusion of transformed mouse, 3T3(U4), and transformed human, W98 VaD, cells. The human, but not the mouse cells, contain effective concentrations of SERF. Following UV-Sendai induced heterokaryon formation, fuzzy-type SV40 DNA is released from integration and is activated to replicate by SERF. Superhelical SV40 DNA and fuzzy strain virus are formed in the heterokaryons. It is possible that the genome resident in some of the transformed human (W98 VaD) cells is also released from integration and that virus particles are formed. If so, the virus particles originating from the W98 VaD cells are noninfectious and are not detected.

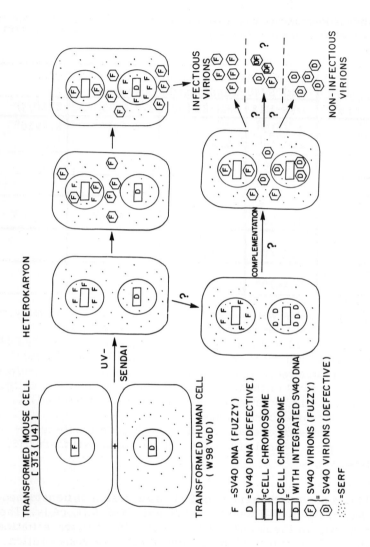

Fig. 6. Postulated events during rescue of SV40 from transformed mouse, 3T3(U4), and transformed human, W98 VaD, cells. The resident genome of W98 VaD cells is assumed to be defective or seldom released from integration. The resident SV40 genome of 3T3(U4) cells was transformed by fuzzy strain SV40(mKS-U4) and yields only fuzzy type virus after fusion with CV-1 monkey kidney cells.

Superinfection of Transformed Monkey and Human Cell Lines with SV40 DNA

Transformed lines of monkey and human cells are often resistant to infection by intact SV40 virions. Experiments by Swetly et al. (36) indicate that this resistance is attributable to the failure of SV40 particles to penetrate and uncoat in the transformed cells, because the transformed human and monkey cells are susceptible to infection by SV40 DNA. However, it was not shown whether virus progeny were derived from the superinfecting SV40 DNA, from the SV40 genome resident in the transformed human cells, or from both.

To answer this question, Kit et al. (35) infected transformed human cell lines, W98 VaD and W98 VaH, with DNA derived from plaque morphology mutants of SV40. Only virus with the plaque type of the infecting DNA was found in extracts from the infected cells. Since the factors essential for replication of nonintegrated SV40 DNA were probably available, this also suggests that the SV40 genomes resident in these transformed human cells were defective or that they were not released from integration.

Spontaneous SV40 Production by Transformed Human Cell Lines

It may be predicted that SERF-positive monkey or human cell lines transformed by nondefective SV40 will occasionally release the SV40 genome from integration and produce infectious virus. Indeed, two transformed human cell lines, WI38 Va13A and W18 Va2(P363), do spontaneously produce small amounts of infectious virus (36). The WI38 Va13A (human embryonic lung) cells have been studied in detail.

Immunofluorescence tests indicated that virtually all of the WI38 Va13A cells were T antigen positive. In any single trial, about 10^{-5} WI38 Va13A cells produced infectious centers on CV-1 monolayers. Infectious center formation was not enhanced by first fusing WI38 Va13A cells with CV-1 cells. The virus produced spontaneously by WI38 Va13A cells replicated normally on CV-1 monolayer cultures.

To determine whether individual cells of the WI38 Va13A population had the capacity to produce virus spontaneously, 12 clonal lines were prepared. Eleven of the 12 clonal lines spontaneously produced small amounts of virus. From 3 of the clonal lines, 17 secondary clones were prepared. Virus was detected in all of the 17 secondary clones (37). From three of the secondary clones, 15 tertiary clones were prepared. At this writing, virus has been detected in 14 of the 16 tertiary clones. Essentially all of the cells of all clones were positive for the SV40 T antigen. It is to be emphasized that from any single clone, virus is not produced at every passage. Thus, trials on repeated passages are often necessary to detect virus production. These findings suggest that spontaneous virus production is a hereditary property of certain SV40-transformed human lines.

Superinfection of Transformed Monkey Cells by SV40 Particles

Many, but not all, cell lines transformed by SV40 and polyoma virus are resistant to superinfection by virus particles. The 3T3 mouse fibroblast line transformed by polyoma virus mutant TS-a was found to be fully susceptible to infection with wild-type polyoma virus (20). A polyoma-transformed mouse line isolated by Benjamin (38), was susceptible to superinfection by polyoma virus, and was used to select host range mutants. Winocour and Sachs (39) found that susceptibility to challenge infection of polyoma virus-induced parotid tumor transplant cells changed during the process of obtaining clonal populations. Twenty-three clones were derived from two polyoma-induced parotid tumor transplant lines, which had been shown to be resistant to challenge with polyoma virus. However, all 23 clonal lines were susceptible to challenge with virions. They suggested that this could have been due either to a change in the physiological state of the cells or to the segregation of new sensitive types.

Two SV40-transformed monkey kidney lines studied in Houston were susceptible to SV40 DNA infection. One line, BSC/SV40 cl 1-1, was highly resistant to SV40 particle infection, but the second line, BSC/SV40(NCM4), consisted of a mixture of virion-resistant and virion-sensitive cells (37).

The BSC/SV40(NCM4) line provides a further example of SERF-positive transformed cells. A postulated mechanism for the isolation of SERF-positive transformed cells is shown in Fig. 7.

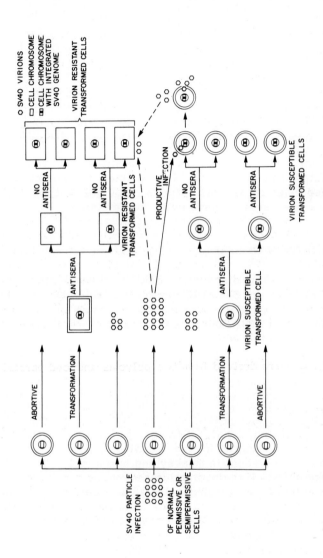

Fig. 7. A postulated mechanism for the isolation of virion-susceptible and virion-resistant transformed cell lines.

Normal monkey or human cells may be infected with either normal or defective SV40 particles. The infections: (i) may be productive, leading to a variable yield of infectious virus per cell; (ii) may be abortive; or (iii) may result in transformation. The transformed cells may be susceptible or may develop resistance to SV40 particle infection. By propagating the cells in SV40 antisera, the virion susceptible cells can be protected against superinfection by virus released from productively infected normal cells and by virus spontaneously produced by some of the transformed cells. Resistance to virion infection entails changes in the cell which reduce virion penetration and uncoating. After resistance has developed, the SERF-positive cells can be propagated without antisera. Clonal lines of susceptible cells may also be grown without antisera, provided that the SV40 genomes integrated in these cells are defective.

THE FUNCTION OF SERF

In the preceding discussion, SERF has been defined as a host factor essential for SV40 DNA replication which is present in monkey and human cells, but lacking in mouse cells. The hypothesis that SERF functions in translation or in transcription will now be discussed.

In abortively infected mouse cells, SV40-specific transplantation and T antigens are made, the activities of several enzymes are enhanced, and cellular DNA synthesis is stimulated. This suggests that ribosomes and the translation factors concerned with initiation, elongation, and termination of polypeptide chains are all functional. It is possible, however, that species-specific tRNAs required for the decoding of SV40 messenger RNA sequences are present in monkey and human cells, but deficient in mouse cells. If the nucleotide sequences coded for a viral protein required for superhelical SV40 DNA replication, then the failure of mouse cells to duplicate SV40 DNA could be explained. Furthermore, replication of parental SV40 DNA seems to be a prerequisite for production of capsid proteins. Failure to synthesize the viral gene product would thus explain the lack of capsid protein formation in abortively infected mouse cells.

One test of the hypothesis that SERF is a primate-specific tRNA would be to supply exogenous tRNAs to abortively infected mouse cells. Uptake of primate tRNA could be enhanced by treating mouse cells with DEAE-dextran. If the tRNAs were taken up in sufficient amounts, SV40 DNA replication might be detected.

An alternative possibility is that SERF is a transcription factor. The present evidence suggests that transcription of SV40 DNA is incomplete in mouse cells. Productively infected monkey cells synthesize two types of SV40-specific RNA, an "early" SV40 RNA that is made prior to the replication of viral DNA and a "late" SV40 RNA, detectable after the onset of viral DNA synthesis (40). The SV40 RNAs can be distinguished by DNA-RNA hybridization-competition experiments and by base ratio analysis. The RNA species made early in infection is also synthesized late in infection.

The size distribution of polyribosome-associated SV40 RNA has been examined at various times after productive infection of monkey cells (41). Early in infection, virus-specific RNA was detected in the 14 to 17S region of a sucrose gradient. At later times, SV40 RNA was detected in more rapidly sedimenting regions (28S). Nuclear SV40 RNA, prepared late in infection, was distributed in regions of the gradient peaking at about 32 to 35S and some nuclear virus-specific RNA could be detected in the 45 to 50S region. During abortive infection of mouse cells, the sedimentation profile of SV40 RNA was very similar to that observed during the early phases of the lytic cycle.

The finding that the size of SV40-specific nuclear RNA is greater than the size of a single strand of viral DNA suggests that processing of the RNA occurs in the nucleus. This raises the possibility that processing of "late" virus-specific RNA may be abnormal in the mouse cells.

An alternative regulatory device may be considered. The transcription of "late" species of SV40 RNA may be controlled by a termination (rho) factor. Effective concentrations of an anti-rho factor might be present in monkey and human cells, but not in mouse cells. Perhaps in mouse cells lacking the anti-termination factor, transcription only progresses through the "early" SV40 gene cistrons. In monkey and human cells containing the anti-rho factor, transcription would proceed through the cistron which controls SV40 DNA replication.

The isolation and purification of the putative anti-rho factor and of the mouse and primate RNA polymerases are required to test the anti-rho factor hypothesis. In a more general sense, the isolation of temperature-sensitive mutants of monkey cells lacking the ability to support SV40 DNA replication would facilitate the identification of SERF.

REFERENCES.
1. E.A.C. Follett and L.V. Crawford. J. Mol. Biol. 34 (1968) 565.
2. K. Yoshiike and A. Furuno. Fed. Proc. 28 (1969) 1899.
3. A.J. Levine, H.S. Kang and F.E. Billheimer. J. Mol. Biol. 50 (1970) 549.
4. H. Kubinski and J.A. Rose. Proc. Nat. Acad. Sci. U.S. 57 (1967) 1720.
5. P. Rüst. Biochem. Biophys. Res. Commun. 39 (1970) 455.
6. M.F. Bourguignon. Biochim. Biophys. Acta 166 (1968) 242.
7. B. Hirt. Proc. Nat. Acad. Sci. U.S. 55 (1966) 997.
8. B. Hirt. J. Mol. Biol. 40 (1969) 141.
9. S. Kit, T. Kurimura, R.A. de Torres and D.R. Dubbs. J. Virol. 3 (1969) 25.
10. S. Kit, R.A. de Torres, D.R. Dubbs and M.L. Salvi. J. Virol. 1 (1967) 738.
11. P. Bourgaux, D. Bourgaux-Ramoisy and R. Dulbecco. Proc. Nat. Acad. Sci. U.S. 64 (1969) 701.
12. B. Hudson, W.B. Upholt, J. Devinny and J. Vinograd. Proc. Nat. Acad. Sci. U.S. 62 (1969) 813.
13. P. Bourgaux, in: Second Lepetit Colloquium on "The Biology of Oncogenic Viruses" (North-Holland, Amsterdam, in press).
14. J.A. Huberman and A.D. Riggs. J. Mol. Biol. 32 (1968) 327.
15. M. Schnös and R.B. Inman. J. Mol. Biol. 51 (1970) 61.
16. W. Goebel and D.R. Helinski. Proc. Nat. Acad. Sci. U.S. 61 (1968) 1406.
17. M.M.K. Nass. Nature 223 (1969) 1124.

18. B.M. Alberts. Fed. Proc. 29 (1970) 1154.
19. M. Fried. Virology 40 (1970) 605.
20. M. Vogt. J. Mol. Biol. 47 (1970) 307.
21. F. Cuzin, M. Vogt, M. Dieckmann and P. Berg. J. Mol. Biol. 47 (1970) 317.
22. W. Eckhart. Virology 38 (1969) 120.
23. D. Trkula, S. Kit, T. Kurimura and K. Nakajima. J. Gen. Virol. 10 (1970) 000.
24. H. Diderholm, B. Stenkvist, J. Ponten and T. Wesslen. Exp. Cell Res. 37 (1965) 452.
25. P.H. Black and W.P. Rowe. Virology 27 (1965) 436.
26. S.A. Aaronson and G.J. Todaro. Science 166 (1969) 390.
27. S.A. Aaronson. J. Virol. 6 (1970) 470.
28. S. Kit, D.R. Dubbs, L.J. Piekarski, R.A. de Tores and J.L. Melnick. Proc. Nat. Acad. Sci. U.S. 56 (1966) 463.
29. S. Kit, K. Nakajima, T. Kurimura, D.R. Dubbs and R. Cassingena. Intern. J. Cancer 5 (1970) 1.
30. F.C. Jensen and H. Koprowski. Virology 37 (1969) 687.
31. D.R. Dubbs, S. Kit, R.A. de Torres and M. Anken. J. Virol. 1 (1967) 968.
32. D.R. Dubbs and S. Kit. J. Virol. 2 (1968) 1272.
33. S. Kit, T. Kurimura, M.L. Salvi and D.R. Dubbs. Proc. Nat. Acad. Sci. U.S. 60 (1968) 1239.
34. G.H. Wever, S. Kit and D.R. Dubbs. J. Virol. 5 (1970) 578.
35. P. Swetly, G. B. Brodano, B. Knowles and H. Koprowski. J. Virol. 4 (1969) 348.
36. S. Kit, T. Kurimura, M. Brown and D.R. Dubbs. J. Virol. 6 (1970) 69.
37. S. Kit and D.R. Dubbs, in: Second Lepetit Colloquium on "The Biology of Oncogenic Viruses" (North-Holland, Amsterdam, in press).

38. T.L. Benjamin. Proc. Nat. Acad. Sci. U.S. 67 (1970) 394.
39. E. Winocour and L. Sachs. Virology 16 (1962) 496.
40. Y. Aloni, E. Winocour and L. Sachs. J. Mol. Biol. 31 (1968) 415.
41. M.A. Martin and J.C. Byrne. J. Virol. 6 (1970) 463.

This investigation was aided by a grant from the Robert A. Welch Foundation (Q-163), and by Public Health Service grants CA-06656-08 and 1-K6-AI 2352 from the National Cancer Institute and the National Institute of Allergy and Infectious Diseases.

We thank Judith Rotbein, Carolyn Smith, and Marjorie Johnson for able technical assistance.

DISCUSSION

Dr. D. Baltimore, Massachusetts Institute of Technology:
In the cells which occasionally release small amounts of virus but will yield more if infectious DNA is added, if you grow up the spontaneously released virus and use it to transform other cells -- does it tend to yield transformed cells which do not yield much virus; or does it have any special properties; or is the fact that it doesn't come out very often seem to be a property of the cell that it's in?

Dr. S. Kit: What we have done, so far, with virus recovered from WI 38 Va 13A cells is to study the properties of the virus in LYTIC infection only; and we have demonstrated that the virus recovered from WI 38 Va 13A cells is perfectly normal in inducing all of the SV40 functions which we have measured. These include T antigen and thymidine kinase induction and the induction of cellular DNA synthesis. Now we have not so far carried out quantitative transformation experiments with that virus.

DISCUSSION TO VIRAL ONCOLOGY
BY S. SPIEGELMAN

Dr. B. Alberts, Princeton University: Do your studies with either the synthetic polymers or the purified polymerase allow you to guess whether, during the course of infection, an RNA-DNA hybrid is formed which then continually liberates single-strand DNA products or whether there is instead a stage in which double-stranded DNA is replicated by the viral polymerase?

Dr. S. Spiegelman: I'm afraid I can't -- we haven't done the right experiments to get the answer that you want. I think perhaps Baltimore may. Dr. Temin would like to know which embryonic tissue has the most activity. Well, that really depends on when you take the sample. Surprisingly, for example, kidney at certain stages of embryonic development has enormous amounts of activity, the highest we've seen. But it may be, we have not seen the others. The lowest has always been the brain, and it's rather interesting that in the course of going through tumors, we have looked at 6 brain tumors. There we were very fortunate because (well, it wasn't fortunate for the patient), they had neighboring normal material along with the tumor, right next to it. The neighboring white matter was completely inactive and the tumor had tremendous amounts of both dC;dG and dT;rA. But in the course of the development of at least the chicken and the mouse, which we had done pretty carefully, and thanks to the abortion laws in New York State, the human, it appears that the organs which lose this activity fastest are things like the brain, muscle; they are the first two that lose it. The last to lose it are the kidney and the liver. In fact, you can actually find activity in hatched chicks in the kidney and the liver.

Dr. J. Hurwitz, Albert Einstein College of Medicine: Sol, one of the questions here which becomes very important is in a comparison of various synthetic polynucleotides, and especially in the system, I think a number of people have seen very similar results with regards to the type of priming which you reported and one of the important features here is that 1) most of the reactions are repair reactions which you are measuring; therefore, with dI:dC, with rA:dT the ratio of input of the two homopolymers becomes very, very critical. Equally critical is the relative susceptibility of these strands to nuclease action. Therefore, the question which really becomes very important in any comparison of various tissues and embryonic tissues, is the susceptibility of these polymers. And, for example, we know that dI:dC will be relatively resistant to a variety of nucleases, (exonucleases?), whereas rA:dT will be very, very susceptible. So, one of the things which really I think would have to be carefully spelled out in such comparisons is the degree of nuclease contamination and specifically under the conditions which most people are working, you're measuring picomole incorporation, and these are relatively low orders of nucleotide incorporation and therefore, the degree of nuclease contamination becomes a very, very important reference point for measuring the efficiency of the different primers that you're using. Well, Jerry, all I can say is I agree completely and the next thing we're trying to do obviously is to purify out these activities. I mean, we recognize they're there and then see what they're really doing in the absence of contamination. We have one great difficulty, and that's a problem of supply. I have been deliberately rather pessimistic about certain aspects of this thing with its relation to, let's say, leukemia problem, and I think we should be. We cannot assume, however, that the activities we are seeing for example in embryonic tissue is identical to what we see in leukemic patients. We have to be able to purify both enzyme systems and then make the comparison when the things are really clean. Our hangup here is one of pure logistics and that is to get enough leukemic cells to work with to do that kind of enzymology, and that's really been one of the big stumbling blocks-I think both for us as well as Dr. Gallo. But I agree, we have to go in the direction you have indicated, otherwise we won't know what these things mean.

Dr. S. Kit, Baylor College of Medicine: You've indicated that different batches of polymers vary in template activity; some batches are good templates, and some are poor templates. I wonder if you can tell us about the physiochemical properties of the templates that you have used - molecular weight, homogeneity, and also whether there are single-stranded gaps in some of the templates.

Dr. Spiegelman: Well, we know most about the dC:dG principally because of the cooperation of Fred Bullum, who knows how to make these compounds, and he characterizes them. Some of the dC:dG's we used have had no gaps detectable, and there's a very easy way of finding this out. You can always make a comparison between a template in which you deliberately have gaps; in other words, you prepare something of the following nature: we've done it with the cooperation of Dr. Nussbaum at LaRoche. You take dT and now you put triplets, no, you take rA and then you put triplets say of dT on it or formers of fibers and so on and then compare the ability of the enzyme to respond to those with or without gaps. Now, in the human case it's quite clear that if it needs gaps, it doesn't need very many; I mean, it does not, as you fill the molecule up, lose activity. Even if you go to completion. Now, the same thing has been done by comparing a variety of dC:dG templates, some with deliberately made gaps and others with none at all where there are virtually no gaps. These same dC:dG's, when given to the Bollum enzyme, do not work at all. But remember, no conclusion can be drawn here until we get our enzyme pure enough so that we know what's in it.

Dr. Kit: What about the size of your templates?

Dr. Spiegelman: The size doesn't seem to make a lot of difference over the range that we have looked at, but it hasn't been very wide.

Dr. G. D. Novelli, Oak Ridge National Laboratory: You showed a lot of data, I might have missed this point, Sol, but you used four templates on the patient's enzyme. Did you see any remarkable differences of the response of individual patients to these templates and would this cause you to think that the enzyme is different in the different patients?

Dr. Spiegelman: No. We do find differences in the relative activities we see on dC:dG compared the dT:rA compared to rA:iU. Now, I don't think this means that the patients have different enzymes. I think it's more likely that the patients have different mixtures of several enzymes. That's a hunch.

Dr. U. Z. Littauer, Weizmann Institute of Science: Did you try polythio U or some other thiolated polynucleotide because these are supposed to be much more resistant to nucleases degradation and induce better interferon response?

Dr. Spiegelman: No, I'd love to try some.

Dr. Littauer: They are made in Gottingen, Germany.

Dr. Spiegelman: Thank you.

Dr. J. Willis, P-L Biochemicals: I feel compelled to say something about the supply of poly I poly C. We have recently completed a contract with NIH for fairly large quantities of this material for use in interferon induction studies and could supply your needs and possibly resolve the problem you spoke of.

Dr. Spiegelman: Well, it's not merely the size, but how good is it as a template. Because Miles has piles of that stuff which is no good for us at all, and they don't know why.

Dr. Willis: Would you like a sample?

Dr. Spiegelman: Yes, I'd be very happy to test it. It would take us about twenty minutes and we'll find out.

Dr. A. Tomasz, The Rockefeller University: Does the enzyme, which you purify to homogeneity, have ligase activity?

Dr. Spiegelman: We have not checked that.

Dr. Julius Schultz, Papanicolaou Cancer Research Institute: Are you going to encourage the wide use of this test before you answer Dr. Hurwitz's question in the laboratory?

Dr. Spiegelman: First of all, I don't want to encourage anybody to do anything and I don't need to encourage people to try to use this test. The information we have, I think, says that even without complete understanding we may get some pragmatically useful information by piling up more information. It is conceivable that by accumulating enough of this information, we could make prognostic judgments as to how a patient is going to respond to a particular course of chemotherapy. And that's just by empirical observation of what happens if you start out with a very high level of enzyme, does such a patient have a better chance of going into remission or not and so on, with one particular course of therapy or not. So long as the data are consistent and give you the same result such pragmatic utility can come out of it even without understanding. Medicine is filled with such.

Dr. J. Warren, Nova University: You mentioned, Dr. Spiegelman, that you first added known amounts of virus to the human plasma. How did the activity compare when you added known amounts of virus to the activities you found in unknowns, such as, Burritt Lymphoma, or HeLa cells. I think, also, somebody should say that this is beautiful and elegant and exciting work.

Dr. Spiegelman: Well, let me put it this way. From the amount of activity that we see in leukemic cells, if they had been due to virions, we would have had no difficulty seeing the virions.

Dr. R. E. Parks, Brown University: If the RNA-dependent DNA polymerase is of viral origin, it might behave as a foreign protein and be antigenic. I wonder if you have had a chance to look for antibodies to the enzyme in patients with malignancies?

Dr. Spiegelman: Look for antibodies in what?

Dr. Parks: In patients with malignancies.

Dr. Spiegelman: Well, lot's of people are doing that.

Dr. Parks: I mean antibodies to the viral RNA-dependent DNA polymerase. For example, the enzyme might be released from broken cells in a patient with leukemia. If the patient develops antibodies to the enzyme, they might be detected.

Dr. Spiegelman: Well, we have not done it. I'm sure that, in fact I think that Darrow is doing something like that, isn't he?

Dr. G. Pieczenik, New York University: I was wondering, have you tried your assay on regenerating systems since you report finding activity in embryonic systems?

Dr. Spiegelman: Regenerating liver. Yes, it's there.